Proceedings in Adaptation, Learning and Optimization

Volume 10

The role of adaptation, learning and optimization are becoming increasingly essential and intertwined. The capability of a system to adapt either through modification of its physiological structure or via some revalidation process of internal mechanisms that directly dictate the response or behavior is crucial in many real world applications. Optimization lies at the heart of most machine learning approaches while learning and optimization are two primary means to effect adaptation in various forms. They usually involve computational processes incorporated within the system that trigger parametric updating and knowledge or model enhancement, giving rise to progressive improvement. This book series serves as a channel to consolidate work related to topics linked to adaptation, learning and optimization in systems and structures. Topics covered under this series include:

- complex adaptive systems including evolutionary computation, memetic computing, swarm intelligence, neural networks, fuzzy systems, tabu search, simulated annealing, etc.
- machine learning, data mining & mathematical programming
- hybridization of techniques that span across artificial intelligence and computational intelligence for synergistic alliance of strategies for problem-solving
- aspects of adaptation in robotics
- agent-based computing
- autonomic/pervasive computing
- dynamic optimization/learning in noisy and uncertain environment
- systemic alliance of stochastic and conventional search techniques
- all aspects of adaptations in man-machine systems.

This book series bridges the dichotomy of modern and conventional mathematical and heuristic/meta-heuristics approaches to bring about effective adaptation, learning and optimization. It propels the maxim that the old and the new can come together and be combined synergistically to scale new heights in problem-solving. To reach such a level, numerous research issues will emerge and researchers will find the book series a convenient medium to track the progresses made.

More information about this series at http://www.springer.com/series/13543

Jiuwen Cao · Chi Man Vong
Yoan Miche · Amaury Lendasse
Editors

Proceedings of ELM-2017

 Springer

Editors
Jiuwen Cao
Institute of Information and Control
Hangzhou Dianzi University
Zhejiang, China

Chi Man Vong
Department of Computer and Information
Science
University of Macau
Macau, China

Yoan Miche
Nokia Bell Labs
Espoo, Finland

Amaury Lendasse
Department of Information and Logistics
College of Technology at the University
of Houston
Houston, TX, USA

ISSN 2363-6084   ISSN 2363-6092   (electronic)
Proceedings in Adaptation, Learning and Optimization
ISBN 978-3-030-13182-1   ISBN 978-3-030-01520-6   (eBook)
https://doi.org/10.1007/978-3-030-01520-6

This Springer imprint is published by the registered company Springer Nature Switzerland AG
The registered company address is: Gewerbestrasse 11, 6330 Cham, Switzerland

Contents

Adaptive Control of Vehicle Yaw Rate with Active Steering System and Extreme Learning Machine - A Pilot Study

Pak Kin Wong[1](✉), Wei Huang[1], Ka In Wong[2], and Chi Man Vong[3]

[1] Department of Electromechanical Engineering,
University of Macau, Macau, China
fstpkw@umac.mo
[2] Institute for the Development and Quality, Macau, China
[3] Department of Computer and Information Science,
University of Macau, Macau, China

Abstract. The active steering system can enhance the vehicle yaw stability, which is essential to road safety. However, control of vehicle yaw rate is very challenging due to the presence of nonlinearity and uncertainties in the vehicle dynamics. To address the problems, an extreme-learning-machine (ELM)-based adaptive control algorithm is proposed, and a vehicle dynamic model is also developed to identify the necessary exogenous variables for control system inputs. To validate the performance of the proposed controller, simulation is conducted with an industry software. Simulation result indicates that the proposed controller achieves superior performance in tracking nominal vehicle yaw rate. A comparison is also carried out with fuzzy logic control. The pilot result shows that the proposed controller outperforms the fuzzy logic control.

Keywords: Adaptive ELM control · Active steering · Vehicle yaw rate control

1 Introduction

In many scenarios, if the driver wants to make a turn when his/her vehicle is running on a road surface with extremely low friction coefficient, the driver may not be capable of maintaining the stability of the vehicle yaw rate. Hence, the vehicle may easily skid and spin out, possibly resulting in a road accident. The major cause for this kind of accident is that the nominal yaw rate motion during turning is difficult to be achieved as a result of small road friction. If there exists a control system that can smartly assist the driver in correcting the vehicle motion according to the driving input, many accidents can be avoided. This research is therefore to develop an intelligent vehicle yaw rate controller to compensate for the driver manipulation.

For modern vehicles, the yaw rate can usually be controlled by adjusting the steering angle with an active front steering (AFS) system. Figure 1 illustrates how an AFS system works. In addition to the steering wheel mechanism, an electromechanical actuator (usually a geared DC motor) is installed on the steering column, which can provide a superposition angle to the steering wheel angle to adjust the actual steering

J. Cao et al. (Eds.): ELM 2017, PALO 10, pp. 1–11, 2019.
https://doi.org/10.1007/978-3-030-01520-6_1

angle. The yaw rate can then be corrected via proper control of the AFS-actuator. In order to guarantee the vehicle yaw stability with the AFS system, the objective of the vehicle yaw rate controller is to determine an appropriate front steering angle for the AFS-actuator to adjust. However, due to the fact that the dynamics of vehicle-road system are highly nonlinear with uncertainties, design of an effective vehicle yaw rate controller with good transient tracking performance is very challenging [1]. The nonlinear characteristics of the vehicle-road system are largely attributed to the complicated tire dynamics, whereas the uncertainties are mainly due to the unmeasurable parameters like the road surface adhesion coefficients and the cornering stiffness, etc. In the existing literature, several control algorithms have been attempted to address these issues. Considering the parametric uncertainties of cornering stiffness and vehicle longitudinal speed, a sliding mode control was recently tried by [2] to control the AFS system for vehicle yaw stability. However, their experimental result indicated that the sliding mode controller cannot avoid the chattering problems [3] because of the slow performance of the steering actuator.

In view of the rapid growth of artificial intelligence in recent years, researchers then started to apply different artificial intelligence methods to deal with the problems of complex nonlinear dynamics and significant uncertainties in yaw rate control. Kirshna et al. [4] and Eski et al. [5] respectively employed fuzzy logic and neural network (NN) for yaw rate control. The advantage of the fuzzy logic control lies in providing a nonlinear control methodology on the basis of human's heuristic knowledge rather than mathematical model, while NN-based prediction control benefits from its capability of modeling the nonlinear input-output relationship of the vehicle-road system from observed dataset without the need of understanding the detailed system dynamics. They were shown to be better than traditional control algorithms [4, 5], but they have several drawbacks for practical use. For fuzzy logic control, the performance greatly relies on how comprehensive and accurate the logic rules are defined. For NN-based prediction control, collection of input-output data for modeling is usually very time-consuming or often unavailable. Furthermore, if the trained network does not cover all the operating

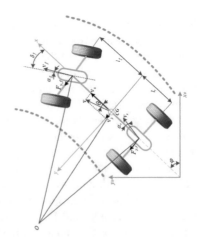

Fig. 1. Working principle of an AFS system   **Fig. 2.** Bicycle model

conditions of the system, or the prediction accuracy of the network is not sufficiently high for some of the range, the resulting control performance is unsatisfactory.

To overcome the difficulties in using fuzzy logic control and NN-based prediction control for practical cases, certain adaptive control algorithms for other applications based on NNs have been proposed [6–8]. These algorithms are categorized as adaptive NN control, and have the advantages that no fuzzy rules or pre-collected data are required in controller development [8]. Nevertheless, the conflict between fast response time, good tracking performance and low computational complexity has been exposed in most of adaptive NN control schemes [8]. Since both fast response time and excellent transient tracking are essential for yaw rate control in order to minimize any potential danger of vehicle skid, existing adaptive neural control algorithms still may not be a good solution. Recently, a new learning algorithm called Extreme Learning Machine (ELM) was proposed by Huang et al. [9]. It has the form of single-hidden-layer feedforward network and requires very low computational cost to compute the network parameters. Many studies [10, 11] showed that ELM can approximate complicated functions faster and more accurate than traditional NNs. These features are very attractive for real-time control. Therefore, ELM could replace traditional neural networks to perform adaptive neural control. Aiming to construct a controller with extremely fast response time and superior transient tracking performance for vehicle yaw rate control, this paper originally proposes an ELM-based adaptive control algorithm. By adaptively tuning the parameters of ELM based on the real-time feedback data, the dynamics of the vehicle-road system can be modeled online efficiently and controlled simultaneously.

2 Description of Vehicle Dynamics

This section illustrates the nonlinear characteristics of vehicle yaw rate under cornering and how to identify the necessary exogenous variables for control system inputs. To describe the vehicle lateral dynamics for vehicle yaw rate control, usually a time-invariant linear bicycle is employed, which assumes that the vehicle tires operate in the linear region and the vehicle longitudinal speed remains unchanged during the cornering process. However, the use of such time-invariant linear bicycle model in the controller may sometimes lead to poor tracking performance as it is inconsistent to the actual nonlinear vehicle-road system. Therefore, the linear bicycle model is still employed to represent the nonlinear vehicle dynamics, but is characterized by time-varying parameters, such as cornering-stiffness and vehicle longitudinal speeds. The following sub-sections present the details of the time-varying linear bicycle model.

2.1 Bicycle Model

Figure 2 shows a bicycle model to describe the vehicle lateral dynamics. The roll and pitch motions are neglected in this linear bicycle model. Derivation of the equations of motion for the simple model can be obtained from the force and moment equilibrium:

$$m_v \cdot \left(\dot{v}_y + v_x \cdot \dot{\varphi} \right) = F_{y,f} + F_{y,r} \tag{1}$$

$$I_z \cdot \ddot{\varphi} = l_f \cdot F_{y,f} - l_r \cdot F_{y,r} \tag{2}$$

Where (x, y) are vehicle body-fixed coordinates; (x_0, y_0) are global coordinates; m_v is vehicle mass; v_x is vehicle longitudinal speed; y_x is vehicle lateral speed; φ is the heading angle of the vehicle; $\dot{\varphi}$ is the yaw rate of the vehicle body; $F_{y,f}$ and $F_{y,r}$ are the front and rear lateral tire forces (the subscripts f and r stand for front and rear respectively); I_z is moment of inertia about a vertical axis through its Center of Gravity (CG); l_f and l_r are the distances from the front and rear axles to the vehicle CG, respectively. The lateral forces at the front and rear are related to slip angles by the varying cornering stiffness of the front and rear tires as

$$F_{y,f} \approx -C_{af} \cdot \alpha_f \tag{3}$$

$$F_{y,r} \approx -C_{ar} \cdot \alpha_r \tag{4}$$

Where C_{af} and C_{ar} are the time-varying cornering stiffness for the front and rear tires, respectively; the slip angles α_f and α_r between the orientation of the tires and the orientation of the speed vector are defined on basis of small angle approximations:

$$\alpha_f = \tan \theta + \frac{l_f \cdot \varphi}{v \cdot \cos \theta} - \delta_f \tag{5}$$

$$\alpha_r = \tan \theta - \frac{l_r \cdot \varphi}{v \cdot \cos \theta} \tag{6}$$

where α_f is the steering angle provided by the driver; θ is the vehicle sideslip angle. Substituting Eqs. (3) to (6) into Eqs. (1) and (2), the vehicle lateral dynamics can be expressed in the discrete representation of state-space form as follows:

$$\begin{bmatrix} v_y(k+1) \\ \dot{\varphi}(k+1) \end{bmatrix} = \begin{bmatrix} -\dfrac{C_{af}(k) + C_{ar}(k)}{m_v \cdot v_x(k)} & -v_x(k) - \dfrac{C_{af}(k) \cdot l_f + C_{ar}(k) \cdot l_r}{m_v \cdot v_x(k)} \\ \dfrac{C_{af}(k) \cdot l_r - C_{ar}(k) \cdot l_f}{I_z \cdot v_x(k)} & -\dfrac{C_{af}(k) \cdot l_f^2 + C_{ar}(k) \cdot l_r^2}{I_z \cdot v_x(k)} \end{bmatrix} \begin{bmatrix} v_y(k) \\ \dot{\varphi}(k) \end{bmatrix} + \begin{bmatrix} \dfrac{C_{af}(k)}{m_v} \\ \dfrac{C_{af}(k) \cdot l_f}{I_z} \end{bmatrix} u(k) \tag{7}$$

$$Y(k) = \begin{bmatrix} A_1 & A_2 \end{bmatrix} \begin{bmatrix} v_y(k) \\ \dot{\varphi}(k) \end{bmatrix} \tag{8}$$

where u is the controlled superposition angle; Y is the output of vehicle performance vector including vehicle lateral speed v_y and yaw rate $\dot{\varphi}$; A_1 and A_2 are constant parameters; k is time step.

2.2 Tire-Road Model

To approximate the nonlinear tire-road friction coefficient, the model in [12] is employed and written as

$$\mu = \rho_1 \cdot e^{-\rho_2 \cdot S} \cdot S^{(\rho_3 \cdot S + \rho_4)} \cdot e^{-\rho_5 \cdot v} \tag{9}$$

where ρ_1, ρ_2, ρ_3, ρ_4 and ρ_5 are constant parameters determined by experimental data; the wheel slip s is defined as:

$$S = \begin{cases} 1 - \frac{R \cdot \omega}{v}, & v \geq R \cdot \omega \\ 1 - \frac{v}{R \cdot \omega}, & v < R \cdot \omega \end{cases} \tag{10}$$

where R is the effective radius of the wheel; ω represents the angular velocity of wheel; and v represents the vehicle speed. To ensure the proposed control system incorporating sufficient exogenous variables, an exact formula [13] is also employed to describe the tire forces as follows:

$$F_f = k_{1,f} \cdot \left\{ k_{2,f} \cdot \tan^{-1} \left(k_{3,f} \cdot \text{abs} \left(1 - k_{4,f} \right) \cdot \alpha_f + k_{4,f} \cdot \tan^{-1} \left(k_{3,f} \cdot \alpha_f \right) \right) \right\} \tag{11}$$

$$F_r = k_{1,r} \cdot \left\{ k_{2,r} \cdot \tan^{-1} \left(k_{3,r} \cdot \text{abs} \left(1 - k_{4,r} \right) \cdot \alpha_f + k_{4,r} \cdot \tan^{-1} \left(k_{3,r} \cdot \alpha_r \right) \right) \right\} \tag{12}$$

where $k_{1,j}$, $k_{2,j}$, $k_{3,j}$, and $k_{4,j}(j = f, r)$ in the models are unknown constant parameters. In the above functions, slip angles α_f and α_r are the only variables which are usually unmeasurable. Combining Eqs. (5), (6), (11) and (12), the tire force $F_f(j = f, r)$ can depend on vehicle slide slip angle θ, yaw rate $\dot{\varphi}$, vehicle speed v and steering angle δ_f.

3 Vehicle Yaw Rate Control

Generally, a vehicle can respond to the steering wheel input well if the road is dry and has a high tire-road friction coefficient. However, if the coefficient of friction is small or if the vehicle speed is too high, then the vehicle should follow the nominal motion requested by the driver. In this case, the yaw control system can partially be useful by making the vehicle yaw rate closer to the desired nominal yaw rate. To enhance the yaw stability control, an ELM-based adaptive control is herein proposed.

3.1 Desired Vehicle Lateral Dynamics

In vehicle yaw stability control, the target control variables are yaw rate and sideslip angle. The desired sideslip angle is zero, namely $\theta_{des}(k) = 0$. Hence, the desired vehicle lateral speed should be $v_{y,des}(k) = 0$ Since the definition of desired yaw rate affects the vehicle stability at high speeds, based on [14], the desired yaw rate $\dot{\varphi}'_{des}$ can be expressed in terms of the longitudinal speed as follows:

$$\dot{\varphi}'_{des}(k) = \frac{v_x(k) \cdot \delta_f(k)}{\left(l_f + l_r\right)\left(1 + K_{us} \cdot v_x^2(k)\right)} \tag{13}$$

where K_{us} is a stability factor. Furthermore, Reference [15] pointed out that the desired yaw rate obtained from Eq. (13) can only show good performance when the road friction coefficient is sufficiently high. A low friction road can result in increasing sideslip angle even though the yaw rate tracks the desired value well. Thus, the lateral acceleration of vehicle cannot exceed the maximum friction coefficient. The desired yaw rate is limited as follows:

$$\text{abs}\left(\dot{\varphi}'_{des}(k)\right) = \min\left\{\text{abs}\,\frac{v_x(k) \cdot \delta_f(k)}{\left(l_f + l_r\right)\left(1 + K_{us} \cdot v_x^2(k)\right)}, \quad \frac{\mu \cdot g}{v_x(k)}\right\} \tag{14}$$

where the tire-road friction coefficient μ is estimated on the basis of Eq. (9), which is related to the angular velocity of wheel ω; g is the gravitational acceleration. The desired vehicle lateral dynamics can then be summarized as:

$$Y_{des}(k) = [A_1 \quad A_2]\begin{bmatrix} v_{y,des}(k) \\ \text{abs}\left(\dot{\varphi}'_{des}(k)\right) \cdot \text{sgn}\left(\dot{\varphi}'_{des}(k)\right) \end{bmatrix} \tag{15}$$

3.2   Control Law

3.2.1   Error Feedback Control

To generate an explicit relationship between the system output Y (vehicle lateral speed and yaw rate) and the control signal to the vehicle $\Delta\delta$ (compensative steering angle), it is necessary to differentiate the output Y based on discrete Eqs. (7) and (8):

$$
\begin{aligned}
Y(k+1) =& \left(A_2 \frac{C_{ar}(k)l_r - C_{af}(k)l_f}{I_z} - A_1 \frac{C_{af}(k)l_r - C_{ar}(k)}{m_v}\right) \tan\theta(k) \\
&+ \left(-A_2 \frac{C_{af}(k)l_f^2 - C_{ar}(k)l_r^2}{I_z \cdot v_x(k)} - A_1\left(v_x(k) + \frac{C_{af}(k)l_f - C_{ar}(k)l_r}{m_v v_x(k)}\right)\right)\dot{\varphi}(k) \\
&+ \left(\frac{C_{af}(k)l_f}{I_z} + \frac{C_{af}(k)}{m_v}\right)u(k)
\end{aligned}
\tag{16}
$$

Then, the tracking error is defined as

$$e(k) = Y_{des}(k) - Y(k) \tag{17}$$

To measure $Y(k)$ in practice, the real-time vehicle lateral speed v_y is measured via an accelerometer while the yaw rate $\dot{\varphi}$ is measured by using a gyroscope. To achieve closed-loop control stability, the target tracking error should be:

$$e(k+1) = -\lambda \cdot e(k) \tag{18}$$

where $0 < \lambda < 1$ is user-defined constant. Substituting Eqs. (16) and (18) into Eq. (17), the control law for the superposition angle to the steering angle can be designed as:

$$u*(k) = \frac{1}{\left(\frac{C_{af}(k)l_f}{I_z} + \frac{C_{af}(k)}{m_v}\right)} \left\{ \dot{Y}_{des}(k) + \left(A_1 \frac{C_{af}(k) - C_{ar}(k)}{m_v} - A_2 \frac{C_{ar}(k)l_r - C_{af}(k)l_f}{I_z} \right) \tan\theta(k) \right.$$
$$\left. \left(A_1 \left(v_x(k) + \frac{C_{af}(k)l_f - C_{ar}(k)l_r}{m_v v_x(k)} \right) + A_2 \frac{C_{af}(k)l_f^2 - C_{ar}(k)l_r^2}{I_z \cdot v_x(k)} \right) \dot{\varphi}(k) - \lambda e(k) \right\} \tag{19}$$

Theoretically, the compensative steering angle $\Delta\delta^*$ provided by AFS system to enhance the vehicle stability can be calculated as:

$$\Delta\delta*(k) = u*(k) - \delta_f(k) \tag{20}$$

However, the vehicle longitudinal speed v_x is an exogenous variable, and the cornering stiffness C_{af} and C_{ar} in Eq. (19) would be fast varying and uncertain that depend nonlinearly on tire slip angles respectively.

3.2.2 ELM-Based Control

To solve the above problems, this study employs ELM to rapidly approximate the unknown control signal $\Delta\delta^*$ in Eq. (20) and to take the multiple inputs into account. In ELM, a single-hidden-layer feedforward network is employed to approximate the target function, which can be described as:

$$f(x) = \sum_{i=1}^{m} \beta_i \cdot G(a_i, b_i, x) = h(x) \cdot \beta \tag{21}$$

Where n is the number of input nodes, m is the number of hidden nodes, $x = [x_1, \cdots, x_n]$ is the input vector, $G(\cdot)$ is the sigmoid function; $a_i = [a_{1i}, \cdots, a_{ni}]^T$ is a randomly chosen input weight vector; bi is a randomly chosen bias term; β_i is the output weight; $h(x) = [G(a_1, b_1, x), \cdots, G(a_m, b_m, x)]$ is the feature mapping vector; $\beta = [\beta_1, \cdots, \beta_m]^T$ is the output weight vector. Therefore, the control law using ELM can be described by:

$$\Delta\delta(k) = h(z(k)) \cdot \beta(k) \tag{22}$$

where $z(k) = \left(e(k), \theta(k), \dot{\varphi}(k), \delta_f(k), v(k), \omega(k) \right)$ is the input vector.

3.3 Adaptation Law

According to the theory of ELM, the feature mapping layer $h(\cdot)$ can be implemented by random hidden nodes, and only the output weights β needed to be adjusted to approximate the target system. So, there exist optimal output weights β^* such that the target system can be approximated with a very small bounded approximation error ς:

$$\Delta\delta^*(k) = h(z(k)) \cdot \beta^* + \varsigma \tag{23}$$

With the control law provided in Eq. (22), the tracking error of the system becomes:

$$e(k+1) = -\lambda \cdot e(k) - \frac{1}{\frac{C_{af}(k)l_f}{I_z} + \frac{C_{af}(k)}{m_v}} \cdot h(z(k)) \cdot \tilde{\beta}(k) + \varsigma \tag{24}$$

where $\tilde{\beta}(k) = \beta^* - \beta(k)$ represents the error of the output weights. Instead of using traditional learning method in ELM, an adaptive weight updating law is designed as:

$$\Delta\beta = \beta(k+1) - \beta(k) = -\eta \cdot h^{\mathrm{T}}(z(k)) \cdot (e(k+1) - \lambda \cdot e(k)) \tag{25}$$

where $0 < \eta < 1$ is the user-defined learning rate.

3.4 Control Scheme

Based on the proposed ELM-based adaptive control law, the overall scheme for vehicle yaw rate control is illustrated in Fig. 3, where ELM is employed to affine the compensative steering angle $\Delta\delta$ and input vector z as expressed in Eq. (22). The adaptation law given in Eq. (25) is used to update the output weight vector β, and the generated control signal $\Delta\delta$ is expected to drive the tracking error e back to zero.

4 Simulation and Discussion of Result

To demonstrate the effectiveness of the proposed adaptive ELM controller, pilot simulation was carried out using the co-simulation between a vehicle simulation software "CarSim" and the Simulink in MATLAB. A B-Class hatchback car was selected from CarSim to be the car model in the simulation, and a double lane change test was used for pilot evaluation. The simulation parameters are listed in Table 1. To evaluate the performance of the controller, the latest fuzzy-logic-based yaw rate controller proposed by Krishna et al. [4] was also implemented for comparison. Figure 4 shows the path of the double lane change test with low road friction coefficient. In the test, the vehicle speed is steadily increased from 60 to 110 km/h in 9 s. The left-hand of Fig. 5 illustrates the steering wheel angle produced by the driver and superposition angle produced by the proposed and fuzzy controllers. The right-hand of Fig. 5 shows that the proposed controller has a good tracking capability to ensure the yaw rate

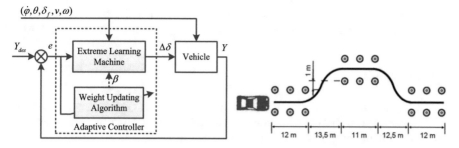

Fig. 3. Scheme of ELM-based adaptive controller

Fig. 4. Double lane change test

Table 1. Simulation parameters

A_1	A_2	K_{us}	l_f(mm)	l_r(mm)	m	λ	η	g (m/s^2)	μ
1	1	0.03	1040	1560	5	0.1	0.001	9.81	0.5

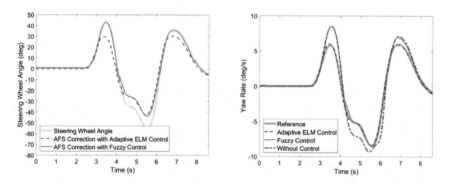

Fig. 5. Simulation result of double lane change test under different control systems

following the reference. In a nutshell, Fig. 5 shows the proposed controller is superior to the fuzzy control and the vehicle without AFS controller.

$$\text{MAR} = \frac{\sum_{r=1}^{P} |E_r|}{P} \tag{26}$$

where E_r is the tracking error; P is the number of samples. Table 2 shows that the proposed controller can reduce 74.4% of MAE as compared to the fuzzy control.

Table 2. Mean of absolute tracking error based on simulation time of 9 s

Control method	Adaptive ELM control	Fuzzy control
Yaw Rate (deg/s)	0.115	0.449

To assess the performance of the proposed controller and fuzzy control, Mean Absolute Error (MAE) is used to measure the tracking error for yaw rate as follows:

5 Conclusion

A new ELM-based adaptive control algorithm is proposed for vehicle yaw rate control. To validate the effectiveness of the proposed controller, simulation was carried out. The performance of the proposed controller was tested in a double lane change. A comparison among the proposed controller, an existing fuzzy logic controller and the vehicle without AFS controller was also carried out. Pilot simulation result indicates that the vehicle yaw stability can be improved with the proposed controller, and the proposed controller outperforms the fuzzy logic controller in terms of tracking ability. So it can be concluded that the proposed controller is effective to improve road safety for modern vehicles. Since this is a pilot study, more simulation and experimental studies as well as controller stability analysis should be carried out in the future.

Acknowledgment. This research is supported by the research grants of the University of Macau under grant numbers MYRG2016-00212-FST and MYRG2017-00135-FST.

References

1. Aripin, M.K., Yahaya, M.S., Danapalasingam, K.A., Peng, K., Hamzah, N., Ismail, M.F.: A review of active yaw control system for vehicle handling and stability enhancement. Int. J. Veh. Technol. **2014**, 1–15 (2014)
2. Guldner, J., Utkin, V., Ackermann, J., Bunte, T.: Sliding mode control for active steering of cars. IFAC Adv. Automot. Control **1**, 61–66 (2016)
3. Utkin, V. and Hoon, L.: Chattering Problem in Sliding Mode Control Systems. In: 2006 International Workshop on Variable Structure Systems, pp. 346–350. IEEE Press, Alghero (2006)
4. Krishna, S., Narayanan, S., Denis Ashok, S.: Fuzzy logic based yaw stability control for active front steering of a vehicle. J. Mech. Sci. Technol. **28**, 5169–5174 (2014)
5. Eski, İ., Temürlenk, A.: Design of neural network-based control systems for active steering system. Nonlinear Dyn. **73**, 1443–1454 (2013)
6. He, W., Chen, Y., Yin, Z.: Adaptive neural network control of an uncertain robot with full-state constraints. IEEE Trans. Cybern. **46**, 620–629 (2016)
7. Tsai, C.C., Huang, H.C., Lin, S.C.: Adaptive neural network control of a self-balancing two-wheeled scooter. IEEE Trans. Industr. Electron. **57**, 1420–1428 (2010)
8. Ge, S.S., Hang, C.C., Lee, T.H., Zhang, T.: Stable Adaptive Neural Network Control. Springer Science & Business Media (2013)
9. Huang, G.B., Zhu, Q.Y., Siew, C.K.: Extreme learning machine: theory and applications. Neurocomputing **70**, 489–501 (2006)
10. Huang, G.B., Dian, H.W., Yuan, L.: Extreme learning machines: a survey. Int. J. Mach. Learn. Cybernet. **2**, 107–122 (2011)
11. Wong, P.K., Wong, K.I., Vong, C.M., Cheung, C.S.: Modeling and optimization of biodiesel engine performance using kernel-based extreme learning machine and cuckoo search. Renew. Energy **74**, 640–647 (2015)

12. Alvarez, L., Yi, J.G.: Adaptive emergency braking control in automated highway systems. In: 38th IEEE Conference on Decision and Control, pp. 3740–3745. IEEE Press, Phoenix (1999)
13. Bakker, E., Pacejka, H.B., Lidner, L.: A New Tire Model with an Application in Vehicle Dynamics Studies. SAE Tech. Paper **98**, 83–95 (1989)
14. Esmailzadeh, E., Goodarzi, A., Vossoughi, G.R.: Optimal yaw moment control law for improved vehicle handling. Mechatronics **13**, 659–675 (2003)
15. Zheng, S.B., Tang, H.J., Han, Z.Z., Zhang, Y.: Controller design for vehicle stability enhancement. Control Eng. Pract. **14**, 1413–1421 (2006)

Sparse Representation Feature for Facial Expression Recognition

Caitong Yue[1], Jing Liang[1(⊠)], Boyang Qu[2], Zhuopei Lu[1],
Baolei Li[3], and Yuhong Han[4]

[1] School of Electrical Engineering, Zhengzhou University,
Zhengzhou, Henan, China
zzuyuecaitong@163.com, liangjing@zzu.edu.cn,
846350826@qq.com
[2] School of Electric and Information Engineering,
Zhongyuan University of Technology, Zhengzhou, Henan, China
qby1984@hotmail.com
[3] Physics & Electronic Engineering College,
Nanyang Normal University, Nanyang, China
bl_li@qq.com
[4] MOE Key Lab of Specially Functional Materials and Institute
of Optical Communication Materials,
South China University of Technology, Guangzhou, China
suthanyuhong@163.com

Abstract. Facial expression recognition is a challenging task, because it is difficult to recognize facial expressions of different persons if they are of diverse races and ages. Extracting distinctive feature from original facial image is a critical step for successful facial expression recognition. This paper proposes sparse representation feature for facial expression recognition. First of all, a dictionary is established using training images. Then sparse representation feature is extracted by sparse representation orthogonal matching pursuit method. Finally the extracted features of different expressions are classified by two-hidden-layer extreme learning machine. Facial expression images of both Cohn-Kanade and JAFFE databases are classified using sparse representation feature. Experimental results show that the sparse representation feature is suitable for facial expression recognition.

Keywords: Facial expression recognition · Sparse representation
Extreme learning machine · TELM

1 Introduction

Facial expression is one of the most immediate ways to express our inner feelings. With the help of facial expression recognition technology, computers or robots can understand the emotions of human beings. During the last few years, facial expression recognition has been widely applied in human-computer interaction, human-robot interaction and virtual reality [1]. According to the Facial Action Coding System (FACS) in [2], facial expressions are classified into six categories, which are anger,

J. Cao et al. (Eds.): ELM 2017, PALO 10, pp. 12–21, 2019.
https://doi.org/10.1007/978-3-030-01520-6_2

disgust, fear, happy, sad and surprise. It is still a challenge work to recognize all the six expressions with a high accuracy regardless of gender, race and age.

Feature extraction is a vital step in facial expression recognition. Many different types of features have been adopted in facial expression recognition. The commonly used facial features include geometric features [3] and appearance features [4]. Geometric features include the angles, lines or locations of facial components, such as mouths and eyes. Appearance features are extracted though image transformation, like Gabor-wavelet transformation, Fourier transformation or Principal Component Analysis (PCA). When extracting geometric features the accurate locations of facial components should be located, which is still a tough task. In addition, the robustness of geometric features is poor. In contrast, appearance features are easy to be extracted and they are relatively robust. Therefore appearance features are widely adopted by researchers. Three different kinds of appearance features based on Discrete Cosine Transform (DCT), Fast Fourier Transform (FFT), and signal value decomposition were employed in [5]. DCT and two-dimensional PCA were compared in [6], and the experimental results in [6] showed that DCT performed better than two-dimensional PCA in respect to recognition rate. Ucar [7] adopted mean, standard deviation and entropy of each region of images as features to classify the facial expression. However, as described in [8] the mentioned feature extraction methods have not made full use of the facial elements and the recognition results are not as good as we expect.

Sparse representation (SR) [9] aims to represent raw data with the linear combinations of atoms in a dictionary. In SR, if the dictionary is overcomplete the linear combination coefficients of atoms are sparse which means that most of the coefficients are zeros. The sparse coefficients contain main information of raw data so that they can be regard as a new kind of feature which is referred to as sparse representation feature in this paper.

In this paper, the training images compose a dictionary and the sparse representation features are obtained through a modified orthogonal matching pursuit (OMP) [10] method. Two-hidden-layer extreme learning machine (TELM) [11] is employed to classify the sparse representation features.

The contributions of this paper are listed as follows. A facial expression dictionary is established with training images. Sparse representation orthogonal matching pursuit method is designed to extract sparse representation features. The extracted features are classified by two-hidden-layer extreme learning machine. Comprehensive experiments are carried out to verify the advantages of sparse representation features in facial expression recognition.

The rest of this paper is organized as follows. Sparse representation feature is introduced in Sect. 2. Section 3 presents sparse representation orthogonal matching pursuit method. A general introduction to two-hidden-layer extreme learning machine is presented in Sect. 4. Experimental results are given in Sect. 5. Finally, conclusion and prospect are presented in Sect. 6.

2 Sparse Representation Feature

Sparse representation [9] has attracted an increasing research interest during the last few years. In sparse representation, signal y can be represented as linear combinations of atoms in an overcomplete-dictionary D, which is described in the following equation.

$$y = Dx \tag{1}$$

In (1), x is the linear combination coefficient. In general, D is redundant and x is sparse. The sparse coefficient x, most elements of which are equal to zero, is calculated according to:

$$x = \arg \min \|x\|_0 \quad \text{s.t.} \quad y = Dx \tag{2}$$

where $\|x\|_0$ represents the number of nonzero components in x.

Recently sparse representation has been adopted in classification in many different manners [12]. Xu [13] designed a discriminative sparse representation method for robust face recognition. A new objective function was given to enhance the distinctiveness of different classes. Then the face images were classified through collaborative representation classification [14]. Quan [15] embedded supervised learning processes into sparse representation and proposed a discriminative sparse coding algorithm.

In this paper, sparse representation coefficients are regarded as sparse representation feature. The general idea of sparse representation feature extraction is described as Fig. 1. Some facial images of different expressions are employed as atoms of dictionary D. Testing sample y is represented by the multiply of dictionary D and sparse represent feature x. The sparse represent feature x is obtained by sparse representation orthogonal matching pursuit method which is introduced in Sect. 3.

The reason why sparse representation coefficients can be employed as sparse representation feature is that the sparse representation coefficients reflect the relationship between testing sample y and atoms in the dictionary D. As to sparse representation coefficients vector x, only indexes corresponding to the relevant facial images

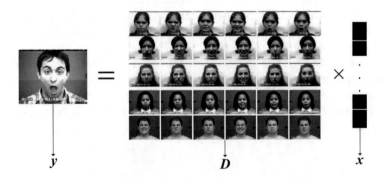

Fig. 1. The general idea of sparse representation feature extraction (y represents testing sample; D is dictionary; x is the sparse representation feature of y).

have nonzero values. Therefore, facial images of the same expression have similar nonzero indexes and values in their sparse representation coefficients vectors, while facial images of the different expression have distinct nonzero indexes and values. Therefore, sparse representation coefficients can be used as facial expression feature.

3   Sparse Representation Orthogonal Matching Pursuit

The orthogonal matching pursuit (OMP) [10] searches optimal solution to problem which is described in (2) in a greedy manner. The general idea of OMP is to gradually maintain the best atom in the measurement matrix and to achieve the sparse coefficient iteratively. In OMP, measurement matrix is often generated through Gaussian random method. In this paper the Gaussian random matrix is replaced by a dictionary D which is composed of training samples. This method can be used exclusively for sparse feature extraction. We refer it to as sparse representation orthogonal matching pursuit (SROMP). The pseudo code of SROMP is shown in **Algorithm 1**. Its task is to obtain sparse representation feature x of testing sample y.

Algorithm1 Sparse Representation Orthogonal Matching Pursuit (SROMP)

Task: Obtain sparse representation feature x according to:

$$x = \arg\min_x \|x\|_0 \text{ s.t. } Dx = y$$

Input: Dictionary D

   Testing sample y

Initialization: Feature $x^0 = 0$

   Residual $r^0 = y - Dx^0 = y$

   Support $S^0 = \varnothing$

   Iteration number $k = 0$

While $\|r\|_2 > \varepsilon$ **do**

   $k = k+1$;

   $z = D'r$

   $S^k = S^{k-1} \cup \{\arg\max_{j \in S^{k-1}} |z(j)|\}$

   $x = A'_{S^k} y$

   $r = y - A_{S^k} x$

End while

Output: Sparse representation feature x

The procedure of SROMP is described as follows. First, initialize feature x, residual r, support S, and iteration number k. Among them, residual r represents the error between testing sample y and Dx. The support S represents the nonzero elements' index in x.

Then, the best matching atom index is chosen according to: $I = \arg\max_{j \notin S^{k-1}} |z(j)|$, where I is the index of the best matching atom; $z = D'r$. The index I is added into support S. Subsequently, the feature x and residual r are updated. The matching atom indexes are found out iteratively until the residual is smaller than ε. ε is set to 1e−3 according to [10]. Finally, the sparse representation feature x is outputted.

In SROMP, the dictionary D is composed of training facial image. In this paper, each training facial image is transformed into one column $p_i = [p_{i1}, p_{i2}, p_{i3}, \ldots, p_{in}]'$ where i represents the ith training image; n is the number of pixels in each image. The mean of p_i is set to zero as following equation:

$$p_{i,j} = p_{i,j} - \frac{1}{n}\sum_{j=1}^{n} p_{i,j} \tag{3}$$

Then p_i is normalized based on l_2 norm according to:

$$p_{i,j} = \frac{p_{i,j}}{\sum\limits_{j=1}^{n} (p_{i,j})^2} \tag{4}$$

The dictionary D is composed of $p_i(i = 1: t)$:

$$D = [p_1; p_2; \ldots; p_t] \tag{5}$$

where t is the number of training images.

4   General Introduction to Two-Hidden-Layer Extreme Learning Machine

Extreme learning machine (ELM) is a feedforward neural network with only one hidden layer. The outstanding character of ELM is that the weights between the input layer and hidden layer together with the bias of hidden neurons are randomized. In addition, the weights between the output layer and the hidden layer are analytically determined though the least-squares method. ELM has been widely applied for classification and regression due to its good generalization performance and fast leaning speed. Based on the ELM, a two-hidden-layer ELM (TELM) [11] is designed by adding a second hidden layer. The structure of TELM is shown in Fig. 2.

TELM has two hidden layers aiming to enhance its classification and regression capabilities. The general procedure of TELM is described as follows. First, the connection weights between the input layer and the first hidden layer and the biases of the first hidden layer are randomly generated. Second, the weights between the second hidden layer and the output layer are estimated with the objective outputs and the outputs of the first hidden layer. Third, the outputs of the second hidden layer can be calculated. Finally the weights between the first hidden layer and the second hidden layer are determined.

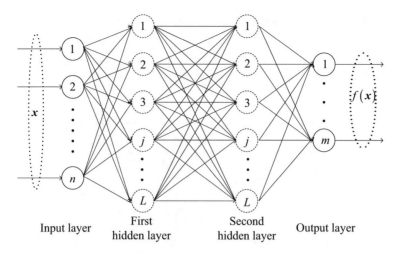

Fig. 2. The structure of TELM

TELM inherits the randomness of ELM to ensure its learning speed. The advantage of TELM is that it can deal with more complex classification and regression problems [11]. Therefore, TELM is employed to classify sparse representation features.

5   Experimental Results

The proposed sparse representation feature is tested on two facial expression databases, JAFFE database [16] and Cohn-Kanade database [17]. The JAFFE database includes 10 females' facial expression images. Each female has two to four images for one expression. In this paper, 180 images including six kinds of expressions (Anger, Disgust, Fear, Happy, Sadness and Surprise) are chosen. In the first part of the experiment, 120 images of them are randomly chosen as training images to construct the dictionary and the other 60 images act as testing images.

Cohn-Kanade database is composed of 2105 images of people whose ages range from 18-40. Approximately 65% of subjects are female and about 35% are male. There are also six different expressions in Cohn-Kanade. In this paper, 1500 images are chosen and 1200 images of them are randomly chosen as training images to establish dictionary and the left 300 images act as testing images in the first part of the experiment.

In the first part of the experiment, the recognition rate of sparse representation (SR) feature is compared with Principal Component Analysis (PCA) feature [18] and Wavelet feature [19]. PCA feature is extracted from more than 90% energy of original images. Wavelet feature is Wavelet coefficient of Wavelet decomposition with Daubechies filter type. All these features are classified by TELM. The number of hidden nodes is set to 2000 and 'Sigmoidal' is chosen as the activated function by experience. Each experiment is carried out 30 times. The means and standard deviations of recognition rates are shown in Table 1. In addition, the confusion matrixes of the three types of features on Cohn-Kanade dataset are shown in Tables 2, 3 and 4.

Table 1. The recognition rates with different features on two databases

	SR	PCA	Wavelet
Cohn-Kanade	**99.41** ± 4.89e−3	95.02 ± 1.11e−2	65.07 ± 3.61e−2
JAFFE	**96.43** ± 8.25e−3	70.30 ± 3.31e−2	75.67 ± 4.15e−2

Table 2. Confusion matrix of SR feature on Cohn dataset

	An* (%)	Di (%)	Fe (%)	Ha (%)	Sa (%)	Su (%)
An	**98.16**	0	0	0.08	0	0
Di	0.04	**100**	0	0.04	0	0
Fe	0	0	**100**	0	0	1.36
Ha	0.04	0	0	**99.84**	0.08	0
Sa	1.76	0	0	0.04	**99.84**	0
Su	0	0	0	0	0.08	**98.64**

*An (Anger), Di (Disgust), Fe (Fear), Ha (Happy), Sa (Sad), Su (Surprise).

Table 3. Confusion matrix of PCA feature on Cohn dataset

	An* (%)	Di (%)	Fe (%)	Ha (%)	Sa (%)	Su (%)
An	**91.72**	0	1.08	3.08	0.36	0.48
Di	0.76	**99.92**	0.64	1.60	1.20	0.36
Fe	1.84	0	**96.32**	1.84	1.04	2.76
Ha	1.52	0	0.84	**89.92**	0.16	0.40
Sa	3	0.04	0.44	2.64	**97.00**	0.76
Su	1.16	0.04	0.68	0.92	0.24	**95.24**

*An (Anger), Di (Disgust), Fe (Fear), Ha (Happy), Sa (Sad), Su (Surprise).

Table 4. Confusion matrix of Wavelet feature on Cohn dataset

	An* (%)	Di (%)	Fe (%)	Ha (%)	Sa (%)	Su (%)
An	**54.84**	4.6	4.96	7.08	5.28	5.20
Di	8.88	**72.44**	5.36	8.56	7.04	5.80
Fe	9.72	5.04	**72.28**	8.36	6.96	9.16
Ha	7.60	6.24	4.72	**57.48**	6.36	6.56
Sa	9.44	5.64	6.32	8.04	**67.04**	6.92
Su	9.52	6.04	6.36	10.48	7.32	**66.36**

*An (Anger), Di (Disgust), Fe (Fear), Ha (Happy), Sa (Sad), Su (Surprise).

In Table 1, the recognition rates with three different types of features on two databases are presented. The reconstruction rates with sparse representation (SR) feature are the highest on both databases, at 99.33% and 96.67% respectively. The performance of

PCA feature is better than Wavelet feature on Cohn-Kanade database. In contrast, Wavelet feature performs better than PCA feature on JAFFE database. In addition, the standard deviations of SR feature are the smallest on both databases which demonstrates that SR feature is more constant than the other two types of features.

In Table 2, 3 and 4, the numbers in bold represent the recognition accuracies in corresponding class. In Table 2, the accuracies for Disgust and Fear expressions are the highest, while the accuracy for Anger expression is the lowest. The reconstruction rate of Disgust expression is the highest in both Tables 3 and 4.

In the second part of the experiment, the relationship between recognition accuracy and the number of training samples are studied. 800 to 1300 training samples are randomly chosen from the facial expression images in Cohn-Kanade database. The experiments are carried out thirty times and the training samples are randomly chosen each time. The error bars of recognition accuracy are shown in Fig. 3. As shown, recognition accuracies of SR and PCA feature are constant while accuracy of Wavelet feature decreases when the number of training samples increases from 800 to 1100 and then increases when the number of training samples increases from 1100 to 1300. Overall, recognition accuracy of SR feature is the highest and the most constant among the three kinds of features.

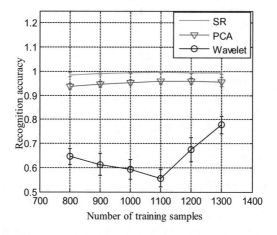

Fig. 3. The recognition accuracy with different number of training samples on Cohn-Kanade database.

6   Conclusion and Prospect

A novel kind of feature, sparse representation feature, is applied to facial expression recognition. This kind of feature aims to make full use of distinctive information in the facial images. Each training image acts as an atom in dictionary after zero-mean and normalization without adopting any extra dictionary learning strategies. Sparse representation coefficients are employed as feature of facial images. Experimental results show that sparse representation feature is effective and constant in facial expression classification.

Sparse representation feature will be applied to other pattern recognition problems in the future. In addition the compressed sparse representation coefficients will be adopted as a new kind of feature for classification in our future work.

Acknowledgements. We acknowledge financial support by National Natural Science Foundation of China (61473266, 61673404, 61305080, and U1304602), China Postdoctoral Science Foundation (No.2014M552013), Project supported by the Research Award Fund for Outstanding Young Teachers in Henan Provincial Institutions of Higher Education of China (2014GGJS-004) and Program for Science and Technology Innovation Talents in Universities of Henan Province in China (16HASTIT041).

References

1. Zhang, Y., Ji, Q.: Facial expression understanding in image sequences using dynamic and active visual information fusion. In: IEEE International Conference on Computer Vision, vol. 2, pp. 113–118 (2003)
2. Ekman, P., Friesen, W.V.: Constant across cultures in the face and emotion. J. Pers. Soc. Psychol. **17**(2), 124–129 (1971)
3. Tian, Y., Kanade, T., Cohn, J.: Handbook of Face Recognition. Springer (2005)
4. Valstar, M., Pantic, M.: Fully automatic facial action unit detection and temporal analysis. In: IEEE Conference on Computer Vision and Pattern Recognition Workshop, pp. 149–156 (2006)
5. Kharat, G.U., Dudul, S.V.: Human emotion recognition system using optimally designed SVM with different facial feature extraction techniques. World Sci. Eng. Acad. Soc. Trans. Comput. **7**(6), 650–659 (2008)
6. Jiang, B., Yang, G.S., Zhang, H.L.: Comparative study of dimension reduction and recognition algorithms of DCT and 2DPCA. In: International Conference on Machine Learning and Cybernetics, pp. 407–410, (2008)
7. Ucar, A., Demir, Y., Güzelis, C.: A new facial expression recognition based on curvelet transform and online sequential extreme learning machine initialized with spherical clustering. Neural Comput. Appl. **27**(1), 131–142 (2016)
8. Qayyum, H., Majid, M., Anwar, S. M.: Facial expression recognition using stationary wavelet transform features. Mathematical Problems in Engineering, 1–9, (2017)
9. Xu, Y., Zhang, Z., Lu, G.: Approximately symmetrical face images for image preprocessing in face recognition and sparse representation based classification. Pattern Recogn. **54**, 68–82 (2016)
10. Tropp, J., Gilbert, A.: Signal recovery from partial information via orthogonal matching pursuit. IEEE Trans. Inf. Theory **53**, 4655–4666 (2006)
11. Qu, B.Y., Lang, B.F., Liang, J.J.: Two-hidden-layer extreme learning machine for regression and classification. Neurocomputing **175**, 826–834 (2016)
12. Zhang, H.M., Wen, H.R., Zhang, X.L.: Sparse representation tracking based on compressed features. J. Zhengzhou Univ. (Eng. Sci.), **37**(3), 21–26 (2016)
13. Xu, Y., Zhong, Z., Yang, J.: A new discriminative sparse representation method for robust face recognition via L_2 regularization. IEEE Trans. Neural Netw. Learn. Syst. **99**, 1–10 (2016)
14. Jia, S., Shen, L., Li, Q.: Gabor feature-based collaborative representation for hyperspectral imagery classification. IEEE Trans. Geosci. Remote Sens. **53**(2), 1118–1129 (2015)

15. Quan, Y., Xu, Y., Sun, Y.: Sparse coding for classification via discrimination ensemble. In: IEEE Conference on Computer Vision and Pattern Recognition, pp. 5839–5847 (2016)
16. http://www.kasrl.org/jaffe_info.html
17. http://www.consortium.ri.cmu.edu/data/ck/
18. Kuncheva, L.I., Faithfull, W.J.: PCA feature extraction for change detection in multidimensional unlabeled data. IEEE Trans. Neural Netw. Learn. Syst. **25**(1), 69–80 (2014)
19. Huang, K., Aviyente, S.: Wavelet feature selection for image classification. IEEE Trans. Neural Netw. Learn. Syst. **17**(9), 1709–1720 (2008)

Protecting User Privacy in Mobile Environment Using ELM-UPP

Yanhui Li[✉], Ye Yuan, and Guoren Wang

School of Computer Science and Engineering,
Northeastern University, Shenyang 110004, Liaoning, China
lyhneu506822328@163.com

Abstract. In this paper, we address the topic of user privacy preservation in mobile environment. Existing techniques mostly rely on structure-based spatial cloaking, but pay little attention to location semantic information. Yet, such information may disclose sensitive information about mobile users. Thus, we propose ELM-UPP, a semantic-awareness privacy preservation framework to protect users privacy from violation. It allows mobile users to explicitly define their preferred privacy requirements in terms of location hiding measures. To provide location semantic protection for mobile users, in our framework, two features are firstly proposed to capture the semantic of locations. Then, a ELM-based unsupervised clustering is leveraged to detect semantic homogeneity locations. Besides, we design cloaking areas that should cover different semantic locations as well as achieving high quality of service. Since the problem of calculating the optimal cloaking area is NP-hard, we design a greedy algorithm that balances quality of service and privacy requirements. Extensive experimental results show the efficiency and effectiveness of our proposed algorithm.

Keywords: Location privacy · l-semantic diversity · k-anonymity

1 Introduction

With rapid advances in mobile communication technologies and continued price reduction of location tracking devices, recent years have witnessed the explosive growth of location-based services (LBS). Examples of these applications include location-based store finder, navigation services, and receiving traffic alerts or notifications. Though LBS offers everyday convenience and new business opportunities, it also presents new threats-the intrusion of location privacy [17].

In order to protect location privacy, previous research has been done using *location k-anonymity* and *location l-diversity*. The key of them is to enlarge the exact position of a mobile user to a cloaking area. These both techniques guarantee some degree of location privacy, nevertheless, they incur a serious drawback. The cloaking area could breach semantic information, which endangers the individuals' privacy. To be specific, the cloaking area could only include

© Springer Nature Switzerland AG 2019
J. Cao et al. (Eds.): ELM 2017, PALO 10, pp. 22–34, 2019.
https://doi.org/10.1007/978-3-030-01520-6_3

semantically similar locations even if it is mixed with other users and locations, and the attacker would be able to infer semantic meanings from the extended area. In other words, it cannot resist the types of *semantic homogeneity attacks*.

Example 1. Figure 1 shows an example of semantic homogeneity attacks. In the Figure, there are 13 mobile users $\{u_1, u_2, ..., u_{13}\}$ and 3 locations $\{S_1, S_2, S_3\}$. S_1 and S_2 are hospitals and S_3 is a club. Consider a scenario, u_1 issues for a LBS request through his GPS-enabled mobile phone with $k = 8$ and $l = 2$. Based on this privacy requirement, the rectangle, $CR_{(k,l)}$ may be a cloaking area. Unfortunately, it is easy for an attacker to infer that u_1 in the hospital, since S_1 and S_2 have the homogeneous semantics, namely hospital. Hence, the mobile user using the cloaking area $CR_{(k,l)}$ for LBS requests would be highly linked to hospitals and can be suspected of having treatment.

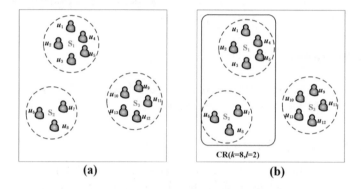

Fig. 1. An example of semantic homogeneity attacks

Although some studies have been done to resist semantic homogeneity attacks, they have different limitations [12]. To tackle the problem in Example 1, we propose *ELM-UPP*, a novel framework which integrates k-anonymity, l-semantic diversity and ELM-based unsupervised clustering, to protect user location privacy. In this framework, the identification of semantic homogeneity locations is a very important task. Generally, the locations in a trajectory are unlabeled. To find the semantic for a location, in this paper, we use two feature to learn location semantics from trajectory data. Further, to cluster similar locations, we focus on the extreme learning machine(ELM for short). It is a powerful clustering model, and it has many advantages such as extremely fast learning and good generalization capacity [6–10].

In ELM-UPP, our primary goal is to protect user privacy while guaranteeing the quality of the LBS (QoS) for snap-shot queries. Our strategy focuses on location k-anonymity and location l-semantic diversity to provide personalized privacy requirements. Besides, it also regards QoS as a critical measure for designing privacy preservation solutions. To achieve this goal, we design a greedy solution, which consists of pre-processing and online cloaking two steps. The pre-processing phase reduces the cloaking space by roughly grouping locations into different clusters, called buckets, while the online cloaking phase generates the desirable cloaking area in each bucket.

In designing our solution to the problem of privacy preservation for mobile users in mobile environment, we thus make a number of contributions, as follows: We develop an <u>ELM</u>-based mobile <u>user</u> <u>privacy</u> <u>preservation</u> (ELM-UPP) framework. In this framework, we first propose using two features to discover the semantic of a location. Then, we adopts ELM-based unsupervised clustering to identify semantic homogeneity locations. Finally, we design a greedy cloaking algorithm based on the identified location semantics. Extensive experiments on real datasets demonstrate the efficiency and effectiveness of our proposed algorithm in providing privacy guarantees.

The reminder of this paper is organized as follows. Section 2 surveys the related work. Section 3 presents the overview of ELM-UPP framework. The proposed algorithm is detailed in Sects. 4 and 5. Experimental evaluation is reported in Sect. 6. Finally, Sect. 7 concludes this paper.

2 Background

Our work relates to two main streams of research, concerning extreme learning machine and location privacy, respectively.

2.1 Brief Introduction to ELM

Extreme learning machine (ELM) was initially proposed for the single hidden-layer feedforward networks(SLFNs) [8–10], and then was extended to general SLFNs where the hidden layer need not be neuron alike [6,7]. It has many advantages such as extremely fast learning and good generalization capacity.

Given N distinct samples $(\boldsymbol{x}_i, \boldsymbol{y}_i)$ where $\boldsymbol{x}_i \in R^{n_i}$ and $\boldsymbol{y}_i \in R^{n_o}$, standard SLFNs with L hidden nodes and activation function g(x) are modeled as

$$\sum_{i=1}^{L} \boldsymbol{\beta}_i g_i(\boldsymbol{x}_j) = \sum_{i=1}^{L} \boldsymbol{\beta}_i \, g(\boldsymbol{w}_i \cdot \boldsymbol{x}_j + b_i) = \boldsymbol{o}_j, j = 1, ..., N \qquad (1)$$

$\boldsymbol{w}_i$ and b_i are the parameters of i-th hidden node, $\boldsymbol{o}_j$ is the output of the jth node, and $\boldsymbol{w}_i$ and $\boldsymbol{\beta}_i$ are the weight vectors connecting the ith hidden node with the input nodes and output nodes, respectively. Besides, b_i is the bias of ith hidden node. If a standard SLFN can approximate these samples with zero errors $\sum_{j=1}^{L} \|\boldsymbol{o}_j - \boldsymbol{t}_j\| = 0$, i.e., there exist $\boldsymbol{w}_i$, b_i and $\boldsymbol{\beta}_i$ satisfying that

$$\sum_{i=1}^{L} \boldsymbol{\beta}_i \, g(\boldsymbol{w}_i \boldsymbol{x}_j + b_i) = \boldsymbol{t}_j, j = 1, ..., n \qquad (2)$$

The above Eq. 2 can be rewritten in compactly as

$$\boldsymbol{H\beta} = \boldsymbol{Y} \qquad (3)$$

where $\boldsymbol{Y} = \left[\boldsymbol{y}_1^T, ..., \boldsymbol{y}_L^T\right]_{n_0 \times N}^T$, $\boldsymbol{\beta} = [\boldsymbol{\beta}_1^T, ..., \boldsymbol{\beta}_L^T]_{n_0 \times L}^T$ and

$$\mathbf{H} = \begin{bmatrix} h(\mathbf{x}_1) \\ \vdots \\ h(\mathbf{x}_N) \end{bmatrix} = \begin{bmatrix} g(\mathbf{w}_1 \cdot \mathbf{x}_1 + b_1) & \cdots & g(\mathbf{w}_L \cdot \mathbf{x}_1 + b_L) \\ \vdots & \cdots & \vdots \\ g(\mathbf{w}_1 \cdot \mathbf{x}_N + b_1) & \cdots & g(\mathbf{w}_L \cdot \mathbf{x}_N + b_L) \end{bmatrix}_{N \times L} \tag{4}$$

In ELM, the parameters $\boldsymbol{w}_i$ and b_i, are chosen randomly without knowing the training data sets. The output weight $\boldsymbol{\beta}$ is then calculated with matrix computation formula $\boldsymbol{\beta} = \boldsymbol{H}^\dagger \boldsymbol{Y}$, where $\boldsymbol{H}^\dagger$ is the *Moore-Penrose generalized inverse* of matrix $\boldsymbol{H}$. To avoid over-fitting, two parameters, $\mathbf{e}_i$ and C, are usually introduced. The first parameter $\mathbf{e}_i$ is the error vector regarding the ith training sample, and the second parameter C is a penalty coefficient on the training errors. In this case, the following equation is used to calculate β.

$$\min_{\beta \in \mathbf{R}^{L \times n_0}} L_{ELM} = \frac{1}{2}\|\beta\|^2 + \frac{C}{2}\|\boldsymbol{Y} - \boldsymbol{H}\beta\|^2 s.t. \ \boldsymbol{H}\beta = \boldsymbol{Y} - e \tag{5}$$

where $\|\cdot\|$ denotes the Euclidean norm and $e = \left[e_1^T, ..., e_L^T\right] \in \mathbf{R}^{N \times n_0}$. The above problem is widely known as the ridge regression or regularized least squares. By setting the gradient of L_{ELM} with respect to $\boldsymbol{\beta}$ to zero, we have

$$\nabla L_{ELM} = \boldsymbol{\beta} + C\boldsymbol{H}^T(\boldsymbol{Y} - \boldsymbol{H}\boldsymbol{\beta}) = 0 \tag{6}$$

If $\boldsymbol{H}$ has more rows than columns and is of full column rank, Eq. (7) is used to calculate $\boldsymbol{\beta}$; Otherwise, $\boldsymbol{\beta}$ is calculated by Eq. (8).

$$\boldsymbol{\beta}^* = (\boldsymbol{H}^T\boldsymbol{H} + \frac{\boldsymbol{I}_L}{C})^{-1}\boldsymbol{H}^T\boldsymbol{Y} \tag{7}$$

$$\boldsymbol{\beta}^* = \boldsymbol{H}^T\boldsymbol{\alpha}^* = \boldsymbol{H}^T(\boldsymbol{H}^T\boldsymbol{H} + \frac{\boldsymbol{I}_N}{C})^{-1}\boldsymbol{Y} \tag{8}$$

where $\boldsymbol{I}_L$ and $\boldsymbol{I}_N$ are the identity matrices of dimensions L and N, respectively.

2.2 Location Privacy

Location anonymization has widely used to protect user location privacy in the LBS. It mainly utilizes *location obfuscation* to hide the exact locations of mobile users. Examples include fake location [20], space transformation [3,16], mix-zones [15], and spatial cloaking. In these diverse anonymization techniques, spatial cloaking is the prominent. it enlarges an user's exact location to a cloaking area until some privacy requirements are satisfied, such as k-anonymity [18] and l-diversity [13].

Location k-anonymity. Gruteser et al. [4] first proposed location privacy technique based on the k-anonymity. After that, a series of studies has improved the computation of a cloaking area under k-anonymity. Casper [14] uses a quadtree-based pyramid data structure for efficient computation of the cloaking area, and CliqueCloak [2] locates a clique in a graph to compute the cloaking area. Probabilistic Cloaking proposed imprecise LBS requests which yield probabilistic results. The HilbertCloak algorithm [11] uses Hilbert space filling curve and its cloaking area is independent of mobile user distribution.

Location l-diversity. The cloaking area generated by location k-anonymity techniques may include only one meaningful location (e.g. a hospital or clinic), Thus, it discloses strong relationships between the mobile users and such a location. To tackle this problem, PrivacyGrid [1] proposed location l-diversity, which extends the cloaking area until at least l-1 different locations are included. Similarly, XSTAR [19] attempted to achieve the optimal balance between high query-processing cost and robust inference attack resilience while considering k-anonymity and l-diversity together. Several researches have identified semantic breach issues, but impractical assumptions were made to resolve these issues, as stated in [12]

3   ELM-UPP: An Overview

3.1   System Architecture

We adopts the classic centralized architecture for supporting anonymous information delivery in a mobile environment, as sketched in Fig. 2. In this architecture, the location anonymizer is a trusted entity that lies between mobile users and LBS service providers (SP). Specifically, the location anonymizer is responsible for (1)receiving the query and the exact position information from the mobile user; (2)anonymizing the location information according to the user's privacy requirements, and relaying it to the SP; (3)extracting the exact answers from the candidate results, and delivering the exact answer to the requester. Besides, the communication between mobile users and the location anonymization server is via establishing an authenticated and encrypted connection.

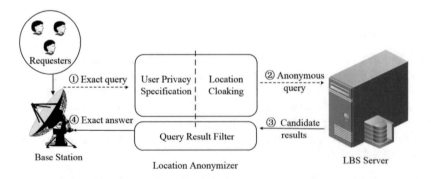

Fig. 2. System architecture

3.2   Privacy Model

Privacy requirement. In ELM-UPP, the privacy protection for mobile users are provided along two orthogonal dimensions: (1) location k-anonymity, which allows a mobile user to control her state of being not identifiable from a set of at least k-1 other users; (2) location l-semantic diversity, which advocates that it should be difficult to link a user with a specific semantic with high certainty. Furthermore, a mobile user may change his privacy preference level (k and l values) as often as required. Thus, such privacy requirements should be customizable and supported on a per query basis.

To guarantee the QoS, e.g., response time, a mobile user u should specify customized QoS requirement. In ELM-UPP, it provides two QoS measures, *maximum spatial tolerance* σ_s and *maximum temporal tolerance* σ_t, to specify this constraint. The parameter σ_s indicates the amount of spatial inaccuracy the user is willing to tolerate to maintain acceptable QoS. The parameter σ_t ensures that the temporal delay introduced for cloaking a request should be within an acceptable time interval. The set $(k, l, \sigma_s, \sigma_t)$ consists of u's service profile. By utilizing these two QoS quality metrics, ELM-UPP aims at devising a cloaking algorithm that will find the smallest possible cloaking area meeting desired privacy levels for each location anonymization request.

3.3   Evaluation Metrics and Algorithmic Framework

Anonymization Success Rate (ASR). The primary objective of the location cloaking algorithm is to maximize the number of messages perturbed successfully while maintaining their anonymization constraints. In this paper, we use the anonymization success rate to capture this goal. Specifically, it is defined as the fraction of messages cloaked successfully by an algorithm with respect to the set of received anonymization requests.

Relative Anonymity (Semantic) Levels. These metrics measures the achieved anonymity and semantic diversity levels for successfully cloaked messages. Intuitively, Relative Anonymity Level (RAL) is the ratio of anonymity level achieved by the location cloaking algorithm to the user specified anonymity level k, i.e., $\frac{k_c}{k}$ and Relative Semantic Level (RSL) provides a similar measure for location l-semantic diversity, i.e., $\frac{l_c}{l}$. k_c and l_c denote the actual values obtained by the location cloaking algorithm.

Cloaking Time (CT). This metrics is used to measure the runtime of the cloaking algorithm. Efficient cloaking implies that the location cloaking algorithm spends less processing time to perturb query requests.

Algorithmic Framework. To achieve privacy preservation, in the location anonymizer, the location cloaking algorithm needs to blur the exact location of each mobile user to a cloaking area that satisfies the user's anonymization requirement. To generate the cloaking area more effectively and efficiently, we devise a location cloaking algorithm consisting of two stages: an offline preprocessing phase and an online cloaking phase. In the offline pre-processing phase, we allocate all locations to different buckets, so that we can perform anonymization in one bucket rather than search the entire spatial space in the cloaking process. In the online cloaking phase, we locate the buckets of query user location and anonymize locations based on user privacy profiles.

4   Location Allocation

This section presents the offline pre-processing phase. It mainly has three steps including mining location semantics (detailed in Sect. 4.1), ELM-based similar locations clustering (detailed in Sect. 4.2), and allocating locations(detailed in Sect. 4.3).

4.1   Mining Location Semantic

To compute semantic-secure cloaking area, we must be aware of the semantics of locations in advance. From the perspective of location privacy, location semantics are interpreted as which type of services are provided at these locations. This interpretation makes sense, because what people want to be private in location privacy is what they did in a location. In the real scenario, most activities of a mobile user are usually performed at where the user stays. For example, we go to restaurants to have food, cinemas to see a film, or schools to attend classes. Due to these different activities, we spend a different amount of time on these locations. Motivated by these observations, we propose two quantitative features for extracting location semantics: staying duration and usage time.

Staying Duration. People spend different amount of time in a location depending on what they do there. The duration of the visiting is called *staying duration*. Intuitively, students usually stay more than six hours at school, whereas people usually take one or two hours to have a meal in a restaurant. Besides, restaurants themselves have different staying duration distributions according to what they actually serve. For instance, eating at a fine dining restaurant takes much longer time than at a fastfood restaurant. Thus, staying duration can captures visiting purposes of each location.

Using Time. Intuitively, each type of locations have their own active times. Obviously, most restaurants are full of customers during lunch or dinner time. In contrast, most bars are full during the night but quiet during the daytime. Inspired by these common sense, locations have differences in the distribution of usage times depending on services provided.

These feature data can be acquired via analyzing trajectory or survey data which contains such information. From the data, each feature's distribution is computed for each location. If two locations exhibit similar visiting patterns, they are possibly to have the same semantic. We thus compute the similarity of two locations based on the two features.

4.2   ELM-Based Location Clustering

In [5], ELMs are extended for both semi-supervised and unsupervised tasks. These two type of learning is built on the following assumptions: (1) all the data $\boldsymbol{X}$ are drawn from the same marginal distribution $\boldsymbol{P_X}$ and (2) if two points $\boldsymbol{x}_i$ and $\boldsymbol{x}_j$ are close to each other, then the conditional probabilities $P(\boldsymbol{y}|\boldsymbol{x}_i)$ and $P(\boldsymbol{y}|\boldsymbol{x}_j)$ should also be similar. Under the manifold regularization framework, it aims to minimize the following cost function:

$$L_m = \frac{1}{2}\sum_{i,j} w_{ij}\|P(\boldsymbol{y}|\boldsymbol{x}_i) - P(\boldsymbol{y}|\boldsymbol{x}_j)\|^2 \tag{9}$$

where w_{ij} is the pair-wise similarity between $\boldsymbol{x}_i$ and $\boldsymbol{x}_j$. It is usually calculated using Gaussian function $exp(\frac{-\|\boldsymbol{x}_i-\boldsymbol{x}_j\|^2}{2\sigma^2})$. Since it is difficult to compute the conditional probability, we use the Eq. (10) to approximate L_m. In this equation, $\hat{\boldsymbol{y}}_i$ and $\hat{\boldsymbol{y}}_j$ are the predictions with respect to $\boldsymbol{x}_i$ and $\boldsymbol{x}_j$, respectively. Besides, it can be simplified in a matrix form (Eq. (11)).

$$\hat{L}_m = \frac{1}{2}\sum_{i,j} w_{ij}\|\hat{\boldsymbol{y}}_i - \hat{\boldsymbol{y}}_j\|^2 \tag{10}$$

$$\hat{L}_m = Tr(\hat{\boldsymbol{Y}}^T\mathbf{L}\hat{\boldsymbol{Y}}) \tag{11}$$

where $Tr(\cdot)$ is the trace of a matrix. In unsupervised setting, the data are unlabeled. The formulation of US-ELM is reduced to $\min\limits_{\beta\in\mathbf{R}^{\mathbf{L}\times\mathbf{n_0}}} \|\boldsymbol{\beta}\|^2 + \lambda Tr(\boldsymbol{\beta}^T\boldsymbol{H}^T\boldsymbol{L}\boldsymbol{H}\boldsymbol{\beta})$. According to the conclusions in [5], if $L \leq N$, the solution to the output weights $\boldsymbol{\beta}$ is given by

$$\boldsymbol{\beta}^* = [\hat{\boldsymbol{v}}_2, \hat{\boldsymbol{v}}_3, ..., \hat{\boldsymbol{v}}_{n_0+1}] \tag{12}$$

where $\hat{\boldsymbol{v}}_i = \boldsymbol{v}_i/\|\boldsymbol{H}\boldsymbol{v}_i\|, i = 2, ..., n_0 + 1$ are the normalized eigenvectors. $\boldsymbol{v}_i$ and γ_i are the i-th smallest eigenvalues and the corresponding eigenvectors for the expression $(\boldsymbol{I}_L + \lambda\boldsymbol{H}^T\boldsymbol{L}\boldsymbol{H})\boldsymbol{v} = \gamma\boldsymbol{H}^T\boldsymbol{H}\boldsymbol{v}$. For the other case, the solution to the output weights $\boldsymbol{\beta}$ is given by

$$\boldsymbol{\beta}^* = \boldsymbol{H}^T[\hat{\boldsymbol{u}}_2, \hat{\boldsymbol{u}}_3, ..., \hat{\boldsymbol{u}}_{n_0+1}] \tag{13}$$

Similarly, $\boldsymbol{u}_i$ is the generalized eigenvectors of the expression $(\boldsymbol{I}_N + \lambda\boldsymbol{L}\boldsymbol{H}^T\boldsymbol{H})\boldsymbol{u} = \gamma\boldsymbol{H}\boldsymbol{H}^T\boldsymbol{u}$, and $\hat{\boldsymbol{u}}_i = \hat{\boldsymbol{u}}_i/\|\mathbf{H}\mathbf{H}^T\hat{\mathbf{u}}_i\|, i = 2, ..., n_0 + 1$ are the normalized eigenvectors.

Based on semantic features, we utilize ELM-based unsupervised learning to cluster locations. The specific algorithm of ELM-based unsupervised clustering is omitted here.

4.3   Location Allocation

In this step, we discuss how to allocate locations into different buckets. The location cloaking algorithm aims at providing location privacy for mobile users.

Meanwhile, it should not deteriorate the QoS. This problem is well-studied in our previous work [12]. To tackle this issue, a greedy solution *EIRank* is proposed. The main idea is that cloaking adjacent locations with different semantic labels provides a compact structure and semantic preference simultaneously. In other words, it prefers cloaking the locations exhibiting *structure similarity* and *semantic dissimilarity*. Generally, it consists of four steps: EI network construction, label clustering, augmented EI Network construction and segment allocation.

Applying EIRank to allocate the locations, it requires some subtle changes. Firstly, in the EI network construction step: each *e-node* represents a location, and two e-nodes are adjacent if their corresponding locations are adjacent. Secondly and more importantly, the dissimilarity metric used in the label clustering is *earth mover's distance*. Once clustered locations are obtained, the semantic of each cluster is updated by computing the average of the distributions in the cluster. By this mean, the semantic of locations in each cluster can be captured more accuracy. In this sense, to evaluate the dissimilarity of location semantics, we need to measure the distances between two distributions. The earth mover's distance is a reasonable choice.

Earth Mover's Distance (EMD). The EMD is based on the minimal cost that must be paid to transform one distribution into the other by moving distribution mass between each other. The cost is the amount of mass moved times the distance by which it is moved. It can be formally defined using the well-studied transportation problem. Let $P = (p_1, p_2, ..., p_m)$, $Q = (q_1, q_2, ..., q_m)$, and d_{ij} and f_{ij} be the ground distance and the flow of mass from element i of P to element j of Q, respectively. Then, the workload, needed to make two distribution the same, is defined as $WORK(P, Q, F) = \sum_{i=1}^{m} \sum_{j=1}^{m} f_{ij} d_{ij}$.

In this expression, $f_{ij} \geq 0$, $1 \leq i, j \leq m$, $p_i - \sum_{i=1}^{m} f_{ij} + \sum_{i=1}^{m} f_{ji} = q_i$, and $\sum_{i=1}^{m} \sum_{j=1}^{m} f_{ij} d_{ij} = \sum_{i=1}^{m} p_i = \sum_{j=1}^{m} q_j = 1$. From this setting, EMD is used to measure the minimum workload, defined as $D_{EMD}(P, Q) = \min_F WORK(P, Q, F)$. The advantage of EMD is in adjusting ground distance which enables us to capture semantic differences. In our approach, the ground distance (d_{ij}) is set to be the normalized difference for two quantitative features (staying duration, using time), i.e., the difference of staying duration or using time divided by the maximum difference. Thus, $0 \leq d_{ij} \leq 1$ for all i and j, which results in $D_{EMD}(P, Q) \in [0, 1]$. In this sense, it can capture (1 hour, 2 hours) pair has a much smaller difference than pair (1 hour, 8 hours) has. Therefore, EMD is a good choice for measuring the semantic difference between locations.

In summary, after deriving the dissimilarity of location semantics, we next perform label clustering and the remaining steps of EIRank to achieve location allocations. The details about EIRank please refer to [12].

5   Online Cloaking Phase

In this phase, the location cloaking algorithm generate a cloaking area according to a user's online request. Generally, it follows the online cloaking framework of EIRank. The only difference lies in the final processing procedure. Specially, it selects neighbor location with largest count to the cloaking area if the cloaking area generated by location l-semantic diversity cannot provide adequate k-anonymity protection. Due to space limit, please refer to the paper [12] for detailed online cloaking. In this paper, to fast and efficient computation of mobile user counts belonging to a location, we introduce the fundamental data structure for location anonymizer.

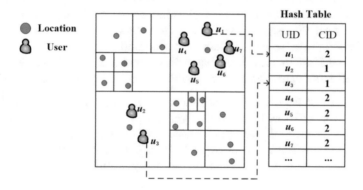

Fig. 3. The location grid

Data Structure. Figure 3 depicts the data structure for the location anonymizer in the ELM-UPP framework. The main idea is to employ a grid-based data structure that recursively decomposes the spatial space into many cells. During the decomposition, we stop decomposing a cell if it only contains one location. Thus, this data structure is named as *location grid.* Each cell in the location grid is represented as (Cid, N) where Cid is the cell identifier while N is the number of mobile users within the cell boundaries. This structure is dynamically maintained to keep track of the current number of mobile users within each cell. In addition to the mapping of each cell to its current count, we also use a hash table HT to keep track of the current location of mobile users. Each entry in HT for each mobile user is the form of (Uid, Cid), where Uid is the mobile user identifier, and Cid is the cell identifier in which the mobile user is located. This structure allows for fast and efficient computation of mobile user counts belonging to a particular cell of grid.

6   Experimental Evaluation

General Settings. Our experiments are based on two real road network datasets: California and Oldenburg road networks. In the experiments, we pick

2000 query points by random from the positions of trajectories. The more detailed description of experiment setting is shown in paper [12]. Besides, we compare our proposed algorithm with XSTAR [19].

Experimental Results. In our experiments, we first examine the effectiveness and efficiency of the proposed location anonymization algorithms. Then, we evaluate the QoS. Since the influence of location k-anonymity on the algorithm performance is similar to that of location l-semantic diversity, we just report the result of location l-semantic diversity.

Effectiveness. In Fig. 4, it can be seen that as l increases, ASR of ELM-UPP decreases sightly and that of XSTAR decreases significantly. Besides, RSL of XSTAR almost remains unchanged and that of ELM-UPP increases. This is because the semantic number of a cloaking area exactly equals to the user-defined semantic diversity for XSTAR. To resist reverse engineering attacks, the lastest cloaking area of each bucket contains more than l semantics for ELM-UPP. In this Figure, it also show that ASL increases as l varies. For a particular k, there will more users involved in cloaking area, and hence, ASL increases.

Efficiency. In Fig. 5(a), we observe that with the increase of semantic diversity l, the time cost of ELM-UPP drops significantly and the time cost of XSTAR increases dramatically. It also can be seen that the cloaking time cost of ELM-UPP is always less than that of XSTAR. This is because, a larger l results in a relative large cloaking area, which increases the possibility to search constructed

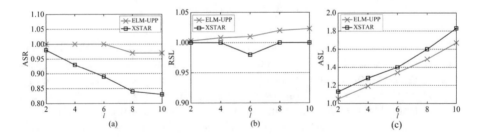

Fig. 4. The effectiveness of the cloaking algorithms

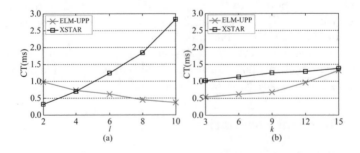

Fig. 5. The efficiency of the cloaking algorithms

area. Thus, it greatly decreases the cloaking time. In contrast, each cloaking area of XSTAR is generated completely dependently. With the increase of l, XSTAR needs to search more region to achieve the cloaking area, which increases the cloaking time.

Figure 5(b) shows the impact of varying k value on the cloaking time for the two algorithms. From the figure, we can see that, the larger k, the more cloaking time. This phenomenon is reasonable. This is intuitive as a larger k imposes a stronger constraint on the mobile users in the cloaking area, thus increasing cloaking time.

Query Processing Cost. Figure 6(a) illustrates the query time of the two algorithms with different values of semantic diversity. Obviously, the query time all increases as the semantic diversity increases. It needs a larger cloaking area for a larger l, and hence the query time increases. Figure 6(b) and (c) show the effect of varying k and k_1 on the query time, respectively. From the Figures, it can be seen that with their increasing, the query costs of two algorithms increase. These phenomena are explained by the following facts: (1) a larger k imposes a stronger constraint on k-anonymity, leading a larger cloaking area. The larger cloaking area, the more query processing time. (2) With k_1 increases, it means for the same cloaking area, we need to search more PoIs. Therefore, it increases the query time.

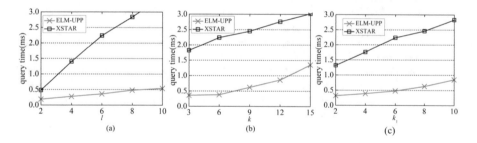

Fig. 6. Query time vs parameters l, k and k_1.

7   Conclusion

Protecting user privacy is a fundamental problem in mobile environment and has attracted intensive interests. However, most of these methods ignore the location semantics information. In this paper, we propose a novel ELM-based user privacy protection framework, which integrates location k-anonymity and l-semantic diversity. In this framework, we propose using two features to capture the semantic of locations. Then, we design a ELM-based unsupervised clustering to detect semantic homogeneity locations. Finally, We design cloaking areas that should cover different semantic locations as well as achieving high quality of service. Extensive experiments on several real datasets verify the efficiency and effectiveness of our proposed algorithm.

References

1. Bamba, B., Liu, L., Pesti, P., Wang, T.: Supporting anonymous location queries in mobile environments with privacygrid. In: Proceedings of WWW, pp. 237–246 (2008)
2. Gedik, B., Liu, L.: Location privacy in mobile systems: a personalized anonymization model. In: Proceedings of ICDCS, pp. 620–629 (2005)
3. Ghinita, G., Kalnis, P., Khoshgozaran, A., Shahabi, C., Tan, K.L.: Private queries in location based services: anonymizers are not necessary. In: Proceedings of SIGMOD, pp. 121–132 (2008)
4. Gruteser, M., Grunwald, D.: Anonymous usage of location-based services through spatial and temporal cloaking. In: Proceedings of Mobile Systems, Applications and Services, pp. 31–42 (2003)
5. Huang, G., Song, S., Gupta, J.N., Wu, C.: Semi-supervised and unsupervised extreme learning machines. Trans. Cybern. 44(12), 2405–2417 (2014)
6. Huang, G.B., Chen, L.: Convex incremental extreme learning machine. Neurocomputing 70(16), 3056–3062 (2007)
7. Huang, G.B., Chen, L.: Enhanced random search based incremental extreme learning machine. Neurocomputing 71(16), 3460–3468 (2008)
8. Huang, G.B., Chen, L., Siew, C.K.: Universal approximation using incremental constructive feedforward networks with random hidden nodes. TNN J. 17(4), 879–892 (2006)
9. Huang, G.B., Zhu, Q.Y., Siew, C.K.: Extreme learning machine: a new learning scheme of feedforward neural networks. Proceedings of International Joint Conference on Neural Networks (IJCNN), vol. 2, pp. 985–990 (2004)
10. Huang, G.B., Zhu, Q.Y., Siew, C.K.: Extreme learning machine: theory and applications. Neurocomputing 70(1), 489–501 (2006)
11. Kalnis, P., Ghinita, G., Mouratidis, K., Papadias, D.: Preventing location-based identity inference in anonymous spatial queries. TKDE J. 19(12), 1719–1733 (2007)
12. Li, Y., Yuan, Y., Wang, G., Chen, L., Li, J.: Semantic-aware location privacy preservation on road networks. In: Proceedings of DASFAA, pp. 314–331 (2016)
13. Machanavajjhala, A., Kifer, D., Gehrke, J.: l-diversity: privacy beyond k-anonymity. TKDD J. 1(1), 3 (2007)
14. Mokbel, M.F., Chow, C.Y., Aref, W.G.: The new casper: query processing for location services without compromising privacy. In: Proceedings of VLDB, pp. 763–774 (2006)
15. Palanisamy, B., Liu, L.: Mobimix: Protecting location privacy with mix-zones over road networks. In: ICDE, pp. 494–505 (2011)
16. Papadopoulos, S., Bakiras, S., Papadias, D.: Nearest neighbor search with strong location privacy. Proc. VLDB 3(1–2), 619–629 (2010)
17. Pedreschi, D., Bonchi, F., Turini, F., Verykios, V.S., Atzori, M., Malin, B., Moelans, B., Saygin, Y.: Privacy protection: regulations and technologies, opportunities and threats. In: Mobility, Data Mining and Privacy, pp. 101–119. Springer (2008)
18. Sweeney, L.: k-anonymity: a model for protecting privacy. UFKBS J. 10(05), 557–570 (2002)
19. Wang, T., Liu, L.: Privacy-aware mobile services over road networks. Proc. VLDB 2(1), 1042–1053 (2009)
20. Yiu, M.L., Jensen, C.S., Huang, X., Lu, H.: Spacetwist: managing the trade-offs among location privacy, query performance, and query accuracy in mobile services. In: Proceedings of ICDE, pp. 366–375 (2008)

Application Study of Extreme Learning Machine in Image Edge Extraction

Xiaoyi Yang[1,2,3(✉)], Xinli Deng[1,2,3(✉)], and Lei Shi[1,2,3(✉)]

[1] School of Education, Chongqing Normal University,
Chongqing 401331, China
495572939@qq.com, 516916990@qq.com, 475794398@qq.com
[2] Chongqing Radio and TV University, Chongqing 400052, China
[3] School of Software Technology, Zhejiang University,
Hangzhou 310058, China

Abstract. Aimed at the puzzle of the edge discontinuity, over segmentation, slow operation speed, and being difficult to select the best parameters existed in image edge detection algorithm caused by noise sensitivity, in order to get better recognition effect and faster execution speed, the paper presented an algorithm of image geometric feature recognition based on extreme learning machine. In order to facilitate the identification of the geometric features of the image, the algorithm first calculates the optimal threshold by means of algorithm model based on ELM (Extreme Learning Machine), in which, the optimal threshold could farthest separate the foreground and background color, then the threshold is used to limit the path cost function so as to narrow the search scope and improve the speed of algorithm execution. A large number of simulation experiments demonstrated that the presented algorithm could obtain a good effect on image processing. The research results show the rationality and validity of the algorithm based on ELM for image geometric feature recognition.

Keywords: Image edge extraction · Geometric feature recognition
Extreme learning machine · Optimal threshold · Path cost function

1 Introduction

Images often have obvious structural features, such as foreground color, background color, contour, edge and so on. In the digital image processing, if these characteristics are fully considered, not only the processing time of can be greatly reduced, but also the ideal processing results can be obtained [1]. At present, there are many methods for image edge extraction [2], commonly used methods of the image edge extraction are the image edge extraction based on fuzzy morphology, the image edge extraction of watershed algorithm and the image edge extraction based on combining the above both methods [3]. However, most of the algorithms are sensitive to noise, easily lead to over segmentation phenomenon, prone to a large number of scattered areas or edge dis- continuities, false edges, over segmentation and so on [4]. Edge extraction algorithm based on morphology and watershed can achieve better image edge extraction effect, however, it is difficult to select the optimal parameters, and in which, there is the

© Springer Nature Switzerland AG 2019
J. Cao et al. (Eds.): ELM 2017, PALO 10, pp. 35–45, 2019.
https://doi.org/10.1007/978-3-030-01520-6_4

puzzles such as the large amount of calculation, the slow execution speed and so on. The algorithm based on ELM presented by Huang [5] can better solve the puzzle of the optimal threshold in edge extraction algorithm based on morphology and watershed, ELM algorithm has many advantages that it has a strong learning ability and can approach the complex nonlinear function, and at the same time, it is fast in learning speed, good in generalization performance, strong in robustness and controllability.

2 Algorithm Model Based on ELM

2.1 Structure Model for ELM

Extreme learning machine is a single hidden layer feedforward neural network, and its typical structure is shown in Fig. 1. The network consists of input layer, hidden layer and output layer, and it is the full connection among the input layer and the hidden layer as well as the hidden layer and the output layer neuron [6, 7]. The weights between the input layer and the hidden layer can be generated randomly, and the algorithm has good generalization performance, relatively fast learning speed and simple network structure [8].

Extreme learning machine can be described as the following.

Input of ELM

- Data, x_i is the sample data, t_j is a category label, N is the number of samples, and it shows as in Eq. (1).

$$\{(x_i, t_j)\}_{j=1}^{N} \subset R^n \times R^m \tag{1}$$

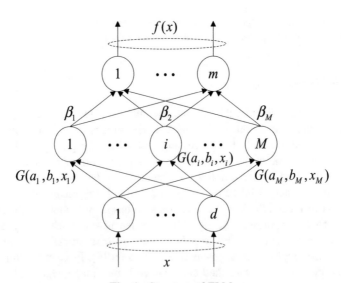

Fig. 1. Structure of ELM.

- The number of hidden layer nodes is M.
- Activation function $G(a_j, b_j, x)$, in which, a_j, b_j is respectively the connection weight values and offset value between the input layer and the hidden layer, and usually the activation function can also select as Sigmoid function.

Output of ELM

The weight matrix between hidden layer and output layer is $\beta_{L \times M}$

2.2 Calculating Steps of Algorithm Based on ELM

The calculating steps are as follows.

Step 1. For all training samples, computing the output matrix H of the hidden layer node, shown as in Eq. (2).

$$H = \begin{bmatrix} G(a_1, b_1, x_1) & \cdots & G(a_L, b_L, x_1) \\ \vdots & \cdots & \vdots \\ G(a_1, b_1, x_N) & \cdots & G(a_L, b_L, x_N) \end{bmatrix} \tag{2}$$

Step 2. The optimal output weight matrix H of hidden layer nodes can be obtained by least square method, shown as in Eq. (3).

$$\|H\hat{\beta} - Y\| = \min_{\beta} \|H\beta - Y\| \tag{3}$$

If the number of N samples is greater than or equal to the number L of hidden layer nodes, then it calculates the Eq. (4).

$$\hat{\beta} = \left(\frac{1}{C} + H^T H\right)^{-1} H^T Y \tag{4}$$

Otherwise, it can be calculated by the Eq. (5).

$$\hat{\beta} = H^T \left(\frac{1}{C} + HH^T\right)^{-1} Y \tag{5}$$

In Eq. (8),

$$Y = \begin{bmatrix} y_1^T \\ \vdots \\ y_N^T \end{bmatrix}_{N \times M}$$

The following makes some explanation for calculating process.

① The theorem has proved that as long as the number L of single hidden layer nodes is much enough, and activation function $G(a_j, b_j, x)$ can achieve infinite differentiability in any interval, then the parameters of the network at this time do not need to be adjusted.

② The optimal output weight matrix of hidden layer nodes has the following characteristics.

- (a) The estimation values of β can makes the algorithm get the minimum training error calculated by means of the least squares method.
- (b) The estimation values of β is the minimal paradigm, therefore, the ELM network has the best generalization ability.
- (c) The estimation values of β has the uniqueness, so the output of the algorithm is a global optimal solution, and not a local optimal solution.

In the given case of input weight w_j and threshold b_j, the extreme learning machine is the same as that of least square solution β of the linear system $H\beta = T$, and that is the following relationship was established.

$$\|H(w_1, \cdots, w_L, b_1, \cdots, b_L)\hat{\beta} - T\| = \min_{\beta}\|H(w_1, \cdots, w_L, b_1, \cdots, b_L)\beta - T\|$$

If the number L of hidden layer nodes is the same as the number N of learning samples $(L = N)$, then the ELM can ideally fit the learning samples. But in practice, it is not a completely zero error approximation, but it is only a nonlinear mapping. There does not necessarily exist $w_j, b_j, \beta_j(i - 1, 2, \cdots, N)$ to make the equation $H\beta = T$ to be set up. The ELM is the minimum training error and the minimum output weight norm, and that is under the condition: *Minimize* : $\|\beta\|$

The presented solution $\hat{\beta} = H^+T$ is as below.

In which, H^+ is the Moore Penrose generalized inverse of hidden layer output matrix H. When $L<N$, $H^+ = (H^TH)^{-1}H^T$

Therefore, in the course of application, the learning machine just sets the number of hidden layer nodes and selects the activation function, without artificial interference in the algorithm implementation process, so it is fast in learning speed, and at the same time, it has the good generalization ability and the advantage of the global optimal solution. Compared with the general EEG analysis, the EEG signals usually need to achieve high recognition rate and low power consumption in the process of analysis and processing, and it is difficult to find a balance between these two algorithms by commonly used recognition algorithm. The advantage of fast learning speed and strong generalization ability of the extreme learning machine can meet the two requirements of EEG data analysis and processing, and therefore, this paper tries to adopt the algorithm based on ELM to extract the edge of the image.

3 Application of ELM in Image Edge Extraction

Based on the gray level features of the image and the ELM algorithm, the optimal segmentation threshold can be calculated so as to separate the foreground image and the background image, then the optimal threshold is used to limit the path cost function in the IFT watershed algorithm, and thus, the foreground and background of the image can be segmented to the greatest extent. The goal of the algorithm based on ELM is to improve the execution speed of the IFT watershed algorithm, and its basic means is to use the optimal threshold to reduce the optimal search path of the original IFT watershed algorithm. The experimental results show that the image processed by this algorithm, it is hardly to appear the phenomenon of over segmentation and false edge, the result of image recognition is softer, and the edge of image is more smooth and continuous.

The optimal threshold that separates the foreground image from the background image can be found by means of the algorithm based on ELM, and the processing results of the original IFT watershed algorithm and the algorithm all are treated as a graph.

Assuming that the gray level t is viewed as the threshold, the image is divided into two types of background and foreground, it is respectively represented as the background class by S1 and the foreground class by S2. The point of the pixel value being not greater than gray level t is the point in the background class, the point of the pixel value being greater than gray level t is the point in the foreground class.

Defines the center of class S_1 to be as in Eq. (6).

$$\omega_1 = \sum_{i=0}^{t} p_i/P_1 \tag{6}$$

Defines the center of class S_2 to be as in Eq. (7).

$$\omega_2 = \sum_{i=t+1}^{L-1} p_i/P_2 \tag{7}$$

In Eqs. (6) and (7), P_1 and P_2 respectively shows the occurrence probability of S_1 and S_2, and $P_1 + P_2 = 1$.

Defines d_1 is the internal spur of class S_1, d_2 is the internal spur of class S_2, and d_1, d_2 is respectively given by Eqs. (8) and (9).

$$\omega_2 = \sum_{i=t+1}^{L-1} p_i/P_2 \tag{8}$$

$$d_2 = \sum_{i=t+1}^{L-1} |i - \omega_2| \frac{p_i}{P_2} \tag{9}$$

Defines D to be the distance between class S_1 and class S_2, shown as in Eq. (10).

$$D = |\omega_1 - \omega_2| \tag{10}$$

From Eq. (10), it can be seen that the smaller the value d_1 and d_2, the smaller the expressed pixel distance between foreground class and background class is, namely the better the performance of the class cohesion. The bigger the D, the bigger the expressed pixel distance between foreground class and background class is, namely the better the classification effect also is. Therefore, the function of classification discriminant can be defined as Eq. (11).

$$H(T) = \frac{P_1 \cdot P_2 \cdot D}{P_1 \cdot d_1 + P_2 \cdot d_2} \tag{11}$$

From Eq. (11) it can be seen that the bigger the $H(t)$, the bigger the expressed distance between S_1 class S_2 and classis, and also the better the separation effect between the foreground and the background is. If a gray level T of the image can be found, and it makes $H(T) = H_{max}(t)$, then T must be as the optimal segmentation threshold. At this time, it continues to have all pixels with a gray value greater than T in the image, namely those belong to all points in the foreground class, and it filters all gray value that is less than T pixels, namely it filters the points in the background.

If f shows the maximum arc path cost function, then it would assign a path cost value $f(\pi)$ to each path, and the maximum of the path cost is ∞. The cost of the path is generally related to the local properties of the image I, the commonly used path cost functions are additive path cost function and maximum arc path function, and here, due to the limitation of space, this paper only describes the maximum arc path function simply, Eqs. (12) and (13) shows the path cost function in the algorithm.

$$f_{max}(<t>) = I(t) \tag{12}$$

$$f_{max}(\pi \cdot <s, t>) = max\{f_{max}(\pi), I(t)\} \tag{13}$$

In Eqs. (12) and (13), $(s, t) \in A$, A is the adjacency relation of each pixel position in the image, s is the end point of the path π, $h(t)$ is the start point, is the initial cost of any point being as the initial point path, and the $I(t)$ is the pixel value of the pixel t.

4   Experimental Simulation Results and Its Analysis

4.1   Experimental Simulation

The flowchart structure of ELM algorithm, which is used for determining the optimal segmentation threshold to separate the foreground image and the background image, is shown in Fig. 2.

Assuming that the gray value in the image has N levels, and in order to better describe the algorithm based on ELM image segmentation threshold, it is necessary to be explained accordingly for the various data structures used in the algorithm. In which, the symbol

I represents the original image that needs to be processed.

L represents the template used in the algorithm.

C represents the value of each node, and the initial value of C is set as infinite.

Q represents the priority queue.

N represents the connection queue between pixels, where $N(p)$ represents the collection of all pixels connected to the pixel p.

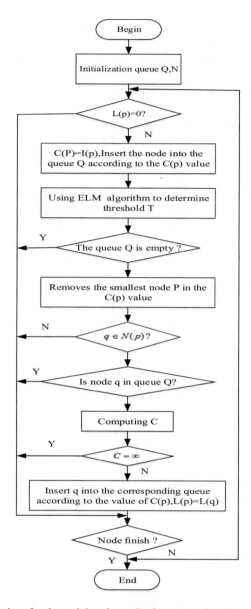

Fig. 2. Flowchart for determining the optimal segmentation threshold of image.

Algorithm analysis

① Reduced the storage space of the algorithm. In the improved algorithm, the queue structure is managed hierarchically because of the limitations of the threshold value. If the image contains N level grayscale, the number of bucket in queue Q can be reduced to $N - t + 1$, the corresponding storage

space becomes $O(n + N - t + 1)$, so this greatly reduces the storage space
of the algorithm.

② Reduced the seed set, reduced from all nodes to the nodes that belong to the
target.

③ The constraint condition of path cost function f_{new} is adjusted based on the
threshold.

④ The efficiency of the algorithm is improved. In the original algorithm, all
nodes need to be searched. In the new algorithm, because the threshold limit
is increased, only the target node can be searched, and so the execution
efficiency of the algorithm is improved.

In order to verify the effectiveness and usability of the algorithm based on ELM and
to get a better comparative recognition effect, the paper takes Matlab 2009 as the
simulation platform. Firstly, the pretreatment of the original image is made by illumination uniformity, and then the optimal threshold value of the image is computed by
the model algorithm based on ELM, and then the image edge is extracted by means of
the optimal threshold. For convenient to compare the processing results of different
algorithm, this paper selects 6 sorts of classical image edge detection algorithms to
detect the edge geometrical feature of the image, and the results of different algorithms
are shown in Fig. 3. Figure 4 is the contour edge of the image detected by the algorithm proposed in this paper, in which, the optimal threshold of the fuzzy morphology
algorithm is selected by algorithm model based on ELM. Compared the Fig. 4 with the
Fig. 3, it can be seen that there does not exist the pseudo edge, edge fracture, over
segmentation in the image edge of the Fig. 4, and the image edge is more smooth, it
can obviously separate the foreground image and the background image clearly.

In the Fig. 3, (a) is the original gray image, (b) is the edge of the image obtained by
Roberts edge detection algorithm, (c) is the edge of the image obtained by Sobel edge
detection algorithm, (d) is the edge of the image obtained by Prewitt edge detection
algorithm, (e) is the edge of the image obtained by Log edge detection algorithm, (f) is
the edge of the image obtained by Canny edge detection algorithm, (g) is the edge of
the image edge obtained by Watershed edge detection algorithm.

4.2 Algorithm Performance Analysis

From the Fig. 4 it can be seen that the image edge positioning is more accurate by
Roberts edge detection algorithm, but the algorithm is more sensitive to noise, the
distinguishing ability is poor on the correct image edge and noise, and it leads to lose
lots of useful edge information the in the recognition result. The image edge detection
algorithm is the first order differential algorithm respectively by Sobel and Prewitt, and
the difference between the both is that the Sobel edge detection algorithm adopts the
average filtering, and Prewitt edge detection algorithm adopts the weighted average
filtering, However, they can not deal with the image that is contained more complex
noise. Therefore, the results of these two algorithms have the phenomenon of multiple
edge fracture. The Log edge operator makes the smooth image treatment, so the
treatment result has a relatively good edge extraction result in some extent, but there are
many pseudo edges.

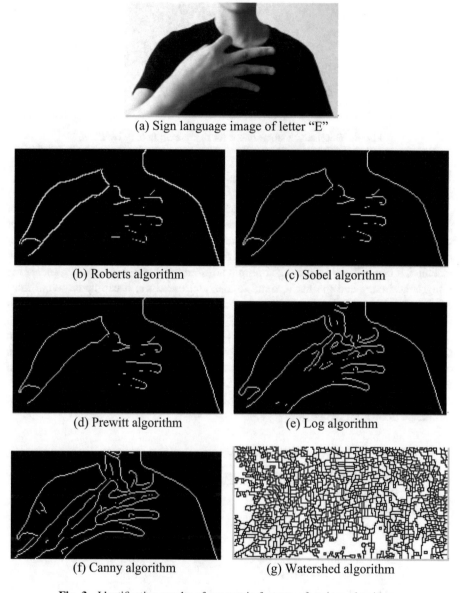

(a) Sign language image of letter "E"

(b) Roberts algorithm

(c) Sobel algorithm

(d) Prewitt algorithm

(e) Log algorithm

(f) Canny algorithm

(g) Watershed algorithm

Fig. 3. Identification results of geometric features of various algorithms.

Canny edge detection algorithm has a better detection effect than the previous algorithms, because the Canny algorithm is based on the first derivative as the criterion to judge the edge points, and the denoising ability is stronger than that of several operators mentioned above. However, when the algorithm is used to smooth the image, it is easy to filter out the edge information which can represent the image feature, and therefore, the results of the algorithm have a large fracture. Although the results of

Fig. 4. Geometric feature based on fuzzy morphology by ELM.

watershed algorithm can identify the image contour, but there are a lot of over segmentation phenomenon.

In a word, for the classical algorithm mentioned above, there are varying degrees of problems, such as edge fracture, false edge, over segmentation or a large amount of calculation and other shortcomings. The image feature recognition algorithm is proposed in this paper, and by means of algorithm model based on extreme learning machine, it can calculate the image optimal threshold that could optimally separate the foreground image and the background image, and therefore, it can better avoid false edges, over segmentation, and other shortcomings, and makes good in identification result.

Table 1 shows the comparative analysis of correct identification rate and execution time for the image recognition algorithm based on fuzzy morphology and several kinds of traditional classical identification algorithm on image geometric feature recognition as well as proposed algorithm in this paper, and the results of the correct identification rate and the execution time are shown as in Table 1 for the size of 512×512 pixels of the image geometric features.

Table 1. Correct recognition rate and execution time of typical algorithm

Fusion algorithm	Correct recognition rate (%)	Execution time (s)
Roberts algorithm	67.564	0.7312
Sobel algorithm	82.655	0.6482
Prewitt algorithm	85.874	0.6311
Log algorithm	88.647	0.5456
Canny algorithm	90.564	0.6785
Watershed algorithm	83.573	0.5678
This paper algorithm	95.564	0.2288

From Table 1, it can be seen that it is the two or three times faster than the classical geometric features of the image identification algorithm in execution speed for the proposed image feature identification algorithm based on fuzzy morphology, and it is the higher in correct recognition rate than the traditional classical algorithm of the image.

5 Conclusions

By means of the algorithm based on ELM, it can better analyze and process the images. Firstly, the image is processed by illumination unification, and then the optimal threshold value of the image is determined, which can be used to identify the geometric characteristics of the image, and to reproduce the image edge. The algorithm enriches the theory of image processing, it has a certain universality and can be applied to other fields, and therefore it has certain engineering practical value. However, there are still some puzzles such as the slow operation speed of the algorithm, the difficulty of choosing the best parameters of the membership function, and so on, which needs further study so as to get better recognition effect of image geometric feature and faster execution speed.

Acknowledgements. This work is supported by Technology Project of Chongqing Municipal Education Commission of China (Grant KJ1603811) and Doctoral Fund Project of Chongqing Normal University of China (Grant 15XLB015).

References

1. Hernandez, S., Barner, E.: Joint region merging criteria for watershed-based image segmentation. In: Proceedings of International Conference on Image Processing, vol. 2, pp. 108–111 (2002)
2. Yang, X.: Study on sign language recognition fusion algorithm using FNN. Adv. Intell. Soft Comput. **78**, 617–626 (2014)
3. Alexandre, X., Jorge Stolfi, F.: The image foresting transform: theory, algorithms, and applications. IEEE Trans. Pattern Anal. Mach. Intell. **26**(1), 364–370 (2014)
4. Yang, X., Qian, W., Zhou, Q.: Method study for facial character recognition based on improved LBP. J. Inf. Comput. Sci. **10**(9), 2519–2528 (2013)
5. Huang, G.B., Ding, X., Zhou, H.: Optimization method based extreme learning machine for classification. Neurocomputing, **74**(1), 155–163 (2010)
6. Deng, W.Y., Zheng, Q.H., Lian, S.G., et al.: Ordinal extreme learning machine. Neurocomputing, **74**(1–3), 447–456 (2010)
7. Huang, G.B., Song, S., Gupta, J.N.D., et al.: Semi-supervised and unsupervised extreme learning machines. IEEE Trans. Cybern. **44**(12), 2405–2417 (2014)
8. Yang, X.: Study on sign language recognition fusion algorithm using FNN. In: Advances in Intelligent and Soft Computing, vol. 78, pp. 617–626. Springer, Heidelberg (2014)

A Normalized Mutual Information Estimator Compensating Variance Fluctuations for Motion Detection

Kun Qin[1(✉)], Lei Sun[1(✉)], Shengmin Zhou[2],
Badong Chen[3], Beom-Seok Oh[4], and Zhiping Lin[4]

[1] School of Information and Electronics, Beijing Institute of Technology,
Beijing 100081, People's Republic of China
2567540768@qq.com, sunlei@bit.edu.cn
[2] Zhejiang HANMA Photoelectric Equipment Co., Ltd.,
Hangzhou 311251, People's Republic of China
[3] Institute of Artificial Intelligence and Robotics, Xi'an Jiaotong University,
Xi'an 710049, People's Republic of China
[4] School of Electrical and Electronic Engineering, Nanyang Technological University,
Singapore 639798, Singapore

Abstract. In motion event detection, an information measure based score function is sensitive to variance fluctuations between sample intervals. In this work, a score function based on a normalization of mutual information measure is proposed to tackle the problem of variance fluctuations. The mutual information is normalized by the maximum entropy which is related to the sample variances in comparison. An estimator using the normalized mutual information measure is implemented by neural networks with random setting of hidden neuron parameters. This estimator is tested by change point detection and motion event detection in experiments. Experimental results show that the normalization scheme in the estimator improves the sensitivity of event detection.

Keywords: Normalized mutual information estimation
Motion detection · Change-point detection
Extreme learning machine · Density ratio approximation

1 Introduction

Motion event detection via an information measure from continuous video frames is an active research area [1–3]. The information measure based approach formulates a score function based on the statistic of motion events between past and present frame intervals. Based on the score function, a motion event or a change point is detected if two interval statistics are significantly different. Various information measures have been proposed for the statistical evaluation of motion events where Shannon mutual information [4], being as the measure of events or as a benchmark, has been extensively discussed in the literature

© Springer Nature Switzerland AG 2019
J. Cao et al. (Eds.): ELM 2017, PALO 10, pp. 46–57, 2019.
https://doi.org/10.1007/978-3-030-01520-6_5

[5,6]. Essentially, the Shannon mutual information evaluates the relationship in sense of statistical independence [7] between the joint probability and the product of marginal probabilities of two random variables. In practical applications, however, the estimation of these probabilities is generally nontrivial or even intractable [8,9].

To alleviate the difficulty of PDF estimation in mutual information estimation, our group recently proposed a neural network based mutual information estimator for change point detection. In that work [10], instead of the estimation of the joint and marginal probabilities, a density ratio estimator [11] was implemented via a random setting scheme of neuron parameters. The random setting of neuron parameters simplified the computation for estimation, meanwhile the obtained estimator showed its effectiveness in the application of change-point detection. However, we noticed that the event score was highly related to the dynamics of the input data. In the application of change point detection, for example, the distributions of two sample intervals are compared by the score based on their mutual information. The larger the score is, the more likely a change point occurs in that interval [12]. However, large variance fluctuations could occur between different sample intervals and lead to fault detection due to the deviation of score baseline caused by the fluctuations.

From the view point of information theory, Shannon mutual information is determined by the contrastive ratio between variances of two random variables [7, Sect. 12]. In the case where samples with large variance fluctuations, the scores obtained from mutual information will largely fluctuate. To tackle this problem, a possible solution is to normalize the mutual information with respect to interval variances. For example, by constraining the information measure to a fixed region, it was shown that the constrained information measure helped to the interpretation and comparison across different conditions [13,14]. In contrast, unconstrained information measures were difficult to be compared [14]. Another example is seen in a clustering problem where the normalized mutual information measure improved the robustness to noises in clustering [15].

Motivated by these observations, in this work, we study a normalization scheme for the neural network based mutual information estimator. In particular, a parameter tuning-free normalized mutual information (NMI) estimator is proposed to tackle the problem of variance fluctuations and is used for a motion detection application.

In the following, the neural network based mutual information estimator is briefly reviewed in Sect. 2. This is followed by a discussion of the score comparison problem caused by variance fluctuations between sample intervals. In Sect. 3, a normalized mutual information estimator is proposed. The proposed NMI is applied for motion detection in Sect. 4. We finally conclude our work in Sect. 5.

2   ELM-MI for Change-Point Detection

2.1   Extreme Learning Machine based Mutual Information Estimator

Let $\mathbf{X}$ and $\mathbf{Y}$ denote two random variables with a set of n instantiations $\{(\mathbf{x}_i, \mathbf{y}_i)|\mathbf{x}_i \in \mathcal{D}_{\mathbf{X}}, \mathbf{y}_i \in \mathcal{D}_{\mathbf{Y}}\}_{i=1}^n$ where $\mathcal{D}_{\mathbf{X}} \subset \mathbb{R}^{d_X}$ and $\mathcal{D}_{\mathbf{Y}} \subset \mathbb{R}^{d_Y}$ are with dimension sizes d_X and d_Y, respectively. Assume the sample pairs are independent and identically distributed (*i.i.d.*). For the two random variables $\mathbf{X}$ and $\mathbf{Y}$, Shannon mutual information is defined over the joint probability density function (PDF) $p_{XY}(\mathbf{x}, \mathbf{y})$ and the product of marginal PDFs $p_X(\mathbf{x})$ and $p_Y(\mathbf{y})$, written as [7]

$$I(\mathbf{X}; \mathbf{Y}) = \iint p_{XY}(\mathbf{x}, \mathbf{y}) \log \frac{p_{XY}(\mathbf{x}, \mathbf{y})}{p_X(\mathbf{x})p_Y(\mathbf{y})} \mathrm{d}\mathbf{x}\mathrm{d}\mathbf{y}. \tag{1}$$

In (1), it is noted that the argument in the logarithmic term is a ratio: $r(\mathbf{x}, \mathbf{y}) = \dfrac{p_{XY}(\mathbf{x}, \mathbf{y})}{p_X(\mathbf{x})p_Y(\mathbf{y})}$. The direct ratio approximation was proposed for the estimation of Shannon mutual information to avoid the difficulty of PDF estimations in real applications [11]. To further reduce computations, a random setting scheme of neuron parameters in Extreme Learning Machine (ELM) [16–18] was adopted and a parameter tuning-free mutual information estimator (ELM-MI) was proposed in [10]. Intuitively, the key idea of ELM-MI is to directly approximate the density ratio by appealing to the random setting of parameters in ELM. When Radial Basis Function (RBF) is adopted in L-length of hidden neurons, the ELM-MI estimator is expressed as [10]:

$$g(\boldsymbol{\mathcal{X}}, \boldsymbol{\mathcal{Y}}) = \sum_{l=1}^{L} \beta_l \phi_l(\boldsymbol{\mu}_l^x, \boldsymbol{\mu}_l^y, \sigma_l, \boldsymbol{\mathcal{X}}, \boldsymbol{\mathcal{Y}}), \tag{2}$$

where RBF kernels $\phi_l(\boldsymbol{\mu}_l^x, \boldsymbol{\mu}_l^y, \sigma, \boldsymbol{\mathcal{X}}, \boldsymbol{\mathcal{Y}})$, $l = 1, 2, \cdots, L$ form the hidden nodes. The two matrices $\boldsymbol{\mathcal{X}} = [\mathbf{x}_1, \mathbf{x}_2, \cdots, \mathbf{x}_n]^T \in \mathbb{R}^{n \times d_X}$ and $\boldsymbol{\mathcal{Y}} = [\mathbf{y}_1, \mathbf{y}_2, \cdots, \mathbf{y}_n]^T \in \mathbb{R}^{n \times d_Y}$ are formed by n sample pairs generated from $\mathbf{X}$ and $\mathbf{Y}$. β_l is a weight value connecting the ith kernel and the output node. The Gaussian kernel is an universal Mercer kernel [8] and is widely adopted for kernel methods and Information Theoretic Learning (ITL) [19,20]. In ELM-MI, Gaussian kernels $\phi_l(\boldsymbol{\mu}_l^x, \boldsymbol{\mu}_l^y, \sigma_l, \boldsymbol{\mathcal{X}}, \boldsymbol{\mathcal{Y}}) = \exp\left(-\frac{\|\boldsymbol{\mathcal{X}} - \boldsymbol{\mu}_l^x\|^2}{2\sigma_l^2}\right)\exp\left(-\frac{\|\boldsymbol{\mathcal{Y}} - \boldsymbol{\mu}_l^y\|^2}{2\sigma_l^2}\right)$ are deployed spontaneously. $\boldsymbol{\mu}_l^x = [\check{\mathbf{x}}_l, \check{\mathbf{x}}_l, \cdots, \check{\mathbf{x}}_l]^T \in \mathbb{R}^{n \times d_X}$ and $\boldsymbol{\mu}_l^y = [\check{\mathbf{y}}_l, \check{\mathbf{y}}_l, \cdots, \check{\mathbf{y}}_l]^T \in \mathbb{R}^{n \times d_Y}$ respectively indicate the lth kernel's centers where $(\check{\mathbf{x}}_l, \check{\mathbf{y}}_l)$ is randomly chosen from $\{\mathbf{x}_j, \mathbf{y}_j\}_{j=1}^n$. $\sigma_l \in \mathbb{R}$ indicates the kernel width which is determined randomly.

The weight vector $\boldsymbol{\beta}$ in (2) is computed by [10]:

$$\widehat{\boldsymbol{\beta}} = \left(\widehat{\mathbf{H}} + \lambda\mathbf{I}\right)^{-1}\widehat{\mathbf{h}}, \tag{3}$$

where

$$\widehat{\mathbf{H}} = \frac{1}{n^2} \sum_{i,j=1}^{n} \boldsymbol{\varphi}_{i,j} \boldsymbol{\varphi}_{i,j}^{T} \in \mathbb{R}^{L \times L}, \quad \widehat{\mathbf{h}} = \frac{1}{n} \sum_{i=1}^{n} \boldsymbol{\varphi}_{i,i} \in \mathbb{R}^{L},$$

$$\boldsymbol{\varphi}_{i,j} = [\phi_1(\check{\mathbf{x}}_1, \check{\mathbf{y}}_1, \sigma_1, \mathbf{x}_i, \mathbf{y}_j), \cdots, \phi_L(\check{\mathbf{x}}_L, \check{\mathbf{y}}_L, \sigma_L, \mathbf{x}_i, \mathbf{y}_j)]^T$$

holds, λ indicates a regularization factor, and $\mathbf{I}$ is an identity matrix with the same dimension size as $\widehat{\mathbf{H}}$. Based on (1)–(3), the Shannon mutual information is calculated as [21]:

$$\widehat{I}(\mathbf{X}; \mathbf{Y}) = \frac{1}{n} \sum_{i=1}^{n} \log \widehat{\boldsymbol{\beta}}^T \boldsymbol{\varphi}_{i,i}. \tag{4}$$

2.2 Impact of Fluctuations on Change-Point Detection Using ELM-MI

In this subsection, we use an application of change-point detection from a time sequence to illustrate the impact of variance fluctuation on change point detection.

Given a time sequence $y_t \in \mathbb{R}, t = 1, \cdots, n$, let τ in the subscript of $y_{t-\tau}$ denote a delay of τ with respect to y_t, the time-delay mutual information (TDMI) is defined as:

$$I(Y_{t-\tau}; Y_t) = \iint p(y_{t-\tau}, y_t) \log \left(\frac{p(y_{t-\tau}, y_t)}{p(y_{t-\tau}) p(y_t)} \right) dy_{t-\tau} dy_t, \tag{5}$$

where $p(y_{t-\tau}, y_t)$ denotes the joint PDF associated with the time-series y_t and its delayed version $y_{t-\tau}$, $p(y_{t-\tau})$ and $p(y_t)$ denote the marginal PDFs, respectively.

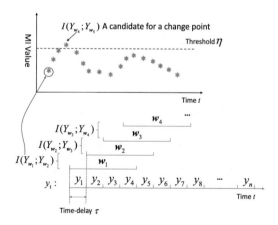

Fig. 1. TDMI based change-point detection framework

TDMI in (5) is adopted for the application of change-point detection and is illustrated in Fig. 1. In this figure, several sample intervals are drawn from a time sequence. Such sample interval is denoted by $\mathbf{w}_t \in \mathbb{R}^k$ where each interval consists of k samples. The mutual information of two overlapped time intervals with a time delay τ are calculated with respect to increment of time. A change-point is detected if the associated score goes beyond a predefined threshold η: $I(Y_{w_{t-1}}, Y_{w_t}) > \eta$.

With the i.i.d. assumption of samples, it is reasonable to assume that the observed time sequence is stationary and a fixed predefined threshold η is sufficient to detect the changes in the time sequence. However, in practice, this assumption can be violated. As is shown in Fig. 2, PDFs of intervals $\mathbf{w}_1$ and $\mathbf{w}_5$ are totally different. Suppose the score of $I(Y_{w_4}, Y_{w_5})$ is within the region $[0, I_c]$ where I_c is determined by the marginal PDFs $p(y_{w_4})$ and $p(y_{w_5})$ and joint PDF $p(y_{w_4}, y_{w_5})$. In contrast, the score $I(Y_{w_1}, Y_{w_2})$ of intervals 1 and 2 is within the ranges from 0 to I_{uc} where I_{uc} is determined by the PDFs with respect to samples of intervals 1 and 2. Obviously, $I_c \neq I_{uc}$ if the distributions of the groups of intervals are different. No doubt that it is hard to determine the fixed η when there are large variance fluctuations between sample intervals where the score regions are different.

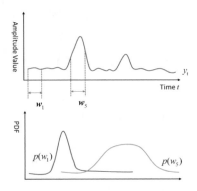

Fig. 2. PDF of two intervals

3   A Normalized Mutual Information Estimator

3.1   Normalization of the Mutual Information

For discrete random variables, Shannon mutual information is limited to the entropies of the random variables shown as [7]:

$$I(\mathbf{X}; \mathbf{Y}) = H(\mathbf{X}) - H(\mathbf{X}|\mathbf{Y}) \leq H(\mathbf{X}). \tag{6}$$

Therefore, it is reasonable to constraint the mutual information to a region by including the maximum value of entropies as an normalizer

$$I_{dnmi} = \frac{I(\mathbf{X}; \mathbf{Y})}{\max\{H(\mathbf{X}), H(\mathbf{Y})\}}, \tag{7}$$

which is defined as the normalized mutual information (NMI), or a variant form [22]: $d_{max} = 1 - I_{dnmi}$.

For the case of continuous random variables, however, the maximum entropy requires a constraint of moment function from observations. For example, when variance is known from observations, the maximum entropy is limited to the entropy of Gaussian distribution with the observed variance. This is known as the maximum entropy principle [23]. In general, the maximum entropy depends on the prior knowledge of the observations. In this work, we consider the case with prior knowledge about variance. In this case, the maximum entropy is

$$h_G(\mathbf{X}) = \frac{1}{2}\log(2\pi e)^{d_X} |\det\mathbf{C}|, \tag{8}$$

where the elements in $\mathbf{X}$ are with the assumption of mutually independent. $\mathbf{C}$ is a diagonal matrix: $|\det\mathbf{C}| = \prod_{i=1}^{d_X} \sigma_i^2$, where σ_i are the standard deviations of elements in $\mathbf{X}$. Based on $h_G(\mathbf{X})$, a modified NMI is given by:

$$I_n(\mathbf{X};\mathbf{Y}) = \frac{I(\mathbf{X};\mathbf{Y})}{\max\{h_G(\mathbf{X}), h_G(\mathbf{Y})\}}. \tag{9}$$

In the following, if not mentioned otherwise, NMI refers to the modified version in (9).

Similar to the ELM-MI described in Sect. 2, based on (4), the estimated NMI is calculated as

$$\text{ELM-NMI} = \frac{\sum_{i=1}^{n} \log\widehat{\boldsymbol{\beta}}^T \boldsymbol{\varphi}_{i,i}}{n \cdot \max\{h_G(\mathbf{X}), h_G(\mathbf{Y})\}}, \tag{10}$$

which is illustrated in Fig. 3.

Here, for practical applications, we propose to include a mapping function to get the normalized information distance (NID):

$$\widetilde{d}_{max}(\mathbf{X};\mathbf{Y}) = 1 - f(I_n) \tag{11}$$

where $f(\cdot)$ denotes the mapping function. This mapping function is to stretch the region of concerned samples. In particular,

$$f(x) = \sqrt{1 - \exp(-2x)}. \tag{12}$$

whose derivative satisfies $f'(x) > 0$ and $f''(x) < 0$ when $x \in (0,1)$ holds, it indicates that $f(x)$ can map an interval with narrow width around 0 into a wider one. Shortly, this mapping function is to be used in a motion detection problem. In motion detection, if the observations with larger change degree are interested, then a finer description on the area with larger $\widetilde{d}_{max}$ (smaller $\widehat{I}_n$) is expected. In this case, function like (12) is an appropriate option.

Finally, NID is estimated by substituting (10) into (11), which is an ELM based NID estimator namely ELM-NID. Note that to overcome the different estimated values in different trials, which is caused by the random settings for parameters, M ELM-NMI is combined and the average value is expected as the estimated result [10].

The normalized mutual information (NMI) has been well discussed in the literature [5,24] and applications can be seen in [14,25]. The two estimators in (9) and (11) are essentially the same. In the following, we shall use (9) for a change-point detection problem and (11) for a motion detection problem, respectively.

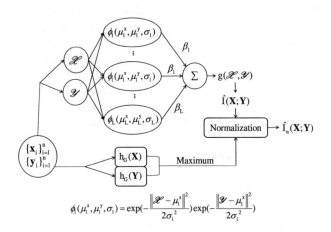

Fig. 3. ELM-NMI model

3.2   A Comparison Between ELM-MI and the Proposed ELM-NMI

ELM-NMI is an improved version of ELM-MI by including a normalization term. To show the effectiveness of the proposed ELM-NMI for adaptation to variance fluctuations, in this subsection, The two methods are compared based on a change-point detection experiment. The data set for change-point detection was used in [10]. The same data set is adopted in this experiment.

There are three model parameters namely: the number of hidden nodes N, the regularization factor λ and the ensemble size M in ELM-MI. In experiments, these model parameters are set at $\lambda = 10^{-3}$, $M = 10$ and $N = 100$ for ELM-MI, as was used in [10]. In ELM-NMI, the three parameters are kept the same as for ELM-MI. For the two estimators, kernel width are randomly selected from a fixed region. In experiments, the kernel width region is fixed at $(0.4, 0.6)$.

Experimental results are shown in Fig. 4. The original signal is shown in the upper part of Fig. 4. In this experiment, both ELM-MI and ELM-NMI value can portray the change of the tested time-series. The score sequence based on ELM-MI is shown in the middle part of Fig. 4, whose bottom part displays the scores described with ELM-NMI. Note that ELM-NMI has a stronger ability in detecting the change-points that are visually small. The change-points marked with pulse $1 \sim 5$ (as shown in Fig. 4) are enhanced in value if the ELM-NMI measure is chosen.

The comparison in Fig. 4 shows that the proposed ELM-NMI can effectively detect small changes which are not detected by ELM-MI. It is noted that those small changes are with small variance fluctuations between sample intervals.

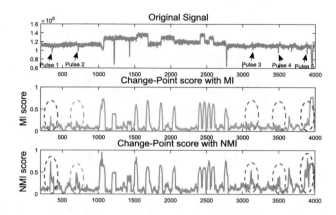

Fig. 4. Change-point detection with MI score and NMI score

Based on the same settings of parameters, it is reasonable to conclude the detection improvement is due to the included normalization in ELM-NMI. In other words, without normalization of mutual information, the scores of these intervals are small and will fail to indicate the changes.

4 Motion Event Detection Experiments

4.1 Motion Event Detection Algorithm

Motion event detection in this paper is to detect event (changes) from continuous video frames. The proposed ELM-NMI is used to detect motion events in an experiment based on a synthetic videos. In the synthetic experiment, one spinning wheel that is simultaneously expanding departing from the wheel center is involved. That is, each point on this expanding wheel are with two speeds with different directions: one is in line with the radius direction and the other is orthogonal to the radius direction. The two speeds are denoted as $\mathbf{v}_1$ and $\mathbf{v}_2$ in Fig. 6(a), respectively. In summary, the spin event and the expansion event are expected to be detected from these video frames.

Based on the proposed ELM-NMI, the motion event detection procedure is illustrated as follows. Sub-frames with length a and width b are intercepted from frames to be processed. The length of overlap between two adjacent sub-frames is defined as k. Let $\{x_l\}_{l=1}^{ab}$ denotes a set of gray values of pixels in sub-frame $\mathbf{X}_j$. These gray values can be treated as samples drawn from a random variable $\mathbf{X}_j$. Similarly, gray values of pixels in sub-frame $\mathbf{Y}_j$ can be considered as samples drawn from a random variable $\mathbf{Y}_j$. Then the distance score S_j for the corresponding sub-frames, which are shown as $\mathbf{X}_j$ and $\mathbf{Y}_j$ in Fig. 5, is characterized through NMI based measure. For a same object between adjacent frames, deviation in its gray values will occur, which is caused by the video quality. Seeing that the existence of such deviations, human's capability of distinguishing two pixels whose gray difference is more than 8 is as a reference in a definition of

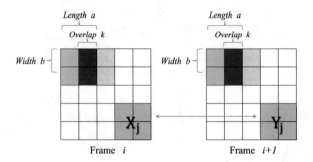

Fig. 5. Motion detection framework

Algorithm 1. NMI based motion detection framework

Input: Given Frames i and $i + 1$, sub-frame pairs $\{\mathbf{X}_j, \mathbf{Y}_j\}_{j=1}^{N_j}$ with length a, width b and overlap k are obtained from corresponding frame. A threshold η is prefixed to determine the possible changing sub-frames.

1: **for** i from 1 to $N_i - 1$ **do**
2:      **for** j from 1 to N_j **do**
3:          **if** $\sum_{l=1}^{ab} f_c(8 - |x_l - y_l|) > \alpha \cdot ab$ **then**
4:              $S_j = 0$
5:          **else**
6:              $S_j = \widetilde{d}_{max}(\mathbf{X}_j; \mathbf{Y}_j)$
7:          **end if**
8:          **if** $S_j > \eta$ **then**
9:              A candidate for a change sub-frame
10:         **end if**
11:     **end for**
12: **end for**

---

invariability of sub-frames [26]. Two sub-frames are regarded as the same, which means $S=0$, if the following **tolerance** holds:

$$\sum_{l=1}^{ab} f_c(8 - |x_l - y_l|) > \alpha \cdot ab \tag{13}$$

where parameter α that ranges from 0 to 1 is related to the video quality, $|\cdot|$ represents absolute value operation. The function $f_c(\cdot)$ is defined as:

$$f_c(x) = \begin{cases} 1 & x > 0 \\ 0 & \text{otherwise} \end{cases}. \tag{14}$$

If two sub-frames fail to satisfy the **tolerance**, their distance score is calculated through:

$$S_j = 1 - f(I_n(\mathbf{X}_j; \mathbf{Y}_j)) \tag{15}$$

which actually is $\widetilde{d}_{max}(\mathbf{X}_j; \mathbf{Y}_j)$. If S_j is a peak value over a predefined threshold η, it can be a candidate for a 'change-point' [10]. In summary, the motion event detection algorithm is listed in Algorithm 1.

4.2 Settings

In the experiment, sub-frames with 10×10 pixels, implying $a = 10$ and $b = 10$, are considered each time. $\alpha = 0.8$ holds for the **tolerance**. The mapping function is chosen as (12). To observe the effect overlap length k has on the detection performance, different values $\{0, 5, 9\}$ are assigned to k. Simply, only two adjacent frames are processed for the motion event detection.

4.3 Results and Discussions

Two speeds $\mathbf{v}_1$ and $\mathbf{v}_2$ with different directions are interested in the synthetic experiment. When comparing these two speeds, it makes no sense if we only consider the gray difference that is actually the first order statistic, meaning changes in the gray value of pixels. As a matter of fact, comparison is hard to made due to the fuzzy outer edges caused by the quality of the video, which is shown in Fig. 6(c). While conclusion is simply made that the wheel spins much faster than it expands by using ELM-NMI estimating the distance scores, for S within the wheel is much larger than that on its edge as observed in Fig. 6f. In this comparison, ELM-NMI can describe changes with higher order statistics features and the proposed model is not appreciably affected by fuzzy edges.

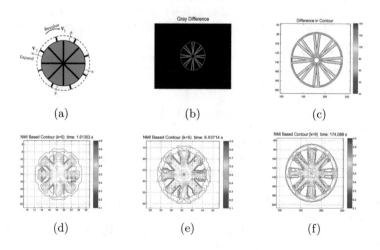

Fig. 6. Results of experiment 2

As are shown in Fig. 6, detection performance with finer description is achieved with larger overlap length k, yet with more time cost. $k = 5$ is a compromise option for its relative fast speed and fine general view of changes in frames. Determination of k is our ongoing work.

5 Conclusion

A mutual information estimator based on normalization of mutual information by maximum entropy of random variables was proposed in this work. The estimator was implemented by neural networks where the hidden neuron parameters were randomly set without a particular determination procedure. This estimator was tested by two experiments. One is change point detection and the other is motion event detection. Both the detection of small variance changes in the change point detection experiment and the particular motion events detected in the motion event detection experiment showed that the proposed estimator is capable of detecting motion events adaptively to variance fluctuations.

Acknowledgment. This work was supported by National NSF of China (No. 61673059, No. 61372152 and No. 91648208) and 973 Program (No. 2015CB351703).

References

1. Guralnik, V., Srivastava, J.: Event detection from time series data. In: Proceedings of the Fifth ACM SIGKDD International Conference on Knowledge Discovery and Data Mining, pp. 33–42. ACM (1999)
2. Ke, Y., Sukthankar, R., Hebert, M.: Event detection in crowded videos. In: IEEE 11th International Conference on Computer Vision, ICCV 2007, pp. 1–8. IEEE (2007)
3. Yamagiwa, S., Kawahara, Y., Tabuchi, N., Watanabe, Y., Naruo, T.: Skill grouping method: mining and clustering skill differences from body movement bigdata. In: 2015 IEEE International Conference on Big Data (Big Data), pp. 2525–2534. IEEE (2015)
4. Zhou, M., Bao, G., Pahlavan, K.: Measurement of motion detection of wireless capsule endoscope inside large intestine. In: 2014 36th Annual International Conference of the IEEE Engineering in Medicine and Biology Society (EMBC), pp. 5591–5594. IEEE (2014)
5. Kraskov, A., Stögbauer, H., Grassberger, P.: Estimating mutual information. Phys. Rev. E **69**(6), 066138 (2004)
6. Sugiyama, M., Liu, S., Marthinus Christoffel, D., Plessis, M.Y., Yamada, M., Suzuki, T., Kanamori, T.: Direct divergence approximation between probability distributions and its applications in machine learning. J. Comput. Sci. Eng. **7**(2), 99–111 (2013)
7. Cover, T.M., Thomas, J.A.: Elements of Information Theory. Wiley-Interscience, New York (2006)
8. Principe, J.C.: Information Theoretic Learning: Rényi's Entropy and Kernel Perspectives. Springer, New York (2010)
9. Sun, L., Chen, B., Ton, K.-A., Lin, Z.: A parameter-free Cauchy-Schwartz information measure for independent component analysis. In: 2016 IEEE International Conference on Acoustics, Speech and Signal Processing (ICASSP), pp. 2524–2528. IEEE (2016)
10. Beom-Seok, O., et al.: Extreme learning machine based mutual information estimation with application to time-series change-points detection. Neurocomputing (2017). https://doi.org/10.1016/j.neucom.2015.11.138

11. Sugiyama, M., Nakajima, S., Kashima, H., Von Buenau, P., Kawanabe, M.: Direct importance estimation with model selection and its application to covariate shift adaptation. Adv. Neural Inf. Process. Syst. **20**, 1433–1440 (2008)
12. Liu, S., Yamada, M., Collier, N., Sugiyama, M.: Change-point detection in time-series data by relative density-ratio estimation. Neural Netw. **43**, 72–83 (2013)
13. Strehl, A., Ghosh, J.: Cluster ensembles—a knowledge reuse framework for combining multiple partitions. J. Mach. Learn. Res. **3**, 583–617 (2002)
14. Luo, P., Xiong, H., Zhan, G., Junjie, W., Shi, Z.: Information-theoretic distance measures for clustering validation: generalization and normalization. IEEE Trans. Knowl. Data Eng. **21**(9), 1249–1262 (2009)
15. Wu, J., Xiong, H., Chen, J.: Adapting the right measures for k-means clustering. In: Proceedings of the 15th ACM SIGKDD International Conference on Knowledge Discovery and Data Mining, pp. 877–886. ACM (2009)
16. Huang, G.-B., Zhu, Q.-Y., Siew, C.-K.: Extreme learning machine: theory and applications. Neurocomputing **70**(1), 489–501 (2006)
17. Sun, L., Chen, B., Toh, K.-A., Lin, Z.: Sequential extreme learning machine incorporating survival error potential. Neurocomputing **155**, 194–204 (2015)
18. Cao, J., Zhang, K., Luo, M., Yin, C., Lai, X.: Extreme learning machine and adaptive sparse representation for image classification. Neural Netw. **81**, 91–102 (2016)
19. Chen, B., Zhu, Y., Hu, J., Principe, J.C.: System parameter identification: information criteria and algorithms. In: Newnes (2013)
20. Chen, B., Zhao, S., Zhu, P., Príncipe, J.C.: Quantized kernel least mean square algorithm. IEEE Trans. Neural Netw. Learn. Syst. **23**(1), 22–32 (2012)
21. Suzuki, T., Sugiyama, M., Sese, J., Kanamori, T.: A least-squares approach to mutual information estimation with application in variable selection. In: Proceedings of the 3rd Workshop on New Challenges for Feature Selection in Data Mining and Knowledge Discovery (FSDM 2008), Antwerp, Belgium (2008)
22. Kvalseth, T.O.: Entropy and correlation: some comments. IEEE Trans. Syst. Man Cybern. **17**(3), 517–519 (1987)
23. Jaynes, E.T.: Information theory and statistical mechanics. Phys. Rev. **106**(4), 620 (1957)
24. Vinh, N.X., Epps, J., Bailey, J.: Information theoretic measures for clusterings comparison: variants, properties, normalization and correction for chance. J. Mach. Learn. Res. **11**, 2837–2854 (2010)
25. Kraskov, A., Stögbauer, H., Andrzejak, R.G., Grassberger, P.: Hierarchical clustering using mutual information. EPL (Europhys. Lett.) **70**(2), 278 (2005)
26. Tianhe, Y., Jing, J.: New technology of infrared image contrast enhancement based on human visual properties. Infrared Laser Eng. **6**(37), 951–954 (2008)

Reconstructing Bifurcation Diagrams of Induction Motor Drives Using an Extreme Learning Machine

Yoshitaka Itoh[✉] and Masaharu Adachi

Department of Electrical and Electronic Engineering,
Tokyo Denki University, 5 Senju-Asahicho Adachi-ku, Tokyo, Japan
16ude01@ms.dendai.ac.jp

Abstract. We describe a method to reconstruct the bifurcation diagrams of a mathematical model of induction motor drives using extreme learning machines. The reconstruction of a bifurcation diagram estimates the oscillatory patterns of a target system when its parameters are changed. These patterns can be periodic, quasi-periodic, or chaotic. Examples of the parameters are electric current and power, temperature, pressure, and concentration. We also estimate Lyapunov exponents so that the reconstructed bifurcation diagram can be quantitatively evaluated. We use typical chaotic systems as target for the reconstruction of bifurcation diagrams, such as logistic and Hénon maps, as well as Rössler equations. Using numerical simulations, we show that the reconstruction of bifurcation diagrams can be useful for real-world systems, such as induction motor drives. For applying the method to the real-world systems, we evaluate robustness for observation noise. Therefore, we attempt to reconstruct the bifurcation diagram of the induction motor drives using noisy time-series datasets.

Keywords: Extreme learning machine · Chaos
Reconstruction of bifurcation diagrams · Time-series prediction
Induction motor drive

1 Introduction

The reconstruction of bifurcation diagrams (BDs) was proposed in 1994 by Tokunaga *et al.* (1994). The reconstruction of BDs estimates the oscillatory patterns of a target system when its parameters are changed. These patterns can be periodic, quasi-periodic, or chaotic. The method makes no assumption on the number of bifurcation parameters and their values, and it requires only a finite number of time-series datasets, generated by the dynamical system with different parameter values.

We have previously proposed a method for the reconstruction of BDs with Lyapunov exponents using extreme learning machines (ELMs) (Huang *et al.* 2006; Itoh 2017a, b, c). While various time-series predictors can be used for the reconstruction of BDs (Ogawa *et al.* 1996; Bagarinao 1999a, b, 2000; Langer and Parlitz 2004), we believe ELM is the best for most cases of reconstruction of BDs because the synaptic weights are trained to model several time-series datasets for the reconstruction of BDs. It is therefore important to consider ELM as a top choice for the reconstruction of BDs.

© Springer Nature Switzerland AG 2019
J. Cao et al. (Eds.): ELM 2017, PALO 10, pp. 58–69, 2019.
https://doi.org/10.1007/978-3-030-01520-6_6

In addition, a good generalization capability is also important to estimate the oscillatory patterns. Moreover, the time-series predictors must have a simple enough structure to be able to estimate the Lyapunov exponents using the Jacobian matrix of each time-series predictor.

In this paper, we attempt to reconstruct the BD of an engineering system, in the form of a mathematical model of induction motor drives, as proposed in 2014 by Chakrabarty and Kar (2014). In addition, we reconstruct the BD of the induction motor drives using noisy time-series datasets in order to evaluate robustness for observation noise.

It is important for engineering systems to adjust the parameters to a stable oscillatory pattern. We expect that the reconstructed BD will be helpful to easily adjust the optimal parameter values.

The rest of the paper is organized as follows. In Sect. 2, we explain the method for reconstructing BDs using ELM (Itoh *et al.* 2017c). In Sects. 3 to 5, we present the results of our numerical simulations. Finally, we draw our conclusions in Sect. 6.

2 Reconstructing Bifurcation Diagrams Using Extreme Learning Machines

In this section, we give a summary of the method for the reconstruction of BDs using ELMs. The method uses only a finite number of time-series datasets $S_1, \cdots, S_K$, where K is the number of time-series datasets used for the reconstruction of the BD, and assumes that the number of bifurcation parameters and their values are unknown.

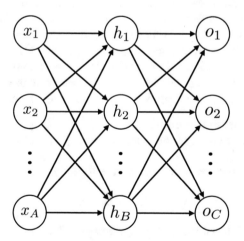

Fig. 1. Structure of an ELM

2.1 Modeling Time-Series Datasets

First, we model the first time-series dataset S_i by using an ELM. The structure of the ELM is displayed in Fig. 1. The output vector $o \in \mathbb{R}^C$ for the output neuron of the ELM is given by

$$o = E\left(\boldsymbol{\beta}_1^T, \boldsymbol{x}\right) = \boldsymbol{\beta}_1^T \cdot sig(W\boldsymbol{x} + \boldsymbol{b}) \tag{1}$$

where $E(\cdot, \cdot)$ is an input-output function of the ELM, $\boldsymbol{\beta}_1 \in \mathbb{R}^{C \times B}$ is the synaptic weight matrix of the output neurons for the first time-series dataset, $W \in \mathbb{R}^{B \times A}$ and $\boldsymbol{b} \in \mathbb{R}^B$ are respectively the synaptic weight matrix and the bias vector of the hidden neurons, $\boldsymbol{x} \in \mathbb{R}^X$ is an input vector, and $sig(\cdot)$ is a sigmoid function. In this paper, the sigmoid function is given by

$$sig(\chi) = \frac{\sigma}{1 + \exp(-\zeta \cdot \chi)} \tag{2}$$

where parameters σ, ζ and ϵ are used to adjust the range of the sigmoid function to one of the target time-series datasets.

The synaptic weights and biases of the hidden neurons are randomly generated. The synaptic weight matrix of the output neuron $\boldsymbol{\beta}_1$ is trained using

$$\boldsymbol{\beta}_1 = H^\dagger D_1 \tag{3}$$

where $H^\dagger$ is the pseudo-inverse matrix of the output matrix of hidden neurons $H = [\boldsymbol{h}(1) \cdots \boldsymbol{h}(L)]^T$ and $D_1 \in \mathbb{R}^{L \times C}$ is the desired output matrix of the output neurons for the first time-series dataset. Here, $\boldsymbol{h} = [h_1 \cdots h_B]$ is the output of the hidden neuron and L is the length of the training data for the target time-series dataset.

Second, we model the remaining time-series datasets $S_i(i = 2, \cdots, K)$. Since we need to obtain trained synaptic weights of output neurons that are close to those already obtained, the synaptic weights and biases of the hidden neurons are not regenerated but are set to the same values used to model the time-series S_i. The procedure is as follows.

1. Calculate an output time-series dataset $O \in \mathbb{R}^{L \times C}$ which is an output of the ELM with the synaptic weights of the output neurons $\boldsymbol{\beta}_{i-1}$ when we input the ith time-series dataset.

2. Calculate the ith difference time-series ΔD_i for the ith desired time-series dataset:

$$\Delta D_i = D_i - O. \tag{4}$$

3. Calculate the difference synaptic weight matrix $\Delta \boldsymbol{\beta}_i$:

$$\Delta \beta_i = H^\dagger D_i. \tag{5}$$

4. Add the ith difference synaptic weight to the $i - 1$th synaptic weight of the output neuron:

$$\beta_i = \beta_{i-1} + \Delta\beta_i. \tag{6}$$

5. Repeat steps 2–5 to model the time-series datasets S_2 to S_K.

2.2 Estimating the Number of Significant Parameters by Principal Component Analysis

First, we prepare a deviation vector $\delta\beta_i$ of the trained synaptic weights using

$$\delta\beta_i = \beta_i - \beta_0 \tag{7}$$

$$\beta_0 = \frac{\sum_{i=1}^{K} \beta_i}{K} \tag{8}$$

where β_0 is the mean vector of the synaptic weights of the output neurons. Second, the eigenvalues and eigenvectors are obtained by applying a principal component analysis to the variance-covariance matrix of the deviation vectors. We estimate the number of parameters for the target system by calculating a contribution ratio and a cumulative contribution ratio from the eigenvalues.

The fth contribution ratio and the fth cumulative contribution ratio are calculated as

$$CR_f = \frac{\lambda_f}{\sum_{j=1}^{F} \lambda_j} \times 100(\%), (1 \leq f \leq F) \tag{9}$$

$$CCR_f = \frac{\sum_{j=1}^{f} \lambda_j}{\sum_{j=1}^{F} \lambda_j} \times 100(\%), (1 \leq f \leq F) \tag{10}$$

where λ_i is the ith eigenvalue and F is the number of trained synaptic weights. We determine that the estimated dimension ρ is f when the fth cumulative contribution ratio is larger than 80%.

2.3 Reconstructing the Bifurcation Diagram

We reconstruct the BD using eigenvectors $V \in \mathbb{R}^{F \times \rho}$ corresponding to the first ρ eigenvalues. The new synaptic weights $\tilde{\beta}$ to reconstruct the BD are given by

$$\tilde{\beta} = V\gamma + \beta_0 \tag{11}$$

where γ is the estimated parameter vector as bifurcation parameters. We generate the time-series dataset using the ELM with the new synaptic weights using

$$o = \tilde{\beta}^T \cdot sig(Wx + b) \tag{12}$$

and we repeat this procedure to reconstruct the BD while changing the estimated parameters.

2.4   Estimating Lyapunov Exponents for the Reconstructed BD

We estimate the Lyapunov exponents of the reconstructed BD by using a Jacobian matrix of the ELM (Itoh *et al.* 2017c). Then, we apply a QR decomposition to the Jacobian matrix using

$$JE(\tilde{\beta}, x(t))Q(t) = Q(t+1)R(t+1), (t = 0, \cdots, L) \tag{13}$$

where $JE(\tilde{\beta}, x(t))$ is the Jacobian matrix of the ELM, $Q(t)$ is an orthogonal matrix and $R(t)$ is an upper triangular matrix. The Lyapunov exponents of the reconstructed BD are given by

$$\mu_j = \frac{1}{L} \sum_{t=1}^{L} \log r_{jj}(t), (j = 1, \cdots C) \tag{14}$$

where $r_{jj}(t)$ is the jth diagonal component of $R(t)$.

3   Numerical Simulations for Reconstruction of BD with Load Torque from Noise Free Time-Series

In this paper, we present our results in reconstructing the BD of the mathematical model of induction motor drives proposed by Chakrabarty and Kar (2014). In this section, we show the results for reconstructing BD using noise free time-series datasets generated with different value of the load Torque.

3.1   A Mathematical Model of Induction Motor Drives

The equations of the induction motor drives are given by

$$\dot{z}_1 = -c_1 \cdot z_1 + c_2 \cdot x_4 - \frac{kc_1}{u_2^0} \cdot z_2 \cdot z_4 \tag{15}$$

$$\dot{z}_2 = -c_1 \cdot z_2 + c_2 \cdot u_2^0 - \frac{kc_1}{u_2^0} \cdot z_1 \cdot z_4 \tag{16}$$

$$\dot{z}_3 = -c_3 \cdot z_3 - c_4 \left[c_5 (z_2 \cdot z_4 - x_1 \cdot u_2^0) - T_l - \frac{c_3}{c_4} \cdot \omega^* \right] \tag{17}$$

$$\dot{z}_4 = (k_i - k_p \cdot c_3) \cdot z_3 - k_p \cdot c_4 \left[c_5 (z_2 \cdot z_4 - x_1 \cdot u_2^0) - T_l - \frac{c_3}{c_4} \cdot \omega^* \right] \tag{18}$$

where z_1 and z_2 are the quadrature and direct axis flux linkage of the rotor, z_3 is the difference between the rotor angular speed and the reference speed and z_4 is the quadrature axis stator current.

The parameters are set as follows: $c_1 = 13.67$, $c_2 = 1.56$, $c_3 = 0.59$, $c_4 = 1176$, $c_5 = 2.86$, $k_p = 0.001$, $k_i = 0.5$, $\omega^* = 181.1$ and $u_2^0 = 4.1$. In this paper, we use the load torque T_l and the ratio between estimated and real rotor time constant k as the bifurcation parameters.

3.2 Numerical Simulation Conditions

In order to reconstruct the BD, we generated five time-series datasets, each of length 5000. We generated the time-series datasets S_i by using Eqs. (15)–(18) together with

$$T_l = -0.1 \cdot \cos\left(\frac{2\pi(i-1)}{8}\right) + 0.6, (i = 1, \cdots, 9) \tag{19}$$

and $k = 3.0$ using a third-order Runge-Kutta method where the step time increment was set to 0.01. Here, we assumed that the z_4-component time-series was measured for the reconstruction of BD. Thus, the z_4-component time-series was used as S_i. The number of input, hidden and output neurons of the ELM were 4, 30 and 4, respectively. The values of the parameters σ, ϵ and ζ in the sigmoid function were set to 10, 4 and 0.1, respectively.

3.3 Estimating the Number of Significant Parameters

We estimate the number of significant parameters using the contribution ratio. Figure 2 (a) shows the contribution ratio of the trained ELM with the induction motor drive data. The number of significant parameters is estimated as one, because only the first principal component is large.

3.4 Bifurcation Path and Locus

A bifurcation path is a sequence of points in the parameter space of the target system, and a bifurcation locus is a sequence of points in the space of the principal component coefficients. Therefore, we can estimate that the space of principal component coefficients corresponds to the parameter space of the target system when the relationships between the points in the bifurcation path and locus are similar. The first principal component coefficients are given by

$$\gamma_1 = \mathbf{v}_1^{-1} \cdot \tilde{\boldsymbol{\beta}} \tag{19}$$

where $\mathbf{v}_1$ is the eigenvector of the first principal component. Figures 2(b) and (c) show the bifurcation path and locus, respectively. As the relationships between the points in these figures are similar, we can use the space of the first principal component instead of the parameter space of the target system.

3.5 Reconstruction of the Bifurcation Diagram

Figures 3(a) and (b) show the original BD and the one reconstructed with the Lyapunov exponents, respectively. The top and bottom panels of each figure show the BD and the Lyapunov exponents, respectively. The solid and dashed lines of the bottom panel of each figure show the largest and the second largest Lyapunov exponents, respectively. Comparing these figures, one can see that the reconstructed BD is qualitatively the same as that of the original system. The period-doubling bifurcation and the window of the reconstructed BD correspond indeed to those of the original BD. In addition, a quantitative comparison of the largest Lyapunov exponents shows that the Lyapunov exponents also correspond to the original ones. The largest Lyapunov exponents of the reconstructed BD are almost equivalent to the those of the original BD.

4   Numerical Simulations for Reconstruction of BD with the Ratio Between Estimated and Real Rotor Time Constant from Noise Free Time-Series

In this section, we show the results for the reconstruction of BD using noise free time-series datasets generated with different value of the ratio between estimated and real rotor time constant k.

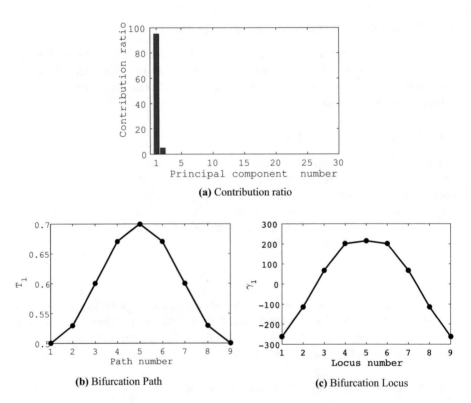

(a) Contribution ratio

(b) Bifurcation Path

(c) Bifurcation Locus

Fig. 2. Contribution ratio, and Bifurcation path and locus when a bifurcation parameter is the load torque

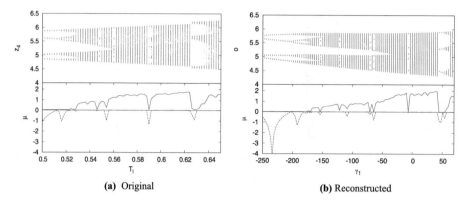

(a) Original

(b) Reconstructed

Fig. 3. Original and reconstructed BD from noise free time-series datasets when a bifurcation parameter is the load torque

4.1 Numerical Simulation Conditions

We generated the time-series datasets S_i by using Eqs. (15)–(18) together with

$$k = -0.1 \cdot \cos\left(\frac{2\pi(i-1)}{8}\right) + 3.1, (i = 1, \cdots, 9) \tag{20}$$

and $T_l = 0.5$. The other conditions are the same as Sect. 3.

4.2 Estimating the Number of Significant Parameters

We show the contribution ratio in Fig. 4(a). The number of significant parameters is estimated as one, because the first principal component is almost 100%.

4.3 Bifurcation Path and Locus

Figures 4(b) and (c) show the bifurcation path and locus, respectively. As the relationships between the points in these figures are similar, although they are in the opposite relationship. Therefore, we see that an opposite space of the first principal component $-\gamma_1$ corresponds to the parameter space k of the target system.

4.4 Reconstruction of the Bifurcation Diagram

Figures 5(a) and (b) show the original BD and the one reconstructed with the Lyapunov exponents, respectively. Comparing these figures, one can see that the reconstructed BD is qualitatively the same as that of the original system. The period-doubling bifurcation and the window of the reconstructed BD correspond indeed to those of the original BD. In addition, a quantitative comparison of the largest Lyapunov exponents shows that the Lyapunov exponents also correspond to the original ones. The largest Lyapunov exponents of the reconstructed BD are almost equivalent to the those of the original BD.

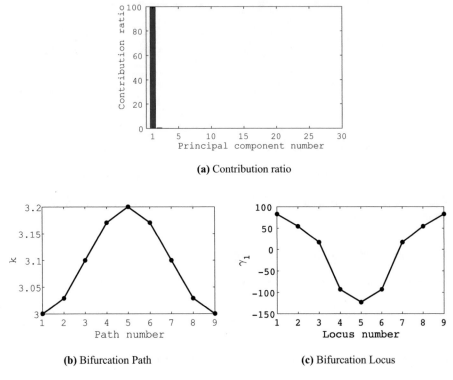

(a) Contribution ratio

(b) Bifurcation Path

(c) Bifurcation Locus

Fig. 4. Contribution ratio, and Bifurcation path and locus when a bifurcation parameter is the ratio between estimated and real rotor time constant

5 Numerical Simulations for Reconstruction of BD from Noisy Time-Series

In this section, we show the results of the reconstruction of BD using noisy time-series datasets. It is important for the real-world system to perform effective noise counter-measures. Therefore, we show the results for reconstructing the BD where we add the noise to the time-series datasets $S_1, \cdots, S_K$. Tokuda *et al.* have reconstructed a BD of Rössler equations using noisy time-series datasets (Tokuda *et al.* 1996). Thus, we referred (Tokuda *et al.* 1996) about noise and filtering in this numerical simulation.

5.1 Noise and Moving Average Filtering

We add a Gaussian noise to time-series datasets generated for reconstructing the BD. Here, mean and standard deviation of the Gaussian noise were set to 0.0 and 0.02, respectively. And we used a moving average filtering for noisy time-series datasets by

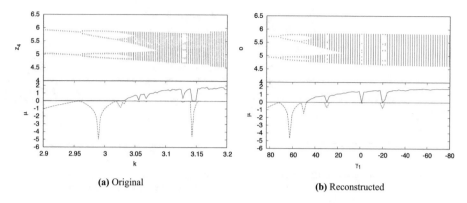

(a) Original   **(b)** Reconstructed

Fig. 5. Original and reconstructed BD from noise free time series datasets when a bifurcation parameter is the ratio between estimated and real rotor time constant

$$\bar{d}(t) = \frac{1}{\xi} \sum_{i=t}^{t+\xi} \hat{d}(i), (t = 1, \cdots, N) \tag{21}$$

where $\hat{d}$ and $\bar{d}$ are the noisy and a filtered time-series data, respectively, and ξ is a window length of moving average. In this numerical simulation, the window length was set to 5.

5.2   Reconstruction of the Bifurcation Diagram

We show the results for reconstructing bifurcation diagram of induction motor drives with the load torque and the ratio between estimated and real rotor time constant as bifurcation parameter. Here, the numerical simulation conditions are the same as Sects. 3 and 4.

In both numerical simulations, the numbers of significant parameters are estimated as one. Because, the contribution ratios of first principal component are more than 90%. And the relationships between the points in bifurcation path and locus are almost similar. Here, the both parameter spaces T_l and k correspond to opposite spaces of the first principal component $-\gamma_1$.

Figures 6(a) and (b) show the results for reconstructing the BD using noisy time-series datasets where the bifurcation parameters in Figs. 6(a) and (b) are the load torque and the ratio between estimated and real rotor time constant, respectively. We see that the reconstructed BDs in Figs. 6(a) and (b) are qualitatively similar bifurcation structure, comparing to the original BDs in Figs. 3(a) and 5(a), respectively. In addition, the Lyapunov exponents also correspond to the original one, though we used noisy time-series datasets.

We show that the BDs of the induction motor drives can be reconstructed using the noisy time-series datasets by applying the moving average filtering.

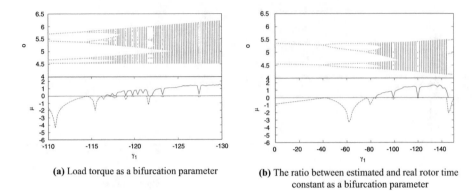

(a) Load torque as a bifurcation parameter

(b) The ratio between estimated and real rotor time constant as a bifurcation parameter

Fig. 6. Reconstructed bifurcation diagram with Lyapunov exponents from noisy time-series datasets

6 Conclusion

We have reconstructed the BD of a mathematical model of induction motor drives using an ELM. By reconstructing the BD of induction motor drives, we have shown that the reconstruction of BDs can be useful for real world systems. In addition, we have shown that the Lyapunov exponents of the reconstructed BD also correspond to those of the original BD, and that the largest Lyapunov exponents of the reconstructed BD are estimated to be almost equivalent to the those of the original BD. Moreover, we have shown that the BD of the induction motor drives can be reconstructed the BD using noisy time-series datasets.

In future work, we will demonstrate the reconstruction of a 2-dimensional BD for a mathematical model of induction motor drives. In addition, we will attempt to reconstruct the BD using real measurement data taken from actual induction motor drives.

Acknowledgement. This work was supported by Tateisi Science and Technology Foundation.

References

Bagarinao, E., Pakdaman, K., Nomura, T., Sato, S.: Reconstructing bifurcation diagrams from noisy time series using nonlinear autoregressive models. Phys. Rev. E **60**(1), 1073–1076 (1999a)

Bagarinao, E., Pakdaman, K., Nomura, T., Sato, S.: Time series-based bifurcation diagram reconstruction. Physica D **130**, 211–231 (1999b)

Bagarinao, E., Pakdaman, K., Nomura, T., Sato, S.: Reconstructing bifurcation diagrams of dynamical systems using measured time series. Meth. Inf. Med. **39**, 146–149 (2000)

Chakrabarty, K., Kar, U.: Bifurcation and control of chaos in induction motor drives. arXiv preprint (2014)

Huang, G.B., Zhu, Q.Y., Siew, C.K.: Extreme learning machine: theory and applications. Neurocomputing **70**, 489–501 (2006)

Itoh, Y., Adachi, M.: Reconstruction of bifurcation diagrams using an extreme learning machine with a pruning algorithm. In: 2017 International Joint Conference on Neural Network, pp. 1809–1816 (2017a)

Itoh, Y., Adachi, M.: A quantitative method for evaluating reconstructed one-dimensional bifurcation diagrams. J. Comput. **13**(3), 271–278 (2017b)

Itoh, Y., Tada, Y., Adachi, M.: Reconstructing bifurcation diagrams with Lyapunov exponents from only time-series data using an extreme learning machine. Nonlinear Theory Appl. IEICE **8**(1), 2–14 (2017c)

Langer, G., Parlitz, U.: Modeling parameter dependence from time series. Phys. Rev. E **70**, 056217 (2004)

Ogawa, S., Ikeguchi, T., Matozaki, T., Aihara, K.: Nonlinear modeling by radial basis function networks. IEICE Trans. Fundam. **E79-A**(10), 1608–1617 (1996)

Tokuda, I., Kajiwara, S., Tokunaga, R., Matsumoto, T.: Recognizing chaotic time-waveforms in terms of a parametrized family of nonlinear predictors. Physica D **95**, 380–395 (1996)

Tokunaga, R., Kajiwara, S., Matsumoto, T.: Reconstructing bifurcation diagrams only from time-waveforms. Physica D **79**, 348–360 (1994)

Ensemble Based Error Minimization Reduction for ELM

Sicheng Yu[1], Xibei Yang[1,2(✉)], Xiangjian Chen[1], and Pingxin Wang[1]

[1] School of Computer Science, Jiangsu University of Science and Technology,
Zhenjiang 212003, Jiangsu, People's Republic of China
yangxibei@hotmail.com

[2] School of Economics and Management, Nanjing University of Science
and Technology, Nanjing 210094, People's Republic of China

Abstract. For better behavior of Extreme Learning Machine (ELM) in the limited condition that the number of training samples less than proper, an error minimization reduction method was employed to ELM. When extends to ensemble strategy, the method does some contribute to learning ability. Experimental results on several UCI data sets show the algorithm we propose effective for promoting learning performance.

Keywords: Extreme Learning Machine · Generalization error
Empirical error · Ensemble learning

1 Introduction

Since the theory of Extreme Learning Machine was proposed [1,2], it has attracted much attention. Profiting from the idea of randomizing hide nodes, the ELM has higher learning speed than conventional single hidden layer feed-forward neural network (SLFN), and provides new perspective of SLFN that reveals the process of biological learning. However, naive ELM has deficiency in practical classification tasks, especially in the face of the data with a small number of samples and complex features at the same time. To fill this gap, ensemble strategy was applied to ELM [3–6].

Ensemble learning which fuses the results from multiple individual learners aims to get a better result. The differences and effectiveness of the base classifiers is usually considered as the key issue to promote the effectiveness of ensemble classifier [7–9]. Generating base classifiers through different feature subsets is one of the efficient ways to make them different. Dimensionality reduction methods can be used to get proper feature subsets.

Dimensionality reduction aims at reducing the dimension of data that is required in various machine learning processes. The main purpose of reduction is optimizing the learning performance through reducing the complexity of model, improving the fitting precision, and so on. In the field of Rough Set, dimensionality reduction is one of the major research fields, for example Hu et al. have

© Springer Nature Switzerland AG 2019
J. Cao et al. (Eds.): ELM 2017, PALO 10, pp. 70–79, 2019.
https://doi.org/10.1007/978-3-030-01520-6_7

proposed a concept of decision error on the neighborhood rough set [10]. The reduct via decision error has effectively promoted the learning performance of neighborhood classifiers. Inspired by the research findings on Rough Set [10–13] and generalization error of ELM [14,15], learning performance of ELM may be elevated via dimensionality reduction according to error minimization. This is what our idea comes from.

The rest of this paper is organized as follows. In Sect. 2, we review the background knowledge that is vital for our research including conditional ELM model and error theory. In Sect. 3, we present the error minimization based reduction method and its application on ELM. In Sect. 4, we propose the reduction based ensemble strategy on ELM. In Sect. 5, the proposed algorithms are tested through several data sets from UCI machine learning repository. Section 6 shows a brief summary of this paper and an outlook.

2 Background Knowledge

2.1 Extreme Learning Machine (ELM)

ELM is a newly proposed learning algorithm based on a SLFN which randomized hide nodes. Compared to conventional learning algorithms based on SLFN, the ELM can be thousands of times faster in learning speed. This is owing to the idea of randomizing hide nodes [1,2].

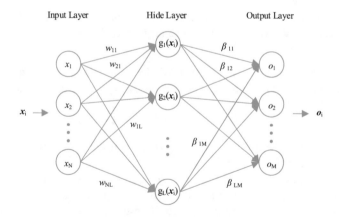

Fig. 1. SLFN based ELM model

Considering a basic ELM model based on SLFN with N input nodes , L hide nodes and M output nodes for S distinct training samples $(\boldsymbol{x}_i, \boldsymbol{t}_i)$ in Fig. 1, its output can be written as:

$$\boldsymbol{o}_i = \sum_{j=1}^{\tilde{N}} \boldsymbol{\beta}_j g_j(\boldsymbol{x}_i), \ i = 1, 2, 3, \ldots, S \tag{1}$$

where $\boldsymbol{\beta}_j = [\beta_{j1}, \beta_{j2}, \ldots, \beta_{jM}]^T$ is the weight vector connecting the jth hide node and the output nodes. $g_j(\boldsymbol{x}_i)$ denotes the output of jth hide node according to the input instance $\boldsymbol{x}_i = [x_1, x_2, \ldots, x_N]^T$ that,

$$g_j(\boldsymbol{x}_i) = f(\boldsymbol{w}_j \cdot \boldsymbol{x}_i + b_j) \tag{2}$$

where $f(\cdot)$ is the aviation function of hide nodes, it can be "Sigmoid", "RBF" and so on, $\boldsymbol{w}_j = [w_{1i}, w_{2J}, \ldots, w_{Nj}]$ is the weight vector connecting the input nodes and the jth hide node, b_j is the threshold of the jth hide node.

In the ELM model $\boldsymbol{w}_j$ and b_j are constant during learning procedure. Hence the algorithm reaches the minimal training error by solving the linear optimization problem,

$$\min_{\beta} \|\boldsymbol{H}\boldsymbol{\beta} - \boldsymbol{T}\| \tag{3}$$

where $\boldsymbol{H}$ is the output matrix of hidden layer, $\boldsymbol{T}$ is the label matrix of training samples, $\boldsymbol{\beta}$ is the weight matrix connecting the hidden layer and output layer.

Further more, Huang et al. gave the solution for ELM [1,2],

$$\boldsymbol{\beta} = \boldsymbol{H}^{\dagger}\boldsymbol{T} \tag{4}$$

where $\boldsymbol{H}^{\dagger}$ is the Moore-Penrose generalized inverse of matrix $\boldsymbol{H}$.

2.2   Error on ELM

Error measures are widely used in order to evaluate the performance of a classifier. We discuss two kinds of error measure on ELM as follow.

Generalization Error is a measure of the ability of a predictor to predict previously unseen data. Formally the generalization error can be described as,

$$\epsilon_h = P_{(x,y)\ D}(h(x) \neq y) \tag{5}$$

where h is the prediction function, (x, y) is the sample produced by a particular distribution D, P is the probability.

When investigating a specific data set or even a part, the universe of discourse limited. Then the generalization error can be approximated by leave-one-out cross-validation error. We employ the leave-one-out cross-validation on ELM, and call the measure "decision error" for convenience. The decision error of the initialized ELM model $elm0$ about the special training set $\hat{U}$ and a set of feature A is denoted by $\text{ELMDR}(elm0, \hat{U}, A)$

While the training samples and test samples are coming from one independent and identically distribution, it is obvious that a smaller value of decision error on training samples will lead to higher classification precision on test samples.

Empirical Error also known as training error, is a typical measure of the model's fitting precision about training data. Formally,

$$\hat{\epsilon}_h = \frac{card(\{x \mid h(x) = y, \forall (x,y) \in \hat{U}\})}{card(\hat{U})} \tag{6}$$

where h is the prediction function, $\hat{U}$ is the set of training samples. With introducing of empirical error, we can decide weather an ELM is underfitting.

3   Error Minimization Reduction on ELM (EM-ELM)

In some classification tasks, a feature subset does well than original feature set because of the following reason.

To solve the problem, we first make assumptions as follows:

1. Due to the limitation of conditions, the structure of ELM may be too simple to represent the inherit prediction function perfectly in practical learning tasks. In this case, the redundant feature would obviously influence the learning performance.

2. As a model based estimator, the training result is usually characterized by a optimal hyper plane. If the training samples have only one dimension, then there are only 2 samples needed for deciding the hyper plane. When the number of dimension turns to two, more than 3 samples will be needed for decision. Sometimes we need to deal with the real world problem that complexity and contains far less proper number of training samples. In this case, lower dimension seems friendlier to model based estimators such as ELM.

According to the description above, we employ the error minimization based reduction on ELM. The definition of an error minimization based reduction on ELM is shown below.

Definition 1. *Given an initialized ELM model elm0, a set of training samples $\hat{U}$ and its feature set AT, A is refer to as a decision error minimization reduction(EM-ELM-D) on elm0 if and only if,*
$A \neq \emptyset$ *and* $\mathrm{ELMDR}(elm0, \hat{U}, A) \leq \mathrm{ELMDR}(elm0, \hat{U}, B), \forall B \subseteq AT$.

After the ELM model is initialized, its hide nodes fixed, the result of training would be stable if training data invariant. In this premise, we introduce the decision error minimization based heuristic reduction to ELM. The algorithm is shown in Algorithm 1.

We can see, in every round of the algorithm the ELM model should be trained independently for N times. Thus, time complexity of EM-ELM-D is $O(M^2 N^4)$, where M is the number of features, N is the number of samples.

Similarly, we can replace the decision error with empirical error to form an empirical error minimization based reduction algorithm (EM-ELM-T). The time complexity of EM-ELM-T is $O(M^2 N^3)$.

Algorithm 1. Decision error minimization based heuristic reduction on elm(EM-ELM-D)

Input: A set of training data $\hat{U}$, A set of features on $\hat{U}$ denoted by AT, an initialized ELM model $elm0$

Output: A minimization error reduct Red

    $Red \leftarrow \emptyset$

    $curDR \leftarrow +\inf$

    $preDR \leftarrow curDR$

    while $card(Red) < card(AT)$ **do**

        $j = \arg\min_{i}\{ \text{ELMDR}(elm0, \hat{U}, Red \cup a_i)) \mid \forall a_i \in (AT - Red)\}$

        $preDR \leftarrow curDR$

        $curDR \leftarrow \text{ELMDR}(elm0, \hat{U}, Red \cup \{a_j\})$

        if $preDR <= curDR$ **then**

            break

        end if

        $Red \leftarrow Red \cup \{a_j\}$

    end while

    return Red

---

4   Ensemble Method on EM-ELM(EEM-ELM)

Based on the algorithm in the previous section. We can propose the ensemble method, and its train and prediction process are shown in Fig. 2.

Differing from conventional voting based ELM, in the proposed method, a reduction procedure is inserted before training the model. While the reduction promotes the efficiency of each base ELM, their voting result will be better. This can be explained as follow.

Considering a binary classification issue. When individually executing the initialization of ELM for n times to get n individual ELM classifiers. The classification precision of classifiers denote as $p_1, \ldots, p_n$. Then the classification precision of the voting result fusing the n ELMs is,

$$P = \underbrace{p_1 \cdots p_n + \bar{p}_1 p_2 \cdots p_n + \cdots + \bar{p}_1 \cdots \bar{p}_{\frac{n}{2}} p_{\frac{n}{2}+1} \cdots p_n}_{\binom{n}{0}+\cdots+\binom{n}{\frac{n}{2}}} \tag{7}$$

assume a better classification is substituted for the classification 1. The new classification precision $p'_1 = p_1 + \alpha \geq p_1, 0 \leq \alpha \leq 1 - p_1$. Then we get the new fusing precision of classification,

$$P_1 = \underbrace{p'_1 \cdots p_n + \bar{p'}_1 p_2 \cdots p_n + \cdots + \bar{p'}_1 \cdots \bar{p}_{\frac{n}{2}} p_{\frac{n}{2}+1} \cdots p_n}_{\binom{n}{0}+\cdots+\binom{n}{\frac{n}{2}}} \tag{8}$$

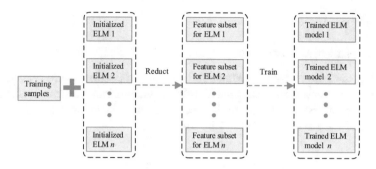

(a) Train process of EEM-ELM

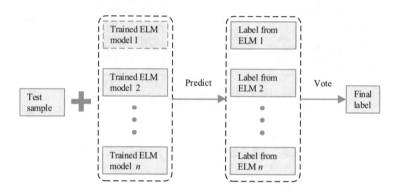

(b) Prediction process of EEM-ELM

Fig. 2. The structure of EEM-ELM

the difference between P_1 and P is,

$$P_1 - P = \alpha\bar{p}_1 \left(\underbrace{\bar{p}_2 \cdots \bar{p}_{\frac{n}{2}}}_{\frac{n}{2}-1} p_{\frac{n}{2}+1} \cdots p_n + \cdots + \underbrace{\bar{p}_{\frac{n}{2}+2} \cdots \bar{p}_n}_{\frac{n}{2}-1} p_1 \cdots p_{\frac{n}{2}+1} \right)$$
$$\underbrace{\phantom{\bar{p}_2 \cdots \bar{p}_{\frac{n}{2}} p_{\frac{n}{2}+1} \cdots p_n + \cdots + \bar{p}_{\frac{n}{2}+2} \cdots \bar{p}_n p_1 \cdots p_{\frac{n}{2}+1}}}_{\binom{n-1}{\frac{n}{2}-1}} \tag{9}$$

$$\geq 0$$

similarly, when replacing the classifier 2 in P_1 by a better one to generate P_2, we have $P_2 \geq P_1 \geq P$, and so on.

Since all the classifiers are replaced by better ones, we finally get a better result than original voting based result.

5   Experiments

5.1   Experimental Data Sets

Several data sets from the UCI Machine Learning Repository were used to validate the performance of the proposed algorithm: Glass Identification, Ionosphere, Libras Movement, Seeds, Wine, and Wisconsin Diagnostic Breast Cancer. The description of data sets is in Table 1 below:

Table 1. Data set description

ID	Name	Samples	Features	Classes
1	Glass Identification	214	9	6
2	Ionosphere	351	34	2
3	Libras Movement	360	90	15
4	Seeds	210	7	3
5	Wine	178	13	3
6	Wisconsin Diagnostic Breast Cancer	569	30	2

we apply leave-one-out cross-validation on each data set in our experiment.

5.2   Experimental Setup

We test the proposed EEM-ELM-D algorithm and EEM-ELM-T algorithm which mentioned in the previous section by comparing with the voting based ELM [6] (vote-ELM). For the vote-ELM, we executed the ELM algorithm many times to generate the ensemble via voting. In the series of experiments, we set the aviation function of hidden layers as Sigmoid function, both algorithms have 50 base estimators, and the number of hide nodes are the same in each independent experiment.

All experiments are run on the PC with 1.1 Ghz CPU and 8 GB RAM via Python3 platform.

5.3   Experimental Results and Discussion

We generate the experimental results, and plot the comparison line charts in Fig. 3. The number of hide node $\tilde{N}$ is varied in the set $\{1, 2, 3, 5, 10, 20, 40, 80\}$.

From the Fig. 3, we can see:

1. The error minimization method tends to reach a lower error rate in the condition of the same number of hide nodes with vote-ELM in most cases. Especially, the simpler is the structure, the more obvious is the advantage of error minimization method.

For example, when employed 3 hide nodes to the base ELM models the leave-one-out cross-validation error rate of EEM-ELM-T and EEM-ELM-D are about

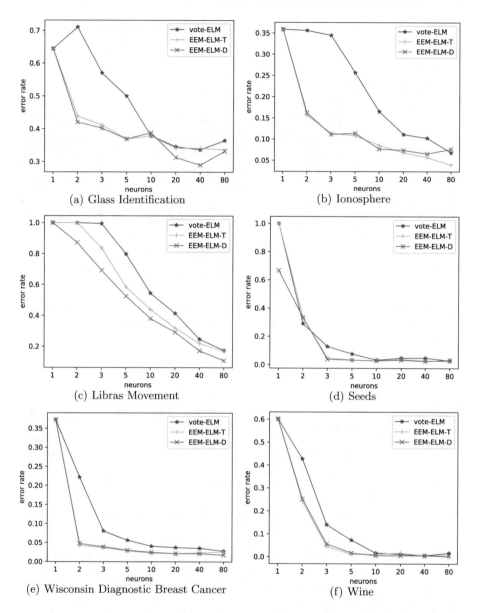

Fig. 3. Leave-one-out cross-validation error for vote-ELM, EEM-ELM-T, and EEM-ELM-D

20% lower than the conventional voting based ELM in the experiment on the Ionosphere data set.

2. When the number of hide nodes becomes large, the resistance of error minimization method to overfitting seems better than conventional vote-ELM.

It is quite evidence the experiment on the Glass Identification data set.

3. In most cases, the proposed EEM-ELM-T and EEM-ELM-D are similar results.

6 Conclusion

In this paper, we investigate the error minimization based reduction on ELM and the relevant ensemble strategy. Our experiments shows that, while random initialized ELMs are internal different from each others, both decision error and empirical error minimization based reduction can promote the learning performance. And both the ensemble methods EEM-ELM-T and EEM-ELM-D lead to better results via voting than conventional voting based method in most cases. However, the training speed of our model is not very desirable, and there is still the defect of instability.

The next step of our work will focus on the following aspects:

1. Develop a fast and stable training algorithm.

2. Ensure the model is efficient in different conditions.

Acknowledgments. This work is supported by the Natural Science Foundation of China(Nos. 61572242, 61503160, 61502211), Qing Lan Project of Jiangsu Province of China.

References

1. Huang, G.B., Zhu, Q.Y., Siew, C.K.: Extreme learning machine: theory and applications. Neurocomputing **70**(1), 489–501 (2006)
2. Huang, G.B., Zhou, H., Ding, X., Zhang, R.: Extreme learning machine for regression and multiclass classification. IEEE Trans. Syst. Man Cybern. Part B (Cybern.) **42**(2), 513–529 (2012)
3. Liu, N., Wang, H.: Ensemble based extreme learning machine. IEEE Signal Process. Lett. **17**(8), 754–757 (2010)
4. Xue, X., Yao, M., Wu, Z., Yang, J.: Genetic ensemble of extreme learning machine. Neurocomputing **129**, 175–184 (2014)
5. Termenon, M., Graña, M., Savio, A., Akusok, A., Miche, Y., Björk, K.M., Lendasse, A.: Brain MRI morphological patterns extraction tool based on extreme learning machine and majority vote classification. Neurocomputing **174**, 344–351 (2016)
6. Cao, J., Lin, Z., Huang, G.B., Liu, N.: Voting based extreme learning machine. Inf. Sci. **185**(1), 66–77 (2012)
7. Wang, X.Z., Xing, H.J., Li, Y., Hua, Q., Dong, C.R., Pedrycz, W.: A study on relationship between generalization abilities and fuzziness of base classifiers in ensemble learning. IEEE Trans. Fuzzy Syst. **23**(5), 1638–1654 (2015)
8. Sun, B., Wang, J., Chen, H., Wang, Y.t.: Diversity measures in ensemble learning. Control Decis. **29**(3), 385–395 (2014)
9. Zhou, Z.H., Yu, Y.: Ensembling local learners through multimodal perturbation. IEEE Trans. Syst. Man Cybern. Part B (Cybern.) **35**(4), 725–735 (2005)
10. Hu, Q., Pedrycz, W., Yu, D., Lang, J.: Selecting discrete and continuous features based on neighborhood decision error minimization. IEEE Trans. Syst. Man Cybern. Part B (Cybern.) **40**(1), 137–150 (2010)

11. Yang, X., Qi, Y., Yu, H., Song, X., Yang, J.: Updating multigranulation rough approximations with increasing of granular structures. Knowl. Based Syst. **64**, 59–69 (2014)
12. Liu, Y., et al.: Maximum relevance, minimum redundancy band selection based on neighborhood rough set for hyperspectral data classification. Measur. Sci. Technol. **27**(12), 125501 (2016)
13. Xu, L., Ding, S., Xu, X., Zhang, N.: Self-adaptive extreme learning machine optimized by rough set theory and affinity propagation clustering. Cogn. Comput. **8**(4), 720–728 (2016)
14. Liu, Q., Yin, J., Leung, V.C., Zhai, J.H., Cai, Z., Lin, J.: Applying a new localized generalization error model to design neural networks trained with extreme learning machine. Neural Comput. Appl. **27**(1), 59–66 (2016)
15. Wang, X.Z., Shao, Q.Y., Miao, Q., Zhai, J.H.: Architecture selection for networks trained with extreme learning machine using localized generalization error model. Neurocomputing **102**, 3–9 (2013)

The Parameter Updating Method Based on Kalman Filter for Online Sequential Extreme Learning Machine

Xiaoming Xu[✉], Chenglin Wen, Weijie Chen, and Siyu Ji

School of Automation, Hangzhou Dianzi University, Hangzhou 310018, China
xuxiaoming0218@163.com, wencl@hdu.edu.cn

Abstract. The recursive least-square algorithm in the on-line sequential extreme learning machine (OS-ELM) is well known for its good convergence property and least square error in stationary system. But the performance in unstable system is not so good due to the matrix singularity and forgetting factor trade-off. In this paper, an improved OS-ELM algorithm is introduced based on kalman filter (KOS-ELM) for online parameter updating through considering the modeling error into the equations to avoid the matrix singularity and randomly walk character into the parameters to handle the non-stationary. The regression experiment which are simulated both in stable and unstable condition system demonstrate that compared with the classic overlay ELM and OS-ELM, the proposed method achieves higher accuracy and stability.

Keywords: Extreme learning machine · OS-ELM · Kalman filter
Parameter estimation

1 Introduction

Feedforward neural network is one of the most popular regression methods which can approximate the complex nonlinear mapping effectively based on the input samples, and has been used in many fields [1–3]. On the other hand, it also has some problems, firstly, the training speed based on the gradient learning algorithm is slow, secondly, all the parameters in the neural network need to be solved which is a time consuming task and is unacceptable for many real-world applications that require fast learning speed [4].

A new neural network learning algorithm, extreme learning machine, is proposed by Huang et al [5] to update the hidden layer output weight only which is different from updating all the parameters in the traditional gradient-based algorithm. According to the literature [6–9], the advantage can be summarized

C. Wen—This work is supported by National Natural Science Foundation(NNSF) of China under Grant U1509203,61490701,61333005,U1664264, and Innovation Fund of Hangzhou Dianzi University(CXJJ2017053).

J. Cao et al. (Eds.): ELM 2017, PALO 10, pp. 80–102, 2019.
https://doi.org/10.1007/978-3-030-01520-6_8

as below: firstly the speed is high which only cost few second for most applications; secondly, has better generalization ability; thirdly, unlike the traditional and classic gradient-based learning algorithms which only work for differentiable activation functions, the ELM algorithm could be used to train with many non-differentiable activation functions.

However, in the actual operation of the system, the data for the system model may not be collected already but obtained one by one or chunk by chunk while the extreme learning machine is an off-line learning algorithm that lack the online updating capability.

In order to solve this problem, some sequential learning algorithms for ELM has been proposed [10–12]. Lim et al [10] proposed recursive complex learning machine which is able to improve the performance especially in case of real valued modulation. Compared with many other sequential learning algorithms, Liang et al [11] introduced the recursive algorithm into the ELM and proposed the online sequential extreme learning machine (OS-ELM), the algorithm can recursively update the hidden layer output weight online based on the previous estimated hidden layer output weight and the new coming information. It has been used in many fields [13–17]. In the manufacturing systems [13], the lack of a reliable and accurate real-time signal processing method for handling massive signal data can be solved by applying the online sequential extreme learning machine that inherits the elegant properties of ELM in terms of extremely fast learning speed and high generalization performance. To cope with the non-linearity, randomness and time-variant of deck-motion caused by sea-wave, tide and wind, an online-sequence extreme-learning- machine is introduced for deck-motion prediction. Sun et al [16] proposed an OS-ELM based ensemble classification framework for distributed classification in a hierarchical P2P work which efficiently decrease prediction error with small network overhead.

As the online sequential extreme learning machine algorithm is widely used in many fields, however, through effectively reading and analyzing the relative works, this still appears to be challenging task with three essential difficulties, the problems and the relative researches [18–25] have been shown as follows:

(1) Firstly, the basic idea of ELM is based on the little approximate which uses the least square method to fit data, due to the limit of the least square itself, the equation belongs to inconsistent equation set due to that the equations are more than the variables, so there exists inconsistence among equation set. As to this problem, Lan [18] proposed to comprise several OS-ELM network and used the average value of the outputs of each OS-ELM in the ensemble for the final measurement of the network performance. Du et al [19] presented an improved OS-ELM algorithm based on L2-regularization which can be used to generate the sparse solutions, then utilize it in the initialization phase of the OS-ELM to reduce the dependency of the number of the hidden nodes. However, it is hard to adjust the parameter lamder of the L2-regularization theoretically.

(2) Secondly, the recursive least-square (RLS) algorithm is well known for its good convergence property and small mean square error (MSE) in stationary environments. However, the RLS using a constant forgetting factor cannot

provide satisfactory performance in time-varying or nonstationary environments. As to this problem, Wang et al [20] introduced the kernals into the online sequential extreme learning machine for nonstationary time series prediction. Nobrega et al [21] used the kalman filter method to adjust the variance of the output weight with factor in order to handle the problem of multicollinearity of the OS-ELM. Zhang et al [22–25] proposed the forgetting factor to balance the origin information and the new obtained information. Lim et al [12] pointed the constant forgetting factor in the OS-ELM cannot provide satisfactory performance in time-varying or nonstationary environment and propose an algorithm with adaptive forgetting factor. However, the method by adding forgeting factor may bring a heavy additional complexity.

In this study, we introduce the kalman filter algorithm which has been widely used in parameter updating [26–28] and combined it with extreme learning algorithm for online sequential updating. Compared with the classic OS-ELM and the other relative improved OS-ELM, the following advantages have been made in this paper.

(1) Firstly, the kalman filter is able to be used for recursive updating. Kalman filter method can use the linear system state equation to estimate the state based on the input and output data, the estimation from time n−1 to time n actually use all the information from time 0 to time n−1 which is contained in covariance.

(2) Secondly, we introduce the modelling error to transform the equations into consistent which can be used as a solution for the singularity of the hidden output matrix.

(3) Thirdly, considering the nonstationary in the system, this can be represented as the randomly walk character of the parameters in the system. And as to the need for updating the model by considering the weight, the kalman filter introduce the process noise which can be used as an forgetting operator of the system.

The rest of this paper is organized as follows: in Sect. 2, we review the basic methods of ELM, and the OS-ELM and relative methods briefly. The online sequential process based on kalman filter and the theoretical reasoning to show the advantage of the kalman filter relative to the recursive least square especially in non-stationary are given in Sect. 3. The simulation experiments are stated in Sect. 4 where we make a comparison between OS-ELM and KOS-ELM. Finally, Sect. 5 gives the conclusion and some interesting future work.

2   Foundations

In this section, we firstly introduce the structure of the single layer feedforward neural network (SLFN), then the basic extreme learning machine algorithm, sequential learning method by sequential filter and recursive least square is also demonstrated respectively, at last, the classic online learning extreme learning machine algorithm is introduced.

2.1   The Introduction of the SLFNs

The typical SLFNs consist of the input layer, the hidden layer and the output layer. The input layer and the hidden layer are connected by neural cell, the same as the hidden layer and the output layer. Besides, the n neural cells in input layer are corresponding to the input dimensions, the l neural cells in hidden layer is set based on the specific circumstance, and the m neural cells in output layer are corresponding to the output dimensions.

Based on the above structure, the parameters of the SLFNs which need to be confirmed contain: (1) the weight between the input layer and hidden layer, (2) the active function of the hidden layer, (3) the bias value of the hidden layer, (4) the weight between the hidden Layer.

Assume w_{ij} stands for the weight between input layer and hidden layer, while i means the ith neural cell in input layer and j means the jth neural cell in hidden layer.

$$w = \begin{bmatrix} w_{11} & w_{12} & \cdots & w_{1n} \\ w_{21} & w_{22} & \cdots & w_{2n} \\ \vdots & \vdots & \vdots & \vdots \\ w_{l1} & w_{l2} & \cdots & w_{ln} \end{bmatrix} \tag{1}$$

Assume β_{ij} stands for the weight between hidden layer and the output layer, while i means the ith neural cell in input layer and j means the jth neural cell in hidden layer.

$$\beta = \begin{bmatrix} \beta_{11} & \beta_{12} & \cdots & \beta_{1m} \\ \beta_{21} & \beta_{22} & \cdots & \beta_{2m} \\ \vdots & \vdots & \vdots & \vdots \\ \beta_{l1} & \beta_{l2} & \cdots & \beta_{lm} \end{bmatrix} \tag{2}$$

Assume b stands for the bias value of the hidden layer:

$$b = \begin{bmatrix} b_1 \\ b_2 \\ \vdots \\ b_l \end{bmatrix} \tag{3}$$

If the trainset contains N samples, then the input matrix and the output matrix is as follows:

$$X = \begin{bmatrix} x_{11} & x_{12} & \cdots & x_{1N} \\ x_{21} & x_{22} & \cdots & x_{2N} \\ \vdots & \vdots & \vdots & \vdots \\ x_{n1} & x_{n2} & \cdots & x_{nN} \end{bmatrix} \tag{4}$$

$$Y = \begin{bmatrix} y_{11} & y_{12} & \cdots & y_{1N} \\ y_{21} & y_{22} & \cdots & y_{2N} \\ \vdots & \vdots & \vdots & \vdots \\ y_{m1} & y_{m2} & \cdots & y_{mN} \end{bmatrix} \tag{5}$$

If the active function in the hidden layer is $g(x)$ then the output T of the SLFNs is:

$$T = \begin{bmatrix} t_1 & t_1 & \ldots & t_N \end{bmatrix}_{1 \times N} \tag{6}$$

$$t_j = \begin{bmatrix} t_{1j} \\ t_{2j} \\ \vdots \\ t_{mj} \end{bmatrix}_{m \times l} = \begin{bmatrix} \sum_{i=1}^{l} \beta_{i1} g(w_i x_j + b_i) \\ \sum_{i=1}^{l} \beta_{i2} g(w_i x_j + b_i) \\ \vdots \\ \sum_{i=1}^{l} \beta_{im} g(w_i x_j + b_i) \end{bmatrix}_{m \times l} \quad (j = 1, 2, \ldots, N) \tag{7}$$

where $w_i = [w_{i1}, w_{i2}, \ldots, w_{in}], x_i = [x_{1j}, x_{2j}, \ldots, x_{nj}]^T$

The formula can also be expressed as:

$$H\beta = T' \tag{8}$$

While T' is the transportation of the matrix T, and H is the output matrix of the hidden layer, the concrete form is:

$$H(w_1, w_2, \ldots, l, b_1, b_2, \ldots, b_l, x_1, x_2, \ldots, x_N) \tag{9}$$

$$= \begin{bmatrix} g(w_1 \cdot x_1 + b_1) g(w_2 \cdot x_1 + b_2) \ldots g(w_l \cdot x_1 + b_l) \\ g(w_1 \cdot x_2 + b_1) g(w_2 \cdot x_2 + b_2) \ldots g(w_l \cdot x_2 + b_l) \\ \vdots \\ g(w_1 \cdot x_N + b_1) g(w_2 \cdot x_N + b_2) \ldots g(w_l \cdot x_N + b_l) \end{bmatrix}_{N_0 \times l} \tag{10}$$

2.2   Extreme Learning Machine

The ELM is mainly based on the two algorithms founded by Huang, here we demonstrate the two principles of ELM [11] and then describe the training process briefly.

Theorem 1.1. Given a standard SLFN with N hidden nodes and activation function $g : R \longrightarrow R$ which is infinitely differentiable in any interval, for N arbitrary distinct samples (x_i, t_i), where $x_i \in R^n$ and $t_i \in R^m$, for any w_i and b_i randomly chosen from any intervals of R_n and R, respectively, according to any continuous probability distribution, then with probability one, the hidden layer output matrix H of the SLFN is invertible and $||H\beta - T|| = 0$.

Theorem 1.2. Given any small positive value $\varepsilon > 0$ and activation functiong : $R \longrightarrow R$ which is infinitely differentiable in any interval, there exists $\tilde{N} \leq N$ such that for N arbitrary distinct samples (x_i, t_i), where $x_i \in R^n$ and $t_i \in R^m$, for any w_i and b_i randomly chosen from any intervals of R_n and R, respectively, according to any continuous probability distribution, then with probability one, $||H_{N \times \tilde{N}} \beta_{\tilde{N} \times m} - T_{N \times m}|| < \varepsilon$.

As described in the above theory, the input weight w and the bias b can be generated randomly, we only need to ascertain the number of the neural cells in the hidden layer and the relatively active function, then we can calculate the weight beta of the hidden layer combined with the output T. The detail procedures are as follows:

Step1: Ascertain the number l of the neural cells in the hidden layer. While the number is equal to the number Q of the trainset samples, it can approximate to the trainset with zero error. However, in most times, the l is less than Q as the Q is too large, but it can still approximate to the trainset with the minimal error.

Step2: Generate the input weight w and bias b in hidden layer randomly.

Step3: Choose a active function which is infinitely differentiable as the active function in the hidden layer, then combined with the above variables to calculate the output matrix H of the hidden layer with formular (10).

Step4: Calculate the output weight β:

$$\beta = H^+ T' \tag{11}$$

where H^+ is the pseudo-inverse of the H.

2.3   Sequential Filter

Sequential filter is a centralized fusion algorithm, compared with the augmented filter which has high complexity, it has the advantage of less calculated amount and better expansibility [31].

The process of sequential filter:

Step1: Assume the original estimation of state and error covariance are $\hat{x}^*_{k-1|k-1}$ and $\hat{P}^*_{k-1|k-1}$, use the state equation to predict the state and error covariance are $\hat{x}^*_{k|k-1}$ and $\hat{P}^*_{k|k-1}$;

Step2: Use the observed value $z_1(k)$ to update $\hat{x}^*_{k-1|k-1}$, the new state $\hat{x}^1_{k|k}$ and error covariance $P^1_{k|k}$ can be calculate based on the following equations

$$\begin{cases} \hat{x}^1_{k|k} = E\{x_k|x_0, z_1^{k-1}, z_1(k)\} = E\{x_k|\hat{X}^*_{k-1|k-1}, z_1(k)\} \\ P^1_{k|k} = E\{\tilde{x}^1_{k|k}(\tilde{x}^1_{k|k})^T\} \end{cases} \tag{12}$$

Here $\tilde{x}^1_{k|k} = x_k - \hat{x}^1_{k|k}$ and new state and error covariance are:

$$\begin{cases} \hat{x}^*_{k|k-1} = \hat{x}^1_{k|k} \\ P^*_{k|k-1} = P^1_{k|k} \end{cases} \tag{13}$$

Step3: Use the next ith observed value to update the state and error covariance based on the step2. Now the state and error covariance are as follows:

$$\begin{cases} \hat{x}^N_{k|k} = E\{x_k|x_0, z_1^k\} = E\{x_k|\hat{X}^*_{k-1|k-1}, z_1(k), \dots, z_N(k)\} \\ P^N_{k|k} = E\{\tilde{x}^{N-1}_{k|k}(\tilde{x}^{N-1}_{k|k})^T\} \end{cases} \tag{14}$$

Then the estimated state based on the global observed values is:

$$\begin{cases} \hat{x}^*_{k|k-1} = \hat{x}^N_{k|k} \\ P^*_{k|k-1} = P^N_{k|k} \end{cases} \tag{15}$$

2.4   The Belief Introduction of the Recursive Least Square

Here we will have a brief introduction to the fundamental principles and the deviations for the recursive least square which is used to compare with the online sequential extreme learning machine based on kalman filter we proposed in this paper, and then we will analyze the relationship and difference between them to explain the advantage of the algorithm we proposed.

Recursive least square is the iterative method of the conventional least-squares for multi-constraint linear system [29, 30]. By taking into account dynamic measurement, it solved the problem of global update when the receiving new measurements every time. The details deviation is as follows.

Furthermore, the process for the recursive least square is shown in Fig. 1.

Assume the input at time t is $X(t) = [x(t), x(t-1), ..., x(t-n+1)]^T$, the $\epsilon(t)$ is the delta between the expection value $y(t)$ and the estimation value $\hat{y}(t)$, and $\varphi(t) = [\varphi_1(t), \varphi_2(t), ..., \varphi_n(t)]^T$ is the optimal solution in the least squares sense, then we can get:

$$\hat{y}(t) = \varphi^T(t)X(t) \tag{16}$$

The evaluated error is $\varepsilon(t) = y(t) - \hat{y}(t) = y(t) - \hat{y}(t)$
the least squares performance evaluation is to minimize J

$$J(t) = \sum_{i=1}^{t} \lambda^{t-i} \varepsilon^2(i) \tag{17}$$

Here the λ is the exponential weighting factor, and take the derivative of the cost function J with respect to φ, then we can get the normal function to calculate the optimal solution in the least squares sense.

$$N(t)\varphi^T(t) = M(t) \tag{18}$$

Here the $n \times n$ correlation matrix N(t) and the $n \times 1$ cross-correlation matrix M(t) is:

$N(t) = \sum_{i=1}^{t} \lambda^{t-i} X(i)X^T(i) = \lambda N(t-1) + X(t)X^T(t)$
$M(t) = \sum_{i=1}^{t} \lambda^{t-i} X(i)y(i) = \lambda M(t-1) + X(t)y(t)$

The $\hat{\varphi}(t)$ can be obtained by the matrix inverse lemma, and it is a recursive form in the process of $t = 1, 2, ...$

$$\hat{\varphi}(t) = \hat{\varphi}(t-1) + K(t)\alpha(t) \tag{19}$$

in which the extension are as follows:

$\alpha(t) = y(t) - X^T(t)\hat{\varphi}(t-1)$
$Z(t) = \lambda^{-1}P(t-1)X(t)$
$K(t) = [1 + X^T(t)Z(t)]^{-1}Z(t)$
$P(t) = \lambda^{-1}P(t-1) - K(t)Z^T(t)$
The initial setting is $P(0) = \delta^{-1}I, \hat{\varphi}(0) = 0, I$ is the unit matrix($t = 1, 2, ...$).

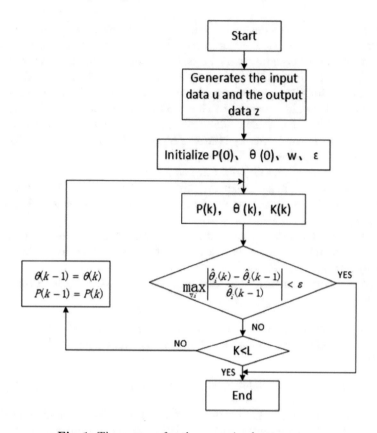

Fig. 1. The process for the recursive least square

2.5 Least Square Based Online Sequential Extreme Learning Machine

As we have known in the introduction, in the actual operating of the system, the data that relative to the system model may not be collected already but obtained one by one or chunk by chunk while the extreme learning machine is an off-line learning algorithm that lack the online update capability. The OS-ELM use the recursive property of the RLS based on the matrix inversion lemma to update the output weights with short time. When the new coming sample arrived, the mathematical model can be modified as below [3]:

$$\begin{bmatrix} H_0 \\ h_{k+1} \end{bmatrix} \hat{\beta} = \begin{bmatrix} T_0 \\ t_{k+1} \end{bmatrix} \tag{20}$$

where H_0 and T_0 are the hidden layer output matrix and output matrix based on the origin samples, and the h_{k+1} and t_{k+1} are the newly generated hidden layer output matrix and relative output based on the new coming sample.

The OS-ELM algorithm includes two phases which are initialization phase and sequential learning phase. In the initialization phase, the algorithm process is almost as the same as the traditional extreme learning machine which is aimed at initializing the net parameters for the single layer feed-forward network, furthermore, the number of the training samples should be more than the number of the hidden nodes according to the theorem 2 to make the training error be approximated to a non-zero decimal, to explain it in mathematical way, the rank of H0 should be equal to the number of the hidden nodes. As to the sequential learning phase, using the recursive property to handle the new coming samples one-by-one or chunk-by-chunk. The procedures of the OS-ELM are can be summarized as follows:

The procedures of the OS-ELM are as follows:

/* Initialization phase: given an initial training dataset for learning, and the process is almost as the same as the traditional extreme learning machine algorithm.

Step1: Ascertain the number l of the neural cells in the hidden layer.

Step2: Assign the input weight w and hidden layer bias b in the hidden layer randomly.

Step3: Calculate the initial hidden layer output matrix H_0 with initial training data;

$$H = \begin{bmatrix} g(w_1 \cdot x_1 + b_1)g(w_2 \cdot x_1 + b_2)\ldots g(w_l \cdot x_1 + b_l) \\ g(w_1 \cdot x_2 + b_1)g(w_2 \cdot x_2 + b_2)\ldots g(w_l \cdot x_2 + b_l) \\ \vdots \\ g(w_1 \cdot x_{N_0} + b_1)g(w_2 \cdot x_{N_0} + b_2)\ldots g(w_l \cdot x_{N_0} + b_l) \end{bmatrix}_{N_0 \times l} \tag{21}$$

Step4: Estimate the initial output weight. Here we use some different description for OS-ELM. Compared with the ELM which minimize the $H_0\beta = T_0$ directly, in OS-ELM, the pseudoinverse of H is expressed by $H^\dagger = (H^t H)^{-1}H^T$, as a result, the object function is given as below:

$$\begin{cases} \min \beta_0 = M_0 H_0^T T_0 \\ M_0 = (H_0^T H_0)^{-1}, T_0 = [t_1, t_2, \ldots, t_N]^T \end{cases} \tag{22}$$

Step5: Set k=0, where k is the index representing the number of the chunks of data presented to the network

/* Sequential learning phase: for each newly arrived sample which is one-by-one or chunk-by chunk, define it as k+1,

Step1: Calculate the partial hidden layer output matrix H_{k+1}

$$H_{k+1} = \begin{bmatrix} g(w_1 \cdot x_{\sum_{j=0}^{k} N_j+1} + b_1)g(w_2 \cdot x_{\sum_{j=0}^{k} N_j+1} + b_2)\ldots g(w_l \cdot x_{\sum_{j=0}^{k} N_j+1} + b_l) \\ g(w_1 \cdot x_{\sum_{j=0}^{k} N_j+2} + b_1)g(w_2 \cdot x_{\sum_{j=0}^{k} N_j+2} + b_2)\ldots g(w_l \cdot x_{\sum_{j=0}^{k} N_j+2} + b_l) \\ \vdots \\ g(w_1 \cdot x_{\sum_{j=0}^{k+1} N_j} + b_1)g(w_2 \cdot x_{\sum_{j=0}^{k+1} N_j} + b_2)\ldots g(w_l \cdot x_{\sum_{j=0}^{k+1} N_j} + b_l) \end{bmatrix}_{N_{k+1} \times l}$$
$$\tag{23}$$

Step2: Update the output weight β_{k+1};

$$\begin{cases} P_{k+1} = P_k - P_k h_{k+1}^T h_{k+1} P_k (1 + h_{k+1} P h_{k+1}^T)^{-1} \\ \beta_{k+1} = \beta_k + P_{k+1} h_{k+1}^T (t_{k+1} - h_{k+1} \beta_k) \end{cases} \tag{24}$$

next, set the T_{k+1} according to the below form:

$$T_{k+1} = \left[t_{(\sum_{j=0}^{k} N_j + 1)} t_{(\sum_{j=0}^{k} N_j + 2)} \cdots t_{(\sum_{j=0}^{k+1} N_j)} \right]_{N_{k+1} \times l} \tag{25}$$

Step3: Make k = k+1 and return to step1 in the sequential learning phase if the new chunk data arrived;

3   Proposed Online Sequential Extreme Learning Machine

In this section, firstly, we review the main idea and the detail process of the kalman filter, then we make a theoretical analysis and an example between the kalman filter and the recursive least square to demonstrate the effectiveness of the kalman filter, at last, the process of the online sequential extreme learning machine algorithm based on kalman filter is introduced to updating the hidden layer output weight.

3.1   Kalman Filter

Kalman filter method can use the linear system state equation to estimate the state based on the input and output data, the estimation from time $n-1$ to time n actually use all the information from time 0 to time $n - 1$ which is contained in covariance, so it can be applied in the estimation of the hidden layer output weights.

In short, the Kalman filter is an optimized autoregressive data processing algorithm. Considering a discrete-time dynamic system, which is represented by the process equation describing the state and the observational equation describing the observation.

Process equation:

$$x(k + 1) = Ax(k) + BU(k) + w(k) \tag{26}$$

Measurement equation:

$$z(k) = Hx(k) + y(k) + v(k) \tag{27}$$

where $x(k)$ is the system state at time k, and A and B are system parameters, $U(k)$ is a known non-random control sequence and constant over the sampling interval; H is the parameter of the measurement system, $y(k)$ is the systematic error term of the observing system, is known and is constant at the sampling interval. For the convenience of analysis, it is usually assumed that the process noise $w(k)$ and the observed noise $v(k)$ are zero-mean white noise processes whose correlation matrices are: $E[w(k)w'(k)] = Q$ and $E[v(k)v'(k)] = R$, and the process noise and the observed noise are not related to each other, $E[w(k)v'(k)] = 0$.

Step 1: We knew observation sequence $z(0)$, $z(1)$, $\cdots$, $z(k)$, required finding the optimal linear estimate of $x(k+1)$.

$$\hat{x}(k+1|k) = E[x(k+1|z(0), z(1), \cdots, z(k))] \tag{28}$$

making the variance $E[\tilde{x}(k+1|k)\tilde{x}'(k+1|k)]$ of the estimated error $\tilde{x}(k+1|k)$ is the smallest, and demanding $\tilde{x}(k+1|k)$ is the linear function of $z(0)$, $z(1)$, $\cdots$, $z(k)$, whose estimation is unbiased, meaning $E[\hat{x}(k+1|k)] = E[x(k+1|k)]$.

We assumed that $\hat{x}(k|k-1)$ is the optimal linear optimal prediction of $x(k)$, then we can prove that $\hat{x}(k+1|k)$ is the optimal prediction of $x(k+1)$.

From the above formula we can obtain:

$$\tilde{x}(k+1|k-1) = x(k+1) - \hat{x}(k+1|k-1) \tag{29}$$

which is also equal to:

$$\tilde{x}(k+1|k-1) = A\tilde{x}(k|k-1) + w(k) \tag{30}$$

Since $\hat{x}(k|k-1)$ is the optimal linear prediction of $x(k)$, the estimate error $\tilde{x}(k|k-1)$ must be orthogonal to $z(0)$, $z(1)$, $\cdots$, $z(k)$ according to the orthogonal theorem, so $A\hat{x}(k|k-1)$ should also be orthogonal to $z(0)$, $z(1)$, $\cdots$, $z(k)$. In addition, $w(k)$ is a white noise sequence independent of $z(0)$, $z(1)$, $\cdots$, $z(k)$, so $w(k)$ is orthogonal to $z(0)$, $z(1)$, $\cdots$, $z(k)$. Hence, when $z(k)$ is unknown, $\hat{x}(k+1|k-1)$ is the optimal prediction of $x(k+1)$.

We can utilize the process model of the system to predict the state of the system. Assuming that the current time of the system is $k+1$, we can get one step prediction

$$\hat{x}(k+1|k) = A\hat{x}(k|k) + BU(k+1) \tag{31}$$

Step2: Defining the error covariance P corresponding to the state $x(k+1|k)$:

$$P(k+1|k) = E[\tilde{x}(k+1|k)\tilde{x}'(k+1|k)] \tag{32}$$

since $w(k)$, $v(k)$ and $\tilde{x}(k+1|k)$ are orthogonal to each other, then we can obtain

$$P(k+1|k) = AP(k|k)A' + Q \tag{33}$$

Step3: Finding the optimal linear estimation of $x(k+1)$, because of

$$\hat{x}(k+1|k+1) = E[x(k+1|z(0), z(1), \cdots, z(k))] \tag{34}$$

if making the variance of the estimated error $x(k+1|k+1)$ is the smallest, and is estimated to be unbiased. In the the optimal linear prediction $\hat{x}(k+1|k-1)$ of $x(k+1)$, we obtain the observed value $\hat{z}(k+1|k)$ at the time of $x(k+1)$, assuming that

$$\hat{x}(k+1|k+1) = \hat{x}(k+1|k) + K(k+1)\tilde{z}(k+1|k) \tag{35}$$

and $K(k+1)$ is the optimal gain matrix, which will be derived in the Step 4.

Step4: To obtain the optimal gain matrix $K(k)$, then we can write the optimal estimation of $x(k)$. We utilize orthogonality theorem to derive the matrix $K(k)$. From the above equations, we know

$$\tilde{x}(k+1|k) = (A - K(k)H)\tilde{x}(k|k-1) + w(k) - K(k)v(k) \tag{36}$$

Utilizing the orthogonality theorem knowing that

$$E[\tilde{x}(k+1|k)z'(k)] = 0 \tag{37}$$

By the above two formulas can be obtained

$$E([A - K(k)H]\tilde{x}(k|k-1) + w(k) - K(k)v(k)z'(k)) =$$
$$E([A - K(k)H]\tilde{x}(k|k-1) + w(k) - K(k)v(k)H(\hat{x}(k|k-1)) + \tilde{x}(k|k-1)) \tag{38}$$

Note that $E[\tilde{x}(k+1|k)\hat{x}(k+1|k)] = 0$, and $v(k)$, $w(k)$ are orthorhombic to $\tilde{x}(k|k-1)$, $v(k)$ and $w(k)$ are white noise with zero mean, then $K(k)$ can be obtain

$$K(k) = (AP(k|k-1)H' + S(k))(HP(k|k-1)H' + R(k)) \tag{39}$$

where $P(k|k-1) = E[\tilde{x}(k|k-1)\tilde{x}'(k|k-1)]$, $S(k) = E[w(k)v'(k)]$ and $R(k) = E[v(k)v'(k)]$. Since $w(k)$ and $v(k)$ are orthogonal to each other, we can get the optimal gain matrix as follow

$$K(k) = P(k|k-1)H'(HP(k|k-1)H' + R(k)) \tag{40}$$

Based on the derivation of the optimal gain matrix $K(k+1)$, we can get the optimal linear estimate of $x(k+1)$

$$\hat{x}(k+1|k+1) = \hat{x}(k+1|k) + K(k+1)(z(k+1) - H\hat{x}(k+1|k)) \tag{41}$$

Step5: The final step is to derive the filter error covariance $P(k+1|k+1)$ of the $\tilde{x}(k+1|k+1)$.

On account of

$$\tilde{x}(k+1|k) = A\hat{x}(k|k) + BU(k) \tag{42}$$

then the optimal prediction error of $x(k+1)$ is

$$\tilde{x}(k+1|k) = A\tilde{x}(k|k) + w(k) \tag{43}$$

And

$$\tilde{x}(k+1|k=1) = (I - K(k+1)H)(A\tilde{x}(k|k) + w(k)) - K(k+1)v(k+1) \tag{44}$$

We know that

$$P(k+1|k+1) = E[\tilde{x}(k+1|k+1)\tilde{x}'(k+1|k+1)] \tag{45}$$

Utilizing the orthogonality among $\hat{x}(k+1|k)$, $\tilde{x}(k+1|k)$, $v(k+1)$, then we can get the $P(k+1|k+1)$ as follow

$$P(k+1|k+1) = P(k+1|k) - P(k+1|k)H'(HP(k+1|k)H'+R) + HP(k+1|k)$$
$$= (I - K(k+1)H)P(k+1k) \qquad (46)$$

Moreover, the process for the kalman filter which include the filter calculating loop and the gain calculating loop are shown in Fig. 2.

3.2 Analyzing between RLS and Kalman filter

While both RLS and Kalman filters have the same format of: NewEstimation = PreviousEstimation + Gain * Innovation. There is still a fundamental difference to the gain and its underlying model. When the system is not dynamic, then the kalman filter will reduce to the recursive least square.

Recursive least square is based on weighted least squares in which while the equations are rearranged in a recursive form. It has several advantages such as less memory, inverting a smaller sized matrix in each step, which is most suitable for the online sequential updating, but this is still the least square.

Kalman introduces a new concept of propagation between the steps that are not precise - each time we propagate the estimation, and we will add some "noise". The kalman filter use the new measurement with all the track history that is packed in the previous estimate. The measurement in itself usually would have a larger covariance with R matrix than the covariance of the all the track history with P matrix, but it is newer information. Therefore when we propagate the P matrix using the linear model, we add the extra covariance of the propagation itself - the process noise defined Q matrix. Q adds a fading memory effect that is dependent on the delta time between the steps and the confidence in the propagation model.

As a result, the recursive least square will only use the observations to update the model when the new arrived samples arrived, but the kalman filter with the propagation steps will provide even more information and hence a more accurate

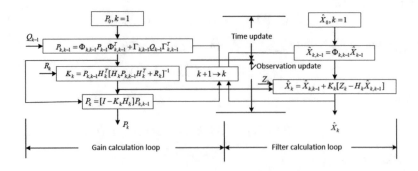

Fig. 2. The process for the kalman filter

estimation. And that is also the reason that we use kalman filter rather than the recursive least square to update the new arrived samples recursively during the online sequential extreme learning machine.

Here we use a simple example to display the difference between kalman filter and recursive least square to demonstrate the effectiveness of kalman filter for parameter updating compare with RLS. As we can see from the Fig. 3, the black real line indicate the inner model we want to follow, the model keeps changing with the time and the observation may exists some noise, during the propagation step, the red real line which stand for the kalman filter always provide even more information and yield a more accurate estimation than the blue imaginary line which stand for recursive least square.

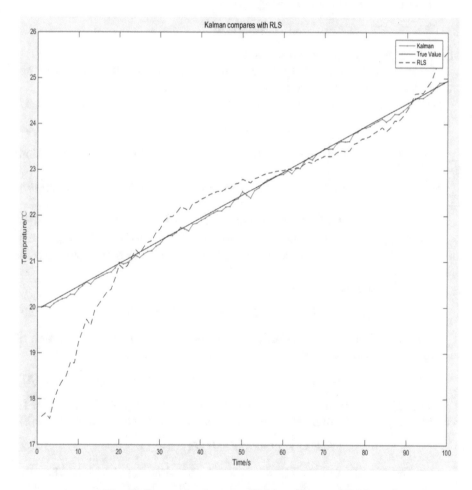

Fig. 3. Comparing between Kalman filter and RLS

3.3   Kalman Filter Based Online Sequential Extreme Learning Machine

The KOSELM algorithm is summarized as bellows. The first part is to calculate the initial β which is the same as the original ELM, the second part is the sequential learning phase by using kalman filter to update the parameter β.

/* Initialization phase: given an initial training dataset for learning,

Step1: Ascertain the number l of the neural cells in the hidden layer.

Step2: Assign the input weight w and hidden layer bias b in the hidden layer randomly.

Step3: Calculate the initial hidden layer output matrix H after choose an active function as formular (20).

Step4: Calculate the output weight β:

$$\beta = H^+T' \tag{47}$$

where H^+ is the pseudo-inverse of the H.

/* Sequential learning phase: for each newly arrived sample which is one-by-one or chunk-by-chunk, define it as k+1.

Step1: Assume the output weight beta is the state x and predict the beta:

$$\beta(k|k-1) = \beta(k-1|k-1) \tag{48}$$

Here, $\beta(k|k-1)$ is the forecast state, $\beta(k-1|k-1)$ is the best state at time $k-1$.

Step2: Next we need to predict the covariance P corresponding to the $\beta(k|k-1)$:

$$P(k|k-1) = AP(k-1|k-1)A' + Q \tag{49}$$

Here $P(k|k-1)$ is the covariance corresponding to $\beta(k|k-1)$, so is $P(k-1|k-1)$ to $\beta(k-1|k-1)$, A' is transposed matrix of A, Q is the covariance of State transfer noise.

Step3: The kalman gain Kg is:

$$Kg(k) = P(k|k-1)H'(HP(k|k-1)H' + R)^{-1} \tag{50}$$

Step4: Based on the prediction of the status, the current best estimation of the state $\beta(k|k)$ can be calculated as follows:

$$\beta(k|k) = \beta(k|k-1) + Kg(k)(Z(k) - H\beta(k|k-1)) \tag{51}$$

Step5: Till now, we have got the best state estimation $\beta(k|k)$, but in order to keep running the kalman filter, we also need to update the covariance of $\beta(k|k)$:

$$P(k|k) = (I - Kg(k)H)P(k|k-1) \tag{52}$$

Here the I is the unit matrix which is 1 to the single model. And when the system turns to the k+1 state, the $P(k|k)$ is the $P(k-1|k-1)$. Then the system can iterate automatically.

Step6: Repeat the process from step1 to step5 till n = N, then assign $\hat{\beta} = \beta_N$ which is the last output weight estimated in the sequential learning step.

4 Performance Evaluation

4.1 Experimental Settings

In this section, the performance of the proposed online sequential extreme learning machine based on kalman filter algorithm (KOS-ELM) is compared with extreme learning machine (ELM) and the online sequential extreme learning machine (OS-ELM). The ELM simply overlay the new coming information to the original train dataset to calculate the beta with the total dataset.

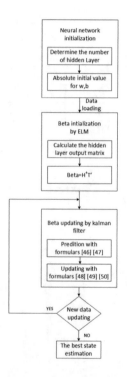

Fig. 4. The process for the OS-ELM based on kalman filter

Table 1. The RMSE and Time cost of the stable system

Nodes	Stable	Overlay ELM		OS-ELM		KOS-ELM	
		Time(/s)	RMSE	Time(/s)	RMSE	Time(/s)	RMSE
20	Sig	2.490	0.041	0.052	0.156	0.025	0.157
20	Sin	2.219	0.037	0.051	0.258	0.024	0.148
50	Sig	5.250	0.035	0.092	0.146	0.117	0.142
50	Sin	4.746	0.037	0.089	0.260	0.094	0.220
100	Sig	13.404	0.041	0.321	0.064	0.319	0.140
100	Sin	13.797	0.039	0.342	0.074	0.319	0.190

The OS-ELM use the origin train dataset to calculate the initial hidden layer output weight beta firstly, then use the recursive least-square method to update the β based on the new coming valuable information [32]. The KOS-ELM use the same method for calculating the initial β as the same as OS-ELM, and then regard the β as the system state and update them with the new coming information based on the kalman filter method (Fig. 4).

Table 2. The RMSE and Time cost of the unstable system

Nodes	Unstable	Overlay ELM		OS-ELM		KOS-ELM	
		Time(/s)	RMSE	Time(/s)	RMSE	Time(/s)	RMSE
20	Sig	2.395	0.505	0.054	0.549	**0.025**	**0.272**
20	Sin	2.085	0.495	0.053	0.566	**0.023**	**0.309**
50	Sig	5.255	0.501	0.092	0.508	0.132	**0.337**
50	Sin	4.619	0.496	0.092	0.499	0.094	**0.327**
100	Sig	12.911	0.493	0.344	0.498	0.355	**0.232**
100	Sin	12.156	0.493	0.333	0.506	0.322	**0.352**

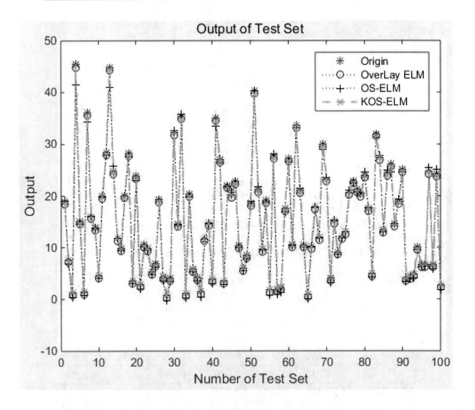

Fig. 5. The output for unstable system based on three methods

As to the initial covariance of the kalman filter, there are some difference with the traditional kalman filter for the initialization of covariance matrix P. Due to the nature of the neural network itself, it is hard to obtain the prior knowledge of the hidden layer output parameters. In general kalman filter algorithm, it is easy to assign an initial value and the influence will be eliminated due to the smoothing parameter, but initialization for the hidden layer output parameters of the extreme learning machine has a much larger range from single digits to the hundreds digits, any small unreasonable covariance may lead to bad results. So in this paper we assign the covariance matrix to be zero and it will change by catching the variation of the system model.

All the three algorithms are used for regression to approximate the function $y = x_1^2 + x_2^2$. As the demand for the problem, we divide the dataset into three parts - Original training dataset, Online training dataset and Test dataset. Furthermore, there are two experiments we take in this paper.

The first experiment is aimed at revealing the adaptive capacity of the three algorithms to the system which may contains noise and new information while operating. Firstly, the original training set is only added by the Gaussian noise which is represented by uniformly randomly distributed in $[-0.2, 0.2]$, secondly, the online training set is added by the Gaussian noise which is as the same as the

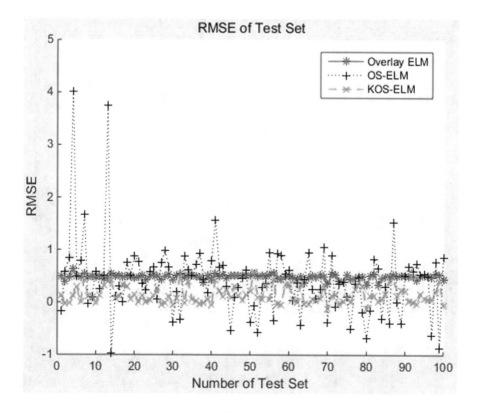

Fig. 6. The RMSE for unstable system based on three methods

original training set and the new information which is represented by the offset 0.5 for the output, thirdly, the test set is only added by the new information which is represented by the offset 0.5.

The second experiment is aimed at exploring the behavior of the three algorithms in the condition of no new information contained while operating. We make the experimental setting which is just like the first experiment described above while without adding the new information that represented by the offset 0.5.

4.2   Statistical Evaluation

The hidden nodes are assigned from 20 to 100 in the ELM for all three algorithms. 50 trials have been conducted for all the algorithms and the average results and standard deviations are shown in Tables 1 and 2.

As to the time cost, it can be seen that both in stationary and non-stationary system, the overlay ELM cost much more time than online sequential algorithm due to that it need to calculate total dataset with every new coming information which increase the computation complexity (Fig. 5).

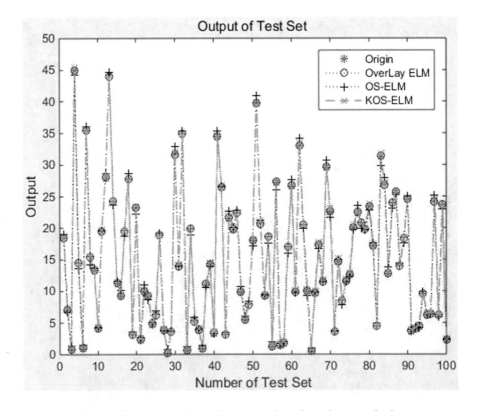

Fig. 7. The output for stable system based on three methods

As to the accuracy which represent by root mean square error (RMSE), in the stable system, the accuracy is good and almost the same with both OS-ELM and KOS-ELM, the reason is that the ELM algorithm can perfectly approximate the model while the hidden nodes is higher than the equation degree of the intrinsic model, as we have chosen more than twenty nodes as the hidden nodes, and it can achieve perfect fit to the model. But in the unstable system, we can see that the KOS-ELM behave better than the overlay ELM and OS-ELM due to that the proposed method can not only treat new coming information with more important weight which increase its real-time performance but also use the random noise to relax the contradiction of the equation set which increase the accuracy.

Figures 6 and 8 show the RMSE for all three algorithms and the true and the approximated function of the three algorithms in the condition of stable and unstable system (Fig. 7).

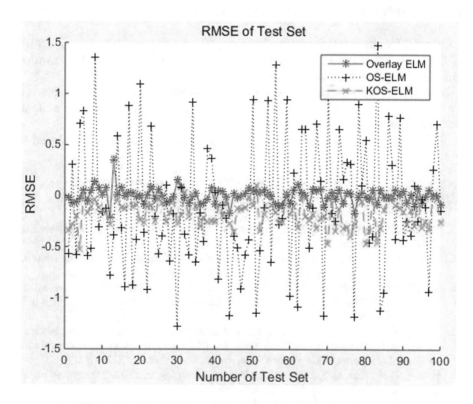

Fig. 8. The RMSE for stable system based on three methods

5   Conclusion

This paper has analyzed the disadvantage of the OS-ELM and proposed a new method combining kalman filter and ELM for online parameter updating. The proposed method can not only learn the models accurately and quickly by ELM but also behave better performance by adjusting the model with the new coming information effectively, as a result, this method would be a robust choice if the system is not sure to be stable or unstable. The simulation based on the regression examples demonstrate the effectiveness of the new method.

Our future work is to exploit the relationship and trendency between the relative initialization of the kalman filter with the extreme learning machine of the SLFN which may be conducive to update the hidden layer output weight more effectively.

6   Compliance with Ethical Standards

Funding: This study was funded by National Natural Science Foundation(NNSF) of China under Grant U1509203,61490701,61333005,U1664264, and Innovation Fund of Hangzhou Dianzi University(CXJJ2017053).

Conflict of Interest: The authors declare that they have no conflict of interest. This article does not contain any studies with human participants or animals performed by any of the authors.

Informed consent was obtained from all individual participants included in the study.

References

1. Ferrari, S., Stengel, R.F.: Smooth function approximation using neural networks. IEEE Trans. Neural Netw. **16**(1), 24–38 (2005)
2. Kwok, T.Y., Yeung, D.Y.: Objective functions for training new hidden units in constructive neural networks. IEEE Trans. Neural Netw. **8**(5), 1131 (1997)
3. Rumelhart, D.E., Hinton, G.E., Williams, R.J.: Learning Internal Representations by Error Propagation. Neurocomputing: Foundations of Research, pp. 318-362. MIT Press (1988)
4. Huang, G.B., Zhu, Q.Y., Siew, C.K.: Extreme learning machine: a new learning scheme of feedforward neural networks. In: IEEE International Joint Conference on Neural Networks, 2004 Proceedings, vol. 2, pp. 985-990. IEEE (2005)
5. Huang, G.B., Zhu, Q.Y., Siew, C.K.: Extreme learning machine: theory and applications. Neurocomputing **70**(1–3), 489–501 (2006)
6. Cao, J., Lin, Z., Huang, G.B.: Self-adaptive evolutionary extreme learning machine. Neural Process. Lett. **36**(3), 285–305 (2012)
7. Wang, W., Vong, C.M., Yang, Y.: Encrypted image classification based on multilayer extreme learning machine. Multidimens. Syst. Signal Process. **28**(3), 851–865 (2017)
8. Zong, W., Huang, G.B., Chen, Y.: Weighted extreme learning machine for imbalance learning. Neurocomputing **101**(3), 229–242 (2013)

9. Huang, G.B., Zhou, H., Ding, X., et al.: Extreme learning machine for regression and multiclass classification. IEEE Trans. Syst. Man Cybern. Part B Cybern. **42**(2), 513-529 (2012). A Publication of the IEEE Systems Man and Cybernetics Society

10. Lim, J., Jeon, J., Lee, S.: Recursive complex extreme learning machine with widely linear processing for nonlinear channel equalizer. In: International Symposium on Neural Networks, pp. 128–134. Springer, Heidelberg (2006)

11. Liang, N.Y., Huang, G.B., Saratchandran, P.: A fast and accurate online sequential learning algorithm for feedforward networks. IEEE Trans. Neural Netw. **17**(6), 1411–23 (2006)

12. Lim, J.S., Lee, S., Pang, H.S.: Low complexity adaptive forgetting factor for online sequential extreme learning machine (OS-ELM) for application to nonstationary system estimations. Neural Comput. Appl. **22**(3–4), 569–576 (2013)

13. Yang, Z., Zhang, P., Chen, L.: RFID-enabled indoor positioning method for a real-time manufacturing execution system using OS-ELM. Neurocomputing **174**(PA), 121-133 (2016)

14. Liu, X.X., Song, Q., Sima, J., Huang, Y.J., Yang, Y.: Deck-motion prediction method for large ship based on online-sequence extreme learning machine. J. Chin. Inert. Technology. **24**(2), 269–274 (2016)

15. Sun, Z.L., Choi, T.M., Au, K.F.: Sales forecasting using extreme learning machine with applications in fashion retailing. Decis. Support. Syst. **46**(1), 411–419 (2009)

16. Sun, Y., Yuan, Y., Wang, G.: An OS-ELM based distributed ensemble classification framework in P2P networks. Neurocomputing **74**(16), 2438–2443 (2011)

17. Budiman, A., Fanany, M.I., Chan, B.: Constructive, robust and adaptive OS-ELM in human action recognition. In: International Conference on Industrial Automation, Information and Communications Technology, pp. 39-45. IEEE (2014)

18. Lan, Y., Soh, Y.C., Huang, G.B.: Ensemble of online sequential extreme learning machine. Neurocomputing **72**(13), 3391–3395 (2009)

19. Du, Z.L., Li, X.M., Zheng, Z.G., Zhang, G.R., Mao, Q.: Extreme learning machine based on regularization and forgetting factor and its application in fault prediction. Chin. J. Sci. Instrum. **36**(7), 1546–1553 (2015)

20. Wang, X., Han, M.: Online sequential extreme learning machine with kernels for nonstationary time series prediction. Neurocomputing **145**(145), 90–97 (2014)

21. Nobrega, J.P., Oliveira, A.L.I.: Kalman filter-based method for online sequential extreme learning machine for regression problems. Eng. Appl. Artif. Intell. **44**(C), 101-110 (2015)

22. Zhang, H., Zhang, S., Yin, Y.: Online sequential ELM algorithm with forgetting factor for real applications. Neurocomputing **261**, 144–152 (2017)

23. Zhao, J., Wang, Z., Dong, S.P.: Online sequential extreme learning machine with forgetting mechanism. Neurocomputing **87**(15), 79–89 (2012)

24. Liu, J.: Adaptive forgetting factor OS-ELM and bootstrap for time series prediction. Int. J. Model. Simul. Sci. Comput. (2017)

25. Schaik, A.V., Tapson, J.: Online and adaptive pseudoinverse solutions for ELM weights. Neurocomputing **149**(PA), 233–238 (2015)

26. Tuan, P.C., Lee, S.C., Hou, W.T.: An efficient on-line thermal input estimation method using kalman filter and recursive least square algorithm. Inverse Probl. Eng. **5**(4), 309–333 (1997)

27. Welch, G., Bishop, G.: An Introduction to the Kalman Filter. University of North Carolina at Chapel Hill (2001)

28. Weill, L.R., Land, D.: The Kalman filter: an introduction to the mathematics of linear least mean square recursive estimation. Int. J. Math. Educ. **17**(3), 347–366 (1986)
29. Xu, Z.Z., Feng, X.L., Wen, C.L.: Sequential fusion filtering for networked multisensor systems based on noise estimation. ACTA ELECTRONICA SINICA **42**(1), 160–168 (2014)
30. Leung, C.S., Young, G.H., Sum, J.: On the regularization of forgetting recursive least square. IEEE Trans. Neural Netw. **10**(6), 1482 (1999)
31. Chi, S.L., Wong, K.W., Sum, P.F.: On-line training and pruning for recursive least square algorithms. Electron. Lett. **32**(23), 2152–2153 (1996)
32. ELM web portal. http://www.ntu.edu.sg/home/egbhuang

Extreme Learning Machine Based Ship Detection Using Synthetic Aperture Radar

Shu-li Jia[1]([✉]), Chong Qu[1], Wenjing Lin[2], Shuhao Cai[2], and Liyong Ma[2]

[1] College of Automation, Shanghai Marine Diesel Engine
Research Institute, Shanghai 201108, China
jiashuli_1019@163.com
[2] School of Information and Electrical Engineering,
Harbin Institue of Technology, Weihai 264209, China

Abstract. Ship detection is an important issue in many aspects, vessel traffic services, fishery management and rescue. Synthetic aperture radar (SAR) can produce real high resolution images with relatively small aperture in sea surfaces. A novel method employing extreme learning machine is proposed to detect ship in SAR. After the image preprocessing, some features including HOG features, geometrical features and texture features are selected as features for ship detection. The experimental results demonstrate that the proposed ship detection method based on extreme learning machine is more efficient than other learning-based methods.

Keywords: Ship recognition · Extreme learning machine
Synthetic aperture radar (SAR)

1 Introduction

Ship detection is an important issue in many aspects, vessel traffic services, fishery management and rescue. Traditional ship detection method such as patrol ships or aircrafts are costly and limited by many circumstances, coverage area and weather condition. Particularly, because of many air cash in recent years, ship detection has become more and more important for ship monitoring and ship searching to save people in time.

Synthetic aperture radar (SAR) can produce images of objects, such as landscapes and sea surfaces. SAR is usually mounted on mobile platforms such as aircrafts. The movement of platform can acquire a lager synthetic antenna and provide better azimuth resolution. Generally speaking, the larger aperture is, the higher image of resolution it will be, no matter the aperture is physical or synthetic. For these reason, SAR can produce a real high resolution image with relatively small aperture. There are many other advantages for us to use SAR images. SAR images can be obtained in many circumstances, regardless of whether it is during day or night, rain or snow. SAR system can be so useful when optical tools can not be used.

© Springer Nature Switzerland AG 2019
J. Cao et al. (Eds.): ELM 2017, PALO 10, pp. 103–113, 2019.
https://doi.org/10.1007/978-3-030-01520-6_9

SAR has been widely used for ship detection. A K-means clustering and land masking method is reported in [1] and proposes a novel method in coastal regions SAR images. A morphological component analysis method is developed in [2] to achieve satisfactory results in complex background SAR images. A new technique using color and texture from spaceborne optical images as a complementary to SAR-based images is employed in [3].

Some ship detection using SAR based on common machine learning algorithms have been reported. Neural network based methods were used in SAR to acquire better results in rough water and false alarm rates in [4]. This method can also use texture features from SAR image to discriminates speckle noise from ship in [5]. A method based on support vector machines (SVM) was combined with grid optimization was employed for false alarm removal in [6]. There are some papers that use deep learning in SAR for ship detection [7]. They present a high network configuration used for ship discrimination.

Extreme learning machine (ELM) is a machine learning algorithm with simple optimization parameters, fast speed and good generalization performance [9–11]. It has been widely used in many applications of image processing and machine vision [12–14]. In comparison with SVM, ELM has a simple implementation and requires less optimization works. Meanwhile, ELM has better learning performance with fast speed for its better generalization ability while SVM can just achieve sub-optimal solutions with higher computational complexity. In this paper, we perform ship recognition in SAR employing ELM method. One of the main contributions of this paper is that an efficient ship recognition method employing ELM with SAR is proposed.

2   Extreme Learning Machine

ELM algorithm includes two steps. The first step is data mapping where input data are mapped into the hidden layer employing random feature mapping or kernel learning approach. The second step is output. The final output can be obtained by multiplying the middle results with their corresponding weights. Different from the traditional learning process of three-layer neural networks where all the parameters are tuned iteratively and severe dependency of the parameters between different layers limits the learning performance, the hidden layer is non-parametric in ELM. This simple policy leads to the smallest training error and the smallest norm of weights, therefore ELM can achieve superior generation performance over other learning approaches for neural networks. ELM is able to improve the performance of single hidden layer neural network which is low and easy to be trapped in local minimums.

Denote the training sample as $(\mathbf{x}_i, \mathbf{t}_i)$, $i = 1, ..., N$, the input feature vectors $\mathbf{x} = [\mathbf{x}_1, \mathbf{x}_2, ..., \mathbf{x}_N]^T \in \mathbb{R}^{D \times N}$, and the label of the supervised sample output $\mathbf{t}_i = [t_{i1}, t_{i2}, ..., t_{iM}]^T \in \mathbb{R}^M$, where N is the sample number, D is the dimension size of the input sample feature vector, and M is the number of network output nodes those are used to solve multi-classification problems.

Denote the number of hidden layer nodes as L, the output of a hidden node indexed by i is

$$g(\mathbf{x}; \mathbf{w}_i, b_i) = g(\mathbf{x} \cdot \mathbf{w}_i + b_i). \tag{1}$$

where $i = 1, ..., N$, $\mathbf{w}_i$ is the input weight vector between the i-th hidden nodes and all input nodes, g is the activation function, and b_i is the bias of this node. A feature mapping function which connects the input layer and the hidden layer is

$$\mathbf{h}(\mathbf{x}) = [g(\mathbf{x}; \mathbf{w}_1, b_1), g(\mathbf{x}; \mathbf{w}_2, b_2), ..., g(\mathbf{x}; \mathbf{w}_L, b_L)]. \tag{2}$$

Denote the output weight between the i-th hidden node and the j-th output node as β_{ij}, where $i = 1, ..., L$, and $j = 1, ..., M$. The value of j-th output node can be obtained by

$$f_j(\mathbf{x}) = \sum_{i=1}^{L} \beta_{ij} \times g(\mathbf{x}; \mathbf{w}_i, b_i)). \tag{3}$$

Thus the output vector of the input sample $\mathbf{x}$ at he hidden layer can be described as

$$\mathbf{f}(\mathbf{x}) = [f_1(\mathbf{x}), f_2(\mathbf{x})..., f_M(\mathbf{x})] = \mathbf{h}(\mathbf{x})\boldsymbol{\beta}, \tag{4}$$

where

$$\boldsymbol{\beta} = \begin{bmatrix} \beta_1 \\ \vdots \\ \beta_L \end{bmatrix} = \begin{bmatrix} \beta_{11} & \cdots & \beta_{1M} \\ \vdots & \ddots & \vdots \\ \beta_{L1} & \cdots & \beta_{LM} \end{bmatrix}. \tag{5}$$

The above ELM calculation can be summed up as following. After randomly select the value of input weights $\mathbf{w}_i$ and the bias of the neural network b_i, we can obtain the output H of hidden layers, therefore the output weights $\boldsymbol{\beta}$ can be obtained. During the training, $\boldsymbol{\beta}$ is obtained based on solving an optimization problem. And during the recognition, the maxim f_j is selected as the ELM output class label.

3   ELM Based Defect Detection Method

A ship detection approach employing ELM classification for SAR images is proposed in this paper. We will discuss image pre-processing, feature extraction and network training of the proposed approach as follows.

3.1   Image Pre-processing

SAR images is calibrated and geocoded for pre-processing, then land areas are removed with segmentation processing as introduced in [3]. The task of ship detection is performed using the ocean areas include ship and ship like objects. In this paper, the ships are distinguished from ship like and ocean areas employing ELM methods.

The edge and texture are important information for the ship detection. To obtain such information, other image processing tasks are employed to extract edge and texture features. To get the ship edge, ships are segmented employing traditional wavelet transform modulus maxima algorithm and morphological operation. To get the texture information, a popular gray level co-occurrence matrix method is employed. The element of the matrix is the coherent distribution for probability of a gray scale in constant distance. The matrix is obtained after scaning all the pixels with the reflection of co-occurrence time probability of related pixels in space. Therefore it can provide the joint probability distribution of the pixels. The gray level co-occurrence matrix is used to obtain entropy.

3.2   Feature Extraction

Due to ships appear with various direction, area and position in SAR, the geometrical feature is important for ships detection. Therefore roundness and roughness are employed as shape feature in our proposed detection method. Roundness is calculated as

$$f_{roundness} = \frac{P^2}{A}, \tag{6}$$

where P is the girth of edge, and A is the area. Roughness can be obtained with

$$f_{roughness} = \frac{1}{N_E} \sum_{i=1}^{N_E} |d(i) - d(i+1)|, \tag{7}$$

where N_E is the number of edge points, $d(m)$ is the m-th standardized radius of the m-th edge point.

Entropy and Contrast. Since texture information is also important for ship detection, we use entropy and contrast as texture feature. Entropy can provide the uniformity and complexity information of the texture, and it is calculated as

$$f_{entropy} = \sum p(i,j) \times [-\ln p(i,j)], \tag{8}$$

where $p(i,j)$ is the element of the position (i,j) in gray level co-occurrence matrix which has been described in Sect. 3.1. For contrast feature, greater contrast means more notable pixels with significant difference compared to average pixel value in the texture. Contrast can be calculated as

$$f_{contrast} = (i-j)^2 \times p(i,j). \tag{9}$$

Histogram of Oriented Gradients. As histogram of oriented gradients (HOG) is efficient for the description of local detailed information [15], HOG is used as local feature in this paper. HOG is obtained as follows. The sample gray image with the size of 30×30 is divided into 36 non-overlapping blocks with the size of 5×5. Two histograms are accumulated in each block. One has seven bins with the direction angle of $0°$–$180°$, and another one has 14 bins with the direction angle of $0°$–$360°$. Each pixel in the block is voted into the corresponding histogram according to its gradient orientation.

3.3   Network Training

In our ship detection method in SAR using ELM network, the input layer is connected to the input feature vector which includes reduced HOG features, geometrical features and texture features as described before. The output layer has one node used to mark ship or not.

After defining the ELM network structure, the network can be used for sample training. The parameters of input weights $\mathbf{w}_i$ and biases b_i in ELM are randomly selected. Consequently the calculation of β is critical for ELM training.

Denote the actual output vector as $\mathbf{Y}$, the input vector $\mathbf{X}$, We can obtain the output vector from (3) as

$$\mathbf{Y} = \mathbf{H}\beta, \tag{10}$$

where

$$\mathbf{H} = \begin{bmatrix} \mathbf{h}(\mathbf{x}_1) \\ \vdots \\ \mathbf{h}(\mathbf{x}_N) \end{bmatrix}$$

$$= \begin{bmatrix} g(\mathbf{x}_1; \mathbf{w}_1, b_1) & \cdots & g(\mathbf{x}_1; \mathbf{w}_L, b_L) \\ \vdots & \ddots & \vdots \\ g(\mathbf{x}_N; \mathbf{w}_1, b_1) & \cdots & g(\mathbf{x}_N; \mathbf{w}_L, b_L) \end{bmatrix}, \tag{11}$$

and

$$\mathbf{Y} = \begin{bmatrix} \mathbf{y}_1 \\ \vdots \\ \mathbf{Y}_N \end{bmatrix} = \begin{bmatrix} y_{11} & \cdots & y_{1M} \\ \vdots & \ddots & \vdots \\ y_{N1} & \cdots & y_{NM} \end{bmatrix}. \tag{12}$$

The object of ELM method is to minimize two errors, they are the training error $||\mathbf{T} - \mathbf{H}\beta||^2$ and the norm of output weight $||\beta||$. This problem can be converted to an optimization problem as below

$$\min \quad \psi(\beta, \boldsymbol{\xi}) = \tfrac{1}{2}||\beta||^2 + \tfrac{C}{2}||\boldsymbol{\xi}||^2 \tag{13}$$

$$\text{s.t.} \quad \mathbf{H}\beta = \mathbf{T} - \boldsymbol{\xi}. \tag{14}$$

where $\boldsymbol{\xi}$ is the output value error between the actual output and the desired output, and C is the regularization factor which is used to improve the training generalization performance with controlling the tradeoff between the closeness to the training data and the smoothness of the decision function. The above optimization problem can be solved employing Lagrange multiplier technique. When the matrix $(\mathbf{I}/C) + \mathbf{H}^T\mathbf{H}$ is not singular, solution β can be calculated as

$$\beta = \left(\frac{\mathbf{I}}{C} + \mathbf{H}^T\mathbf{H} \right)^{-1} \mathbf{H}^T\mathbf{T}. \tag{15}$$

Otherwise, when the matrix $(\mathbf{I}/C) + \mathbf{H}\mathbf{H}^T$ is not singular, solution β can be calculated as

$$\beta = \mathbf{H}^T \left(\frac{\mathbf{I}}{C} + \mathbf{H}\mathbf{H}^T \right)^{-1} \mathbf{T}, \tag{16}$$

where $\mathbf{I}$ is an identity matrix. In practice, when the number of training features of samples is greater than the one of hidden neurons, we use (15) to obtain the output weights, otherwise we use (16).

To improve the stability of ELM in calculating the output weights, an efficient solution is to find high quality mapping between input and hidden layers. RBF function is one of the most efficient mapping functions, and it used in our ELM based bubble defect detection method. $\mathbf{H}^T\mathbf{H}$ in (15) or $\mathbf{H}\mathbf{H}^T$ in (16) is called ELM kernel matrix, and $\mathbf{h}(\mathbf{x}_i) \cdot \mathbf{h}(\mathbf{x}_j)$ is ELM kernel. In our proposed method, the following Gaussian function is selected as the kernel

$$\phi(\mathbf{x}_i, \mathbf{x}_j) = \mathbf{h}(\mathbf{x}_i) \cdot \mathbf{h}(\mathbf{x}_j) = \exp\left(-\frac{||\mathbf{x}_i - \mathbf{x}_j||^2}{\sigma^2}\right). \quad (17)$$

The training process is performed as described above. After the training is finished, the trained ELM can be used for image detection in SAR.

4   Experimental Results

We use a SAR dataset reported in [7] in our experiments. These SAR images acquired from Sentinel-1 and RADARSAT-2 have been radiometrically calibrated. Two sample images are illustrated in Fig. 1.

Some other usually used methods are assessed in our experiments to verify the efficiency of our proposed method. It has been widely validated that neural network based classification has poorer performance than SVM based one. Support Tucker machine (STM) method has been used for object detection in SAR with good performance [8]. So our test methods are K-means (KM) [1], neural network (NN) [4], SVM [6], STM [8], CNN [7] and ELM methods. After testing the different parameters with experiments, we select the optimized parameters for the best performance from the allowed ranges employing optimization search. In our ELM based method, the number of hidden nodes is set to 500.

4.1   Accuracy

The comparisons of classification performance in a variety of sample numbers are performed. The sample number is selected from 200 to 500 with the step of 100. The samples are stochastically selected from the dataset, and half of the sample is selected from the ship images, and others from non-ship ones. Each experiment has been tested 15 times.

The comparison of classification correction rate of different methods is listed in Table 1. And these experimental results are also illustrated in Fig. 2. As shown in the figure, the classification accuracy of the proposed ELM based method is between 99.00% and 100%. With the increasement of training samples, the classification accuracy of our proposed method gradually increases as well. The classification accuracy is more prior to other methods when fewer training samples are employed. Our proposed ELM based method is able to keep the superior performance of accuracy with the different sample numbers. It means that our

Fig. 1. Sample images from the dataset

Table 1. Classification accuracy of different methods

Samples number	Method	Accuracy	Standard deviation
200	KM	0.8900	0.059
	NN	0.9150	0.051
	SVM	0.9450	0.022
	STM	0.9650	0.032
	CNN	0.9750	0.037
	ELM	0.9900	0.012
300	KM	0.9292	0.052
	NN	0.9333	0.051
	SVM	0.9429	0.028
	STM	0.9708	0.013
	CNN	0.9750	0.069
	ELM	0.9917	0.008
400	KM	0.9143	0.014
	NN	0.8964	0.039
	SVM	0.9286	0.028
	STM	0.9607	0.041
	CNN	0.9643	0.033
	ELM	1.0000	0.000
500	KM	0.9188	0.031
	NN	0.8995	0.019
	SVM	0.9438	0.057
	STM	0.9563	0.020
	CNN	0.9875	0.034
	ELM	1.0000	0.000

proposed ELM method has most satisfactory performance with varies of sample number.

4.2 Number of Hidden Nodes

We have tested the proposed ELM based ship detection method with different hidden nodes. Both the numbers of training samples and the ons of test samples are 400, in which 200 samples are positive samples with ships. The experimental results are listed in Table 2. It is clear that the number of nodes in the hidden layer has little influence on the classification accuracy in our proposed method. So the number of hidden nodes is set to 500 in our application. We performed the experiments with Matlab R2013 on a computer with 4 G RAM and i5 CPU of 2.66 GHz. The training time and test time are all listed in the table as well. The

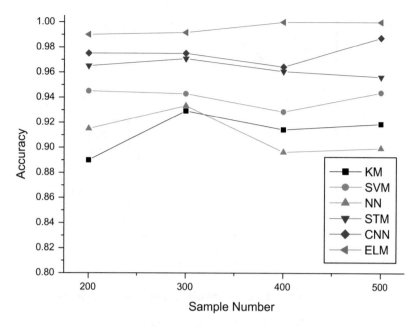

Fig. 2. Comparison of defect detection classification accuracy with different methods

Table 2. Classification results of different number of hidden nodes

Number of hidden nodes	Accuracy	Standard deviation	Training Time (s)	Test Time (s)
300	0.9969	0.003	195.64	9.26
400	0.9969	0.003	263.42	12.83
500	1.0000	0.000	372.56	14.39
600	1.0000	0.000	417.27	15.73

test time of each sample is about 0.023–0.039 s for network test. The proposed ELM algorithm is so fast that it can be employed in real time applications.

4.3 ROC and AUC

Receiver operating characteristics (ROC) curve and Area under the ROC curve (AUC) are widely used for classification performance comparison, and they are also employed in our experiments. We set the threshold varying from maximum to minimum and get the ROC curve. Interpreted as the probability that a classifier is able to distinguish a randomly chosen positive instance from a randomly chosen negative instance, AUC value is greater when the classification has good performance. The detailed ROC curves are illustrated in Fig. 3 in which the number of samples is 200. The curves of our proposed method are all on the above in the figures. It reveals that our proposed method has the best classifica-

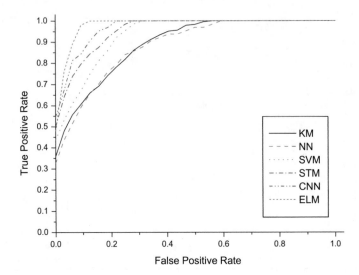

Fig. 3. ROC of different methods (sample number is 200).

Table 3. AUC of different methods

Samples number	KM	SVM	NN	STM	CNN	ELM
200	0.923	0.968	0.919	0.953	0.975	0.983
240	0.938	0.976	0.917	0.964	0.979	0.986
280	0.934	0.978	0.924	0.971	0.982	1.000
320	0.942	0.984	0.936	0.969	0.981	1.000

tion ability than other approaches. AUC value is listed in Table 3. Our proposed method obtain the greatest AUC value. All these experiments have shown that our proposed ELM based method has the best performance.

5   Conclusions

In this work, a novel image processing application is developed to detect ship in SAR images. After the image processing and the feature extraction, ELM based classification is applied to detect ships in SAR images. Compared with other learning based classifiers, the proposed ELM based method can obtain better performance to detect ship. The proposed extreme learning machine based method is potential for object detection in other SAR applications. Research of the more efficient ELM based method with automatic feature selection will be studied in the future.

Acknowledgments. This work was partly supported by National Natural Science Foundation of China (No. 61371045) and Science and Technology Project of Weihai, Shandong China.

References

1. Wang, S., Yang, S., Feng, Z., et al.: Fast ship detection of synthetic aperture radar images via multi-view features and clustering. In: International Joint Conference on Neural Networks, pp. 404–410 (2014)
2. Yang, G., Yu, J., Xiao, C., et al.: Ship wake detection for SAR images with complex backgrounds based on morphological dictionary learning. In: IEEE International Conference on Acoustics, Speech and Signal Processing, pp. 1896–1900. IEEE (2014)
3. Selvi, M.U., Kumar, S.S.: Sea object detection using shape and hybrid color texture classification. Commun. Comput. Inf. Sci. **204**, 19–31 (2011)
4. Martan-De-Nicols, J., Mata-Moya, D., Jarabo-Amores, M.P., et al.: Neural network based solutions for ship detection in SAR images. In: International Conference on Digital Signal Processing, pp. 1–6. IEEE (2013)
5. Khesali, E., Enayati, H., Modiri, M., et al.: Automatic ship detection in Single-Pol SAR images using texture features in artificial neural networks. Int. Arch. Photogramm. Remote. Sens. **XL-1-W5**, 395–399 (2015)
6. Yang, X., Bi, F., Yu, Y., et al.: An effective false-alarm removal method based on OC-SVM for SAR ship detection. In: IET International Radar Conference, pp. 1–4. IET (2015)
7. Schwegmann, C.P., Kleynhans, W., Salmon, B.P., et al.: Very deep learning for ship discrimination in synthetic aperture radar imagery. In: 2016 IEEE International Geoscience and Remote Sensing Symposium, pp. 104–107. IEEE (2016)
8. Ma, L.: Support Tucker machines based marine oil spill detection using SAR images. Indian J. Geo-Mar. Sci. **45**, 1445–1449 (2016)
9. Huang, G.B., Zhu, Q.Y., Siew, C.K.: Extreme learning machine: theory and applications. Neurocomputing **70**, 489–501 (2006)
10. Huang, G., Huang, G.B., Song, S., You, K.: Trends in extreme learning machines: a review. Neural Netw. **61**, 32–46 (2015)
11. Huang, G.B., Zhou, H., Ding, X., Zhang, R.: Extreme learning machine for regression and multiclass classification. IEEE Trans. Syst. Man Cybern. **2**, 513–529 (2016)
12. Wang, S., Deng, C., Lin, W., Huang, G.B.: NMF-based image quality assessment using extreme learning machine. IEEE Trans. Cybern., 255–258 (2016)
13. Yüksel, T.: Intelligent visual servoing with extreme learning machine and fuzzy logic. Expert. Syst. Appl. **47**, 232–243 (2017)
14. Liu, X., Deng, C., Wang, S., Huang, G.B., Zhao, B., Lauren, P.: Fast and accurate spatiotemporal fusion based upon extreme learning machine. IEEE Geosci. Remote. Sens. Lett. **13**, 2039–2043 (2016)
15. Dalal, N., Triggs, B.: Histograms of oriented gradients for human detection. In: Proceedings - 2005 IEEE Computer Society Conference on Computer Vision and Pattern Recognition, San Diego, United States, pp. 886–893 (2005)

Fault Diagnosis on Sliding Shoe Wear of Axial Piston Pump Based on Extreme Learning Machine

Jinwei Hu[1], Yuan Lan[1,2(✉)], Xianghui Zeng[1], Jiahai Huang[1,2],
Bing Wu[1,2], Liwei Yao[1,2], and Jinhong Wei[1,2]

[1] School of Mechanical Engineering, Taiyuan University of Technology,
Taiyuan 030024, China
hujunwei2015@126.com, laynlibiyi@163.com
[2] Key Lab of Advanced Transducers and Intelligent Control System of Ministry
of Education and Shanxi Province, Taiyuan University of Technology,
Taiyuan 030024, China

Abstract. In order to improve the computational efficiency and classification accuracy of fault diagnosis on axial piston pump, a new classifier, namely, Extreme Learning Machine (ELM) is introduced to identify sliding shoe wear on axial piston pump in this paper. The diagnosis procedures can be summarized as follows. Firstly, the vibration signals in normal working condition and fault working condition (i.e. sliding shoe wear with different wear degree) are collected. Secondly, the time domain indexes, wavelet packet energy and results obtained by using Local Mean Decomposition-Singularity Value Decomposition (LMD-SVD) are considered as the features. Finally, we apply ELM and other algorithms to identify the sliding shoe wear of axial piston pump. The results show that ELM has the highest classification accuracy and the fastest diagnosis speed.

Keywords: Fault diagnosis · ELM · Axial piston pump · Sliding shoe wear

1 Introduction

Hydraulic pump is the "heart" of hydraulic systems. As an important part of the mechanical equipment, hydraulic pump can play a decisive role in the work of the mechanical equipment. The operation of the hydraulic pump would directly affect the operation of the whole hydraulic system. An unexpected failure of the hydraulic pump may cause the sudden breakdown of hydraulic system, bringing about enormous financial losses or even personnel casualties. The sliding shoe wear is incipient fault on axial piston pump that could lead to wear failure, which may cause unbalanced

Fund Project: Project supported by the National Natural Science Foundation of China (Grant No. 51405327).Major special science and technology projects in Shanxi province(GJ2016-02).

J. Cao et al. (Eds.): ELM 2017, PALO 10, pp. 114–122, 2019.
https://doi.org/10.1007/978-3-030-01520-6_10

operation of hydraulic pump [1]. It will affect the whole hydraulic system operation, so its fault diagnosis is of great importance.

Essentially, the axial piston pump fault diagnosis is a pattern recognition problem. And pattern recognition can be achieved through some intelligent classification algorithms, some of which such as artificial neural networks (ANN) and support vector machine (SVM) have been used in the axial piston pump fault diagnosis and identification. Wu proposed the multi-source information fault diagnosis method based on combining SVM and D-S evidence theory [2]. Li used neural network for fault diagnosis and recognition [3]. Zhang et al. applied SVN on fault diagnosis and recognition [4]. However, these methods share the common disadvantage like slow computational speed and manual intervention requirement. In this paper, a new classifier, extreme learning machine (ELM) classifier, is introduced and employed to recognize the sliding shoe wear on axial piston pump. ELM [5] is originally proposed by Professor Guang-Bin Huang from Nanyang Technological University at IEEE International Joint Conference Neural Networks in 2004. This method is based on single hidden layer feedforward network (SLFNs) and it is simple and easy to use. The input parameters of ELM are randomly generated, which are independent of the training samples. The method transforms general iterative solution of the single hidden layer neural network into a solution of a linear equation, and accelerate the computation speed. Moreover, the method using least squares solution can produce unique optimal solution [6]. Therefore, ELM have the following advantages: (1) fast learning speed (2) can obtain the global optimal solution.

The rest of paper is organized as follow. Section 2 provides a brief review of ELM. Section 3 describes the experiment datasets preparation and feature extraction. Section 4 presents the results of fault diagnosis by applying ELM classifiers and other classifiers. Finally, Sect. 5 is the conclusion of the paper.

2 Extreme Learning Machine (ELM)

ELM [5, 7, 8] is a learning algorithm that is developed for generalized SLFNs with wide variety of hidden nodes. Consider N arbitrary distinct samples $(X_i, t_i) \in R^n \times R^m$. If a SLFN with L hidden nodes can approximate these N samples with zero error, it then implies that there exist β_i, α_i and b_i such that [9]:

$$f_L(X_j) = \sum_{i=1}^{L} \beta_i G(a_i, b_i, X_j) = t_j, j = 1, \cdots, N \qquad (1)$$

where a_i and b_i are the learning parameters of the hidden nodes, β_i is the output weight, and $G(a_i, b_i, X)$ denotes the output of the i th hidden node with respect to the input X.

Equation (1) can be written compactly as:

$$H\beta = T \qquad (2)$$

where

$$
\begin{aligned}
&H(a_1, \cdots, a_L, b_1, \cdots, b_L, X_1, \cdots, X_N) \\
&= \begin{bmatrix} G(a_1, b_1, X_1) & \cdots & G(a_L, b_L, X_1) \\ \vdots & \cdots & \vdots \\ G(a_1, b_1, X_N) & \cdots & G(a_L, b_L, X_L) \end{bmatrix}_{N \times L}
\end{aligned} \tag{3}
$$

$$
\beta = \begin{bmatrix} \beta_1^T \\ \vdots \\ \beta_L^T \end{bmatrix}_{L \times m} \quad \text{and} \quad T = \begin{bmatrix} t_1^T \\ \vdots \\ t_N^T \end{bmatrix}_{N \times m} \tag{4}
$$

H is called the hidden layer output matrix of the network.

According to ELM theories, all the hidden nodes (a_i, b_i) can be randomly assigned instead of being tuned. The solution of Eq. (2) is estimated as:

$$
\hat{\beta} = H^+ T \tag{5}
$$

where H^+ is the Moore-Penrose generalized inverse of the hidden layer output matrix H. The ELM algorithm which only consists of three steps can then be summarized as:

ELM Algorithm: Given a training set $\aleph = \{(X_i, t_i) | X_i \in R^n, t_i \in R^m, i = 1, \cdots, N.\}$, activation function $g(x)$, and hidden node number L,

1. Initialize input weights a_i and bias values b_i, $i = 1, \cdots, L$.
2. Calculate the hidden layer output matrix H.
3. Calculate the output weight β: $\beta = H^+ T$

Remark: Universal approximation capability of ELM has been analyzed by Huang et al. using incremental methods and it has been shown that single SLFNs with randomly generated (variety of) hidden nodes can universally approximate any continuous target functions [7].

3   Fault Diagnosis on Sliding Shoe Wear of Axial Piston Pump

3.1   Experiment Setup

In order to verify the effectiveness of ELM classifiers on fault diagnosis of sliding shoe wear of axial piston pump, the axial piston pump test bed is designed. Test system diagram is shown in Fig. 1.

In the experiment, the model of axial piston pump is A10VSO45, and the rated speed of the motor is 1480 r/min. The axial piston pump is tested under three different situation (normal, one sliding shoe wear and three sliding shoe wear). In the entire test, the pressure of main hydraulic circuit is maintained at 10 MPa. The axial piston pump shell vibration acceleration signal, flow signal and pressure signal can be collected through the accelerometer, flow meter and pressure transmitter. And then, the signals

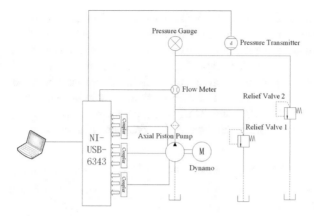

Fig. 1. Axial piston pump test system diagram

are collected by using signal collection card (NI-USB-6343) and it is connected with the labview software in the computer. The sampling frequency of the signal collection card is 45 kHz. A total of 200 sets of data are collected, and each group is sampled for 0.2 s. The locations of three three-axis accelerometer are shown in Fig. 2, and the axes of the accelerometer are shown in Table 1.

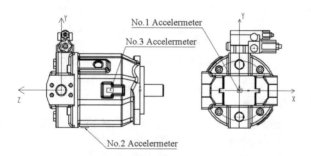

Fig. 2. Locations of three three-axis accelerometer

3.2 Data Acquisition

Follow the above description of the installation of the test system, collect the axial piston pump shell time-domain vibration signal, flow signal and pressure signal under three different situation (normal, one sliding shoe wear and three sliding shoe wear).

Table 1. The direction of the accelerometer

	X-Axis	Y-Axis	Z-Axis
Accelerometer No.1	Main(X-Axis)	Main(Y-Axis)	Main(Z-Axis)
Accelerometer No.2	Main(X-Axis)	Main(Z-Axis)	Main(Y-Axis)
Accelerometer No.3	Main(Z-Axis)	Main(Y-Axis)	Main(X-Axis)

In order to evaluate the ELM classifier and other classifiers, we separate the experiment data sets with six kinds of data sets. For ALL and PALL data sets, it includes all data of three different situations. Data sets OSS and POSS contains data of normal condition and one sliding shoe wear condition. Data sets TSS and PTSS contains data of normal condition and three sliding shoe wear condition. There are 150 training data and 50 testing data in each working condition in the data set ALL, OSS and TSS. For PALL, POSS and PTSS data sets, there are 100 training data and 100 testing data in every working conditions. The data specification is presented in Table 2.

Table 2. Axial piston pump sliding shoe wear data set

Dataset	Training data	Testing data	Sliding shoe wear	Classification label
ALL	150*3	50*3	0	0
			1	1
			3	2
OSS	150*2	50*2	0	0
			1	1
TSS	150*2	50*2	0	0
			3	2
PALL	100*3	100*3	0	0
			1	1
			3	2
POSS	100*2	100*2	0	0
			1	1
PTSS	100*2	100*2	0	0
			3	2

3.3   Feature Extraction

3.3.1   Time Domain Indexes and Wavelet Packet Energy Feature Vector

In the field of fault diagnosis, how to effectively extract the fault sensitive feature vectors is the key to the diagnosis of hydraulic pump. The raw signal collected by the signal collection card (NI-USB-6343) is the modulation signal. And we have to pre-process the raw signal to obtain the useful real envelope signal, thus appropriate to extract the fault feature vector.

In this paper, firstly we use the db1 wavelet to decompose the original signal with two layers of wavelet packet. And then select the result from high frequency band, and use wavelet packet reconstruction algorithm to get the corresponding time-domain signal. Through the threshold de-nosing, we get high frequency time-domain signal by using the band-pass filter. Due to the obtained signals are modulated signal, the Hilbert envelope demodulation process should be performed [10]. In addition, we know that the characteristic frequency of the hydraulic pump fault is below 1 kHz, so the envelope signal is sampled at a sampling frequency of 2 kHz, and the envelope signal frequency is 1 kHz after resampling.

The time-domain signal indexes of the obtained signals can be used as the features, including variance σ^2, standard deviation S, peak metric C_f, pulse index I_f, margin index CL_f, kurtosis index K_v, The formulas are:

$$\sigma^2 = \frac{1}{N} \sum_{i=1}^{N} [x_i - \bar{x}]^2 \tag{6}$$

$$S = \sqrt{\frac{1}{N} \sum_{i=1}^{N} [x_i - \bar{x}]^2} \tag{7}$$

$$C_f = \frac{x_{max}}{x_{rms}} \tag{8}$$

$$I_f = \frac{x_{max}}{|\bar{x}|} \tag{9}$$

$$CL_f = \frac{x_{max}}{x_r} \tag{10}$$

$$Kv = \frac{\frac{1}{N} \sum_{i=1}^{N} x^4}{x_{rms}^4} \tag{11}$$

where $x_{rms} = \sqrt{\frac{1}{N} \sum_{i=1}^{N} x^2}$, $x_r = \left[\frac{1}{N} \sum_{i=1}^{N} \sqrt{|x|} \right]^2$.

Besides, the time-frequency domain index of the envelope signal, i.e. select the largest frequency band energy E from 8 frequency energy generated by three layer wavelet packet decomposition, and the flow rate V of the system are included in the feature vector as well. Hence, the feature vector is with 8 dimensions.

3.3.2 LMD-SVD Feature Vector

In addition to the feature extraction method described above, we have tried to collect features using local mean decomposition (LMD) with singular value decomposition (SVD) [13]. The procedure can be summarized as follows. First, LMD [11, 12] of the collected vibration signal is obtained. LMD decomposition of the original signal contains a large amount of data, and it can't be directly used as a fault feature vector. Therefore, LMD + SVD is applied to prepare the feature vector. Finally, the feature vector is with 6 dimensions. The detailed manipulate procedures are presented in the reference [13]. We have evaluated the fault sensitivity of LMD-SVD feature vectors and compared the results with the feature vectors extracted based on the method described in Sect. 3.3.1.

4   Experiment Results

In this paper, all the evaluations were carried out in the MatlabR2014a environment running. 20 rounds of simulations were conducted and the average values were recorded as the final results.

4.1   The Result of Time Domain Indexes and Wavelet Packet Energy Feature Vector

Data set ALL, includes all data of three different situations. The fault diagnosis results are shown in Table 3. The accuracy of ELM and SVM classifiers is higher than that of the BP neural networks, and exceed 98%, meanwhile, ELM has the fastest diagnosis speed.

The data set OSS contains data of normal condition and one sliding shoe wear condition. The fault diagnosis results are shown in Table 4. The accuracy of ELM is higher than BP neural network and SVM.

The data set TSS contains data of normal condition and three sliding shoe wear condition. The fault diagnosis results are shown in Table 5. The accuracy of the three classifiers is 100%, while the diagnostic speed of ELM is faster than the others.

Table 3. Performance comparison of ELM, SVM and BP neural network on ALL dataset

Dataset	Classifier	Time (s)	Testing accuracy (%)	Testing std
ALL	ELM	0.1032	98	0.0139
ALL	BP neural network	0.7963	97	0.0206
ALL	SVM	0.1125	98	

Table 4. Performance comparison of ELM, SVM and BP neural network on OSS dataset

Dataset	Classifier	Time (s)	Testing accuracy (%)	Testing std
OSS	ELM	0.0547	99.7	0.008
OSS	BP neural network	0.4126	99.5	0.0061
OSS	SVM	0.0062	95	

Table 5. Performance comparison of ELM, SVM and BP neural network on TSS dataset

Dataset	Classifier	Time (s)	Testing accuracy (%)	Testing std
TSS	ELM	0.007	100	0
TSS	BP neural network	0.7869	100	0
TSS	SVM	0.0074	100	

Therefore, it is feasible to use the ELM to diagnose different sliding shoe wear of the axial piston pump. The results show that ELM has relatively the highest accuracy and the fastest diagnosis speed.

Table 6. Comparison of time domain indexes and wavelet packet energy feature vector and LMD-SVD feature vector by using ELM classifiers

Dataset	Feature vector	Time (s)	Testing accuracy (%)	Testing std	Node
PALL	Time domain and wavelet packet energy	0.0031	97.6	0.0166	60
PALL	LMD-SVD	0.0078	91	0.011	58
POSS	Time domain and wavelet packet energy	0.0102	99.8	0.003	45
POSS	LMD-SVD	0.0062	92	0.0225	58
PTSS	Time domain and wavelet packet energy	0.0149	99.9	0.0022	110
PTSS	LMD-SVD	0.0031	95	0.0146	45

4.2 Comparison of Time Domain Indexes and Wavelet Packet Energy Feature Vector and LMD-SVD Feature Vector by Using ElM Classifiers

The results of time domain indexes and wavelet packet energy feature vectors and LMD-SVD feature vectors are shown in Table 6. Although the ELM classifiers using time domain indexes and wavelet packet energy feature vectors require more time than the ones using LMD-SVD in the data set POSS and PTSS, their accuracy in all data sets are far greater than the ones using LMD-SVD. Therefore, it can be concluded that the time domain indexes and wavelet packet energy feature vectors are more suitable for the fault diagnosis of sliding shoe wear of the axial piston pump.

5 Conclusion

(1) In this paper, the time domain indexes and wavelet packets energy feature extraction method can accurately reflect the fault characteristics of sliding shoe wear of the axial piston pump.

(2) ELM can effectively diagnose sliding shoe wear of the axial piston pump. ELM has higher accuracy and diagnosis speed, which can be extended to other fault diagnosis application.

References

1. Liu, S., Wang, C., Yang, M., et al.: Analysis of dynamic characteristics for severe wear process of swash plate axial piston pump slipper pair. Chin. Hydraul. Pneumatics. **1** (2016)
2. Wu, S.: New Fault Diagnosis Methods of Kernel Principal Components Analysis and Evidence Theory on Multi-Domain Feature. Yanshan University (2011)
3. Li, S., Zhang, P., Li, B., et al.: Fault feature selection method for axial piston pump based on quantum genetic algorithm. China Mech. Eng. **12** (2014)
4. Zhang, P., Li, S.: Fault feature selection method based on wavelet packet transform and GA - PLS algorithm. J. Vib. Measurement Diagnosis **34**(2), 385–391 (2014)

5. Huang, G.B., Zhu, Q.Y., Siew, C.K.: Extreme learning machine: a new learning scheme of feedforward neural networks. In: Proceedings of International Joint Conference on Neural Networks (IJCNN 2004), Budapest, Hungary, 25–29 July 2004, vol. 2, pp. 985–990 (2004)
6. Huang, G.B.: An insight into extreme learning machines: random neurons random features and kernels. Cogn. Computation. **6**(3), 376–390 (2014)
7. Huang, G.B., Chen, L., Siew, C.K.: Universal approximation using incremental constructive feedforward networks with random hidden nodes. IEEE Trans. Neural Netw. **17**(4), 879–892 (2006)
8. Huang, G.B., Zhu, Q.Y., Siew, C.K.: Extreme learning machine: theory and applications. Neurocomputing **70**, 489–501 (2006)
9. Lan, Y., Xiong, X., Han, X., et al.: Multifault diagnosis for rolling element bearings based on extreme learning machine. In: Proceedings of ELM-2014, vol. 2, pp. 209–222. Springer (2015)
10. Jiang, W., Liu, S., Zhang, Q.: Intelligent Information Diagnosis and Monitoring of Hydraulic Fault. China Machine Press, Xi'an (2013)
11. Smith, J.S.: The local mean decomposition and its application to EEG perception data. J. R. Soc. Interface **2**(5), 443 (2005)
12. Wang, Y., He, Z., Zi, Y.: A comparative study on the local mean decomposition and empirical mode decomposition and their applications to rotating machinery health diagnosis. J. Vib. Acoust. **132**(2) (2010)
13. Tian, Y., Ma, J., Lu, C., et al.: Rolling bearing fault diagnosis under variable conditions using LMD-SVD and extreme learning machine. Mech. Mach. Theory **90**, 175–186 (2015)

Memristive Extreme Learning Machine: A Neuromorphic Implementation

Lu Zhang[1], Hong Cheng[1(✉)], Huanghuang Liang[1], Yang Zhao[1], Xinqiang Pan[2], Yuansheng Luo[1], Hongliang Guo[1], and Yao Shuai[2]

[1] Center for Robotics, School of Automation Engineering, University of Electronic Science and Technology of China, Chengdu 610054, China
hcheng@uestc.edu.cn
[2] State Key Laboratory of Electronic Thin Films and Integrated Devices, University of Electronic Science and Technology of China, Chengdu 610054, China

Abstract. Neuromorphic computation has been a hot research area over the past few years. Memristor, as one of the neuromorphic computation materials memorizes the conductance value and is able to adapt it according to changing voltages. This paper pioneers a neuromorphic computing paradigm implementation (through memristor) for Extreme Learning Machine (ELM), which is one of most popular machine learning methods. By simulating the biological synapses with memristors and combining the memory property of memristor with high-efficient processing ability in ELM, a three-layer ELM model for classification is constructed. We represent the ELM network weights through memristive conductance values. The conductance values (network weights) are updated through tuning the voltages. Experimental results over the Iris dataset show that the memristor-based ELM achieves the same level performance as the one implemented via traditional software, and exhibits great potential that ELM can be implemented in neromorphic computation paradigms.

1 Introduction

Brain-like computation is one of the hottest research areas in computer society. Researchers have been seeking for neuromorphic chips which mimic human brain functionalities. While one of the most important features inside human brain is the synapse firing process between neurons [6,7,19], memristor, as one of the newly discovered materials, exhibits the most similar property of synapse firing process in human brains [16,23]. Memristor is able to change the conductance value with the input voltages, identical to the synapse firing process, which changes the synapse strength according to the input neuron's activation level. Thus, memristor has the great potential to reach real brain-like computation systems.

On the other hand, with the development of computation hardware and communication technology, artificial intelligence (AI) has been attracting more and

© Springer Nature Switzerland AG 2019
J. Cao et al. (Eds.): ELM 2017, PALO 10, pp. 123–134, 2019.
https://doi.org/10.1007/978-3-030-01520-6_11

more researchers and entrepreneurs and seen great success over a variety of application domains. Extreme learning Machine (ELM), as one of the most popular AI computation algorithms, has seen wide applications across different domains [15]. However, some real-world applications, such as video surveillance, require real time processing of the input data, and cannot rely on remote communications to the backend servers, thus, the on-board implementation of ELM is necessary.

This paper pioneers in using memristor to implement ELM. Compared with traditional on-board implementations of ELM with GPU or FPGA, using memristor to implement ELM is requiring quite little space and do not need electricity to maintain the ELM network weights. Moreover, the synapse firing property of memristor grants its ELM implementation with great potential for future brain-like processing abilities.

As memristor can naturally store the conductance value, which corresponds to the output weight in ELM, and update the conductance values with the input voltages, which is analogous to the external gradient-based update rule in ELM computation, in this paper, we use memristor to replace the weight update process in ELM. Experimental results show that the memristor-based ELM reaches the same performance as the one implemented in normal software, i.e., in Matlab.

The contributions of the paper can be summarized as: (1) memristor-based ELM relays the significant computation process (weight update) to hardware, and hence demonstrates great potential for ELM's on-board implementation; (2) this research work is the first one integrating brain-like computation materials (memristor) with ELM, which grants ELM with brain-like data processing abilities.

The structure of the paper is organized as follows: we give a brief overview of conventional computing platform, ELM and memristor computation and its application in simple neural network in Sect. 2. Section 3 describes the implementation process of memristor-based ELM in detail, followed by experimental results and analysis in Sect. 4. The paper concludes in Sect. 5 with conclusion and future works.

2   Related Works

In this section, we will first review related technologies on memristor-based neuromorphic computation, ELM introduction and literature on hardware accelerated computation.

2.1   Memristor-Based Neuromorphic Computation

Neuromorphic computation [23] is one of the most important research areas in computer society, while memristor, as one of the emerging neuromorphic computation materials has drawn increasing interest from researchers. In this subsection, we focus on reviewing relevant neuromorphic computation implementations with memristors.

The physical model of memristor and its application as simulating synapse functionalities are shown in Fig. 1 [16]. Memristor is a two-terminal device which shows non-volatile resistive switching behaviour, and various physical mechanisms for the non-volatile resistance change have been proposed, such as migration of anions/cations,charge trapping, ferroelectric polarization switching [17,25]. Memristors show nonlinear current-voltage characteristics and continuously adjustable conductivity, which can mimic the typical feature of biological synapses in human brains [6].

Authors in [23] build the neural network that two inputs neurons are connected with an output neuron which is capable of representing associative memory based on memristor. Hopfield neural network realized by memristor is reported in [11], the network can be reconfigured to realize various positive and negative synaptic weights. Although the synaptic operation of memristor by tuning the resistance has been widely demonstrated, implementation of a simple neural network is still the challenge problem. Single memristor has been used as artificial synapse, recent research has been extended to crossbar array of phase-change memristive devices, such as implementation of single-layer perceptron [24].

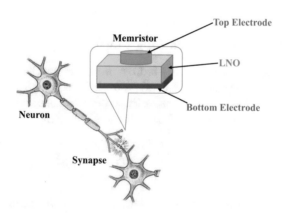

Fig. 1. Memristor's physical model and application as simulating synapse.

An increasing number of experiments show that using memristor as synapse in neuromorphic circuits is able to potentially offer both connectivity and high density as required for efficient computing [6]. Furthermore, memristor has a significant impact on smart chips and smart hardware for machine learning and computational intelligence. Despite much progress in computation of FPGA and GPU, even dedicated for neural network chips, technology development and scientific development still need different technologies to explore neuromorphic computation chips.

2.2 Hardware Accelerated Computation

In this subsection, we review related works on hardware computation implementations. Nowadays, GPU is no longer limited to $3D$ graphics processing, which is able to provide several times or even hundreds of times compared to the performance of CPU with respect to floating-point computation and parallel computation [3,8,9]. OpenCL, CUDA and ATISTREAM are the standards of GPU general computation. Although GPU has strong computing power, high energy consumption is a serious problem over industrial application [1,4,5].

FPGA is another product based on PAL, GAL, CPLD and other programmable device. In general, we need to consider the tradeoff between flexibility and performance when assessing the acceleration of hardware platforms [10,18,20,21]. On the one hand, GPP can promote flexibility and usability, but lacks efficiency. On the other hand, ASIC can provide high performance, but lacks flexibility. Therefore, FPGA is a compromise between the two extremes which offers both the performance benefits of integrated circuits and the flexibility of reconfigurable GPP [26–28]. Existing technologies for the current computing system are approaching its physical limitation, novel device conceptions such as memristive devices are considered as the next generation of computation systems [29].

2.3 Review on Extreme Learning Machine

Extreme Learning Machine (ELM) is first proposed for the single-hidden layer feedforward *neural* networks (SLFNs) and then extended to the generalized single-hidden layer feedforward networks. From the network architecture point of view, the output function of ELM for generalized SLFNs is:

$$f_L(\mathbf{x}) = \sum_{i=1}^{L} \boldsymbol{\beta}_i h_i(\mathbf{x}) = \mathbf{h}(\mathbf{x})\boldsymbol{\beta}, \tag{1}$$

where $\boldsymbol{\beta} = [\boldsymbol{\beta}_1, \cdots, \boldsymbol{\beta}_L]^T$ is the output weight vector between the hidden layer of L nodes to the $m \geq 1$ output nodes, and $\mathbf{h}(\mathbf{x}) = [h_1(\mathbf{x}), \cdots, h_L(\mathbf{x})]$ is the output (row) vector of the hidden layer with respect to the input $\mathbf{x}$. $h_i(\mathbf{x})$ is the output of the ith hidden node output, which may not be unique. According to ELM theories [12–14], $h_i(\mathbf{x})$ can be almost any nonlinear piecewise continuous function. $\mathbf{h}(\mathbf{x})$ is also called ELM feature space, which in most cases can be randomly generated according to any continuous probability distribution function and independent from the training samples.

From the learning point of view, ELM theory aims at reaching the smallest training error but also the smallest norm of output weights [15]:

$$minimize : C\|\boldsymbol{\beta}\|_p^{\sigma_1} + \|\mathbf{H}\boldsymbol{\beta} - \mathbf{T}\|_q^{\sigma_2}, \tag{2}$$

where $\sigma_1 > 0, \sigma_2 > 0, p, q = 1, 2, \cdots, +\infty$, $\mathbf{H}$ is the hidden layer output matrix (*randomized matrix*) and $\mathbf{T}$ is the training data target matrix:

$$\mathbf{H} = \begin{bmatrix} \mathbf{h}(\mathbf{x}_1) \\ \vdots \\ \mathbf{h}(\mathbf{x}_N) \end{bmatrix} = \begin{bmatrix} h_1(\mathbf{x}_1) & \cdots & h_L(\mathbf{x}_1) \\ \vdots & \vdots & \vdots \\ h_1(\mathbf{x}_N) & \vdots & h_L(\mathbf{x}_N) \end{bmatrix}, \tag{3}$$

and $\mathbf{T}$ is the training data target matrix:

$$\mathbf{T} = \begin{bmatrix} \mathbf{t}_1^T \\ \vdots \\ \mathbf{t}_N^T \end{bmatrix} = \begin{bmatrix} t_{11} & \cdots & t_{1m} \\ \vdots & \vdots & \vdots \\ t_{N1} & \cdots & t_{Nm} \end{bmatrix}. \tag{4}$$

The prevailing ELM will set $p = q = \sigma_1 = \sigma_2 = 2$, in this case, one of the ELM analytic solutions is:

$$\boldsymbol{\beta} = (C\boldsymbol{I} + \boldsymbol{H}^\top \boldsymbol{H})^{-1} \boldsymbol{H}^\top \boldsymbol{T}. \tag{5}$$

3   Memristor-Based ELM Implementation

In this section, we first introduce the memristor platform, and then provide details of ELM network for the classification task. The last subsection shows the memristor-based ELM weight update process including the pseudo code showing the memristive ELM implementation process.

3.1   Memristive Platform

The entire structure of the memristors is shown in Fig. 1 above, which is a capacitor-like structure. The vacuum-annealed $LiNbO_3$ thin film is sandwiched between the top electrode of Au and bottom electrode of Pt [22].

3.2   ELM for Classification

Extreme Learning Machine is intrinsically an artificial neural network comprising of single hidden layer. The weights between input layer and hidden layer are randomly projected and weights in the output layer are adapted. In other words, the network only needs to train the weights of output layer. Figure 2 illustrates an example of ELM pattern classification process, where the network has four inputs and each input is connected to a synapse with synaptic weight of w_{ij}^I, the output neuron is expressed as:

$$Y = f(X^T W^I) \cdot W^o, \tag{6}$$

where X is training data, $x_i(i = 1, 2, ..., m)$, W^I is the fixed weight matrix of input layer which is randomly generated beforehand, W^o is the weight matrix of output layer, $w_k^o (k = 1, 2, ..., n)$, are trainable synaptic weights, f is a sigmoid function providing the non-linear mapping functionality.

Such a model is sufficient for performing, for instance, the classification of Iris Flowers data (this is a classical dataset to be found in pattern recognition, the data contains 3 classes of 50 instances each, where each class refers to a type of Iris plant) into three classes, with network inputs corresponding to the features of sepal length and width, petal length and width.

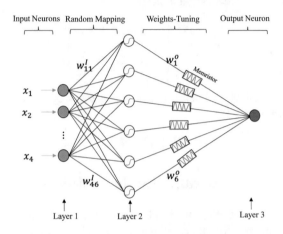

Fig. 2. An illustration of ELM network for pattern classification.

3.3   Memristor-Based ELM Implementation

In this subsection, we use the conductance value of memristors to represent the weights of ELM's output layer. Since it is quite difficult if not impossible to let the memristors update its conductance value along the exact gradient direction, we adopt the coordinate decent method to update the weight values of ELM's output layer. We will present the weight update process in the last subsection.

(1) Challenges and Solutions. The memristors in this work are used as electronic synapses in the simple neural network to realize pattern classification. Due to matrix processing in computer programs, the weights of each layer are tuned simultaneously, however, in the memristor platform, one memristor controls one weight value independently, it is almost impossible to make the memristors' conductance update direction exactly the same as the pre-calculated gradient direction.

In this case, we adopt the coordinate descent strategy, which updates the ELM network's output weight value (memristor's conductance) once at a time. The proposed coordinate decent strategy is able to reduce the training error as the process continues and reaches global optimum. At each iteration, we update the weight value according to $w_{k+1}^{(i)} = w_k^{(i)} + \Delta w^{(i)}$, where $\Delta w^{(i)}$ is the projected gradient information along the corresponding coordinate direction.

(2) Conductance/Weight Update Process. As the scheme shows, the hidden-node outputs z_j (with j = 1,2,...,n) are calculated as sigmoid function: $z_j = 1/(1+ exp^{-I_j})$, where I_j is the matrix-by-vector product components: $I_j = \sum_{i=1}^{m} w_{ji}^I x_i$. Here x_i ($i = 1, ..., m$) are the input signals, and w_{ji}^I are randomly generated parameters of the first layer, moreover, these will be fixed once assigned. Secondly, the outputs Y_k are calculated as: $Y_k = \sum_{j=1}^{n} w_{kj}^o z_j$. In ELM, weights of output layer need to be updated. For each single sample, we define the error of network output and target output as: $Error = 1/2 \sum_{k=1}^{l} (T_k - Y_k)^2$.

In supervised learning, the weights can be obtained by solving the inverse or generalized inverse of matrix. In this work, the weights are updated through the gradient descent method:

$$\frac{\partial Error}{\partial w_{kj}^o} = -(T_k - \sum_{j=1}^{n} w_{kj}^o z_j) z_j. \tag{7}$$

Here T_k is the target value of input sample. At each epoch of the procedure, samples from the dataset are applied incrementally to the networks input, the outputs and weights are updated through following equation:

$$w_{kj}^o(t + 1) = w_{kj}^o(t) + \Delta w_{kj}^o, \tag{8}$$

with $\Delta w_{kj}^o = \frac{\partial error}{\partial w_{kj}^o}$. Equation 8 is realized by the computer program, in the memristive implementation, the synaptic weights are modified as:

$$V = \begin{cases} +\Delta v_m, \ \Delta w > 0 \\ -\Delta v_m, \ \Delta w < 0 \end{cases} \tag{9}$$

Here Δw is given by the program, when $\Delta w > 0$, $+\Delta v_m$ voltage pulse with the width of 100 ms are applied to the memristor, when $\Delta w < 0$, $-\Delta v_m$ voltage pulse with the width of 100 ms is applied to the memristor. The program terminates if the error satisfies the condition of threshold, $error < \Theta$.

Algorithm 1. Memristive ELM implementation

Input: The input dataset: $X = x_1, x_2, x_3, x_4$, x_i refers to the feature of Iris flower;
Output: The target value: $T = 0, 1, 0$ and 1 indicate category labels.
1: Initialize the weight matrixes W^I (layer1 to layer 2) and W^O (layer2 to layer3) are initialized randomly.
2: Calculate hidden layer outputs: $Z = f(X * W^I)$;
3: The weight matrix W^I is fixed, H are the outputs of the hidden layer.
4: **for** each input sample **do**
5:     calculate output of training layer: $Y = W^O * Z$;
6:     **if** $Loss > \theta$ **then**
7:         modify the $W^O = W^O + \delta W$ according to Eq. (7)–(9)
8:     **else**
9:         stop training, obtain the final weight W^O;
10: **return** result

---

(3) Algorithm Flow Process. **Algorithm** 1 illustrates the memristive ELM implementation process. We feed the algorithm with the input data and tune the corresponding output weight values (which is the memristor conductance value) through input voltages.

4   Experimental Results and Analysis

4.1   Memristor System Setup

In the following experiment, we use vacuum-annealed $LiNbO_3$ sandwiched between the top electrode and bottom electrode as a memristor to represent the output weight of ELM, by tuning the input voltage, we adjust the conductance value of the memristor. The conductance value is then linearly transformed to represent the actual weight of ELM network.

4.2   Benchmark Data Set and ELM Algorithm Configuration

We test the proposed memristive ELM with Iris Flower dataset from the UCI Machine Learning Database Repository [2], which contains 150 samples of three categories, and each category contains 50 samples and 4 features. The goal is to classify the unknown samples on the basis of known Iris data object. We choose two classes corresponding to the features of sepal length and width, petal length and width. We select 80 samples (40 for setosa and 40 for versicolor) as training data and use other independent 20 samples as testing data.

In the experiment, the ELM model is constructed with 30 synapses implemented with 6 memristors and 11 neurons. The input neurons $x_i(i = 1, 2, 3, 4)$ correspond to the features of Iris, as the input of the ELM network. w_{ij}^I $((i = 1, ..., 4, j = 1, ...6))$ is the random weight (but fixed during the algorithm's training and testing phase) between the input layer and the hidden layer. w_k^o is the weight between hidden layer and output layer, which is the trainable synaptic weight. Here, the trainable weight is represented by memristor's conducted value, and each weight is implemented with one memristor.

During the memristor implementation process, we first set the input voltage pulse with amplitude $V = -10v$ to activate the memristor [24], then we adopt the pulse with amplitude $V = 5v$ to read the current of memristor. We construct the relationship between the change of weight Δw and current ΔI according to a sequence of experiments, which is defined as: $\Delta w = 1.05 \times 10^5 \Delta I$.

4.3   Performance Evaluation and Results Analysis

To evaluate the performance of the weight-tuning process based on memristor, we compare it with pure MATLAB ELM implementation (we call it 'normal ELM' in the remaining contents of the paper). Figure 3 shows the weight adaptation process comparison between memristive ELM and normal ELM. In Fig. 3, we can see that after around 10 memristor epochs, the two algorithms reach the almost the same value. The time cost of 10 memristor epochs is in the scale of miliseconds, thus can be ignored in most of the applications. It is worth noting that the memristor ELM ends after around 40 epochs as the algorithm satisfies the predefined threshold $error < 0.001$.

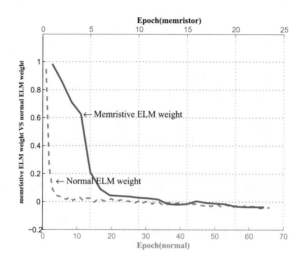

Fig. 3. The weight adaptation process of the memristive ELM VS normal ELM (we use w_1^o as an example, all the other weight adaptation process are of similar patterns). Note the time cost of one epoch in memristor is different from that of one epoch in Matlab (normal ELM), we already scale them to the same time scale in the figure. It also applies to Fig. 4

Figure 4 shows the comparative results of memristive ELM VS. normal ELM. In the figure, we can see that both of the two algorithms will converge and reach very low error. As it is shown in Fig. 4, memristive ELM will exhibit high fluctuation and converge slowly when compared to normal ELM. That is because we cannot use the pure gradient information for memristive ELM implementation, instead, we are using the coordinate descent method which converges slower than gradient-based algorithms. Another difficulty is that we cannot exactly control the weight update value, instead, we can only control the direction of the weight update process. Therefore, the memristive ELM exhibits high fluctuation shown in Fig. 4.

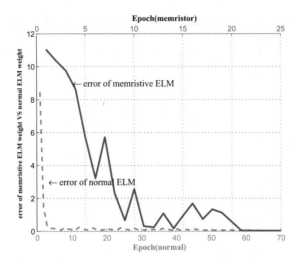

Fig. 4. The error changing process of the memristive ELM VS normal ELM.

5   Conclusion and Future Works

In this paper, we have implemented a prototype of memristive extreme learning machine for the classification task and tested it over the Iris dataset. We use memristors' conductance value to represent ELM network's output weight and adjust it with input voltages. Experimental results show that the adjusted memristor-based ELM weights are similar to the normal ELM weights and the performance in term of mean square errors is of no difference to normal ELM. Memristive ELM has great potential for neuromorphic computation applications.

This paper proposes a new way for neuromorphic hardware implementation of ELM and serves as a pioneer in memristor-based ELM. In the future, we will implement various variations of ELM such as L1-ELM, incremental ELM, unsupervised ELM with memristors and build the real memristive ELM system. We are also trying the extend the memristor-based AI algorithms to other machine learning algorithms like neural networks, support vector machines. Considering the uniqueness of memristor's conductance value preservation ability, we will perform memristor-based spiking neural network implementation in the future.

References

1. Bakkum, P., Skadron, K.: Accelerating SQL database operations on a GPU with CUDA. In: Proceedings of the 3rd Workshop on General-Purpose Computation on Graphics Processing Units, pp. 94–103. ACM (2010)
2. Blake, C., Merz, C.J.: UCI repository of machine learning databases (1998)
3. Bolz, J., Farmer, I., Grinspun, E., Schröoder, P.: Sparse matrix solvers on the GPU: conjugate gradients and multigrid. ACM Trans. Graph. (TOG) **22**(3), 917–924 (2003)

4. Chakravarty, M.M., Keller, G., Lee, S., McDonell, T.L., Grover, V.: Accelerating haskell array codes with multicore GPUs. In: Proceedings of the sixth workshop on Declarative aspects of multicore programming, pp. 3–14. ACM (2011)
5. Diamos, G.F., Wu, H., Lele, A., Wang, J.: Efficient relational algebra algorithms and data structures for GPU. Georgia Institute of Technology, Technical report (2012)
6. Eryilmaz, S.B., Kuzum, D., Jeyasingh, R., Kim, S., BrightSky, M., Lam, C., Wong, H.-S.P.: Brain-like associative learning using a nanoscale non-volatile phase change synaptic device array, arXiv preprint arXiv:1406.4951 (2014)
7. Furber, S., Temple, S.: Neural systems engineering. J. R. Soc. Interface **4**(13), 193–206 (2007)
8. Grauer-Gray, S., Kambhamettu, C., Palaniappan, K.: GPU implementation of belief propagation using CUDA for cloud tracking and reconstruction. In: 2008 IAPR Workshop on Pattern Recognition in Remote Sensing (PRRS 2008), pp. 1–4. IEEE (2008)
9. Hartley, T.D., Catalyurek, U., Ruiz, A., Igual, F., Mayo, R., Ujaldon, M.: Biomedical image analysis on a cooperative cluster of GPUs and multicores. In: ACM International Conference on Supercomputing 25th Anniversary Volume, pp. 413–423. ACM (2014)
10. Hetherington, T.H., Rogers, T.G., Hsu, L., O'Connor, M., Aamodt, T.M.: Characterizing and evaluating a key-value store application on heterogeneous CPU-GPU systems. In: IEEE International Symposium on Performance Analysis of Systems and Software (ISPASS), pp. 88–98. IEEE (2012)
11. Hu, S., Liu, Y., Liu, Z., Chen, T., Wang, J., Yu, Q., Deng, L., Yin, Y., Hosaka, S.: Associative memory realized by a reconfigurable memristive hopfield neural network. Nature commun. **6**, 7522 (2015)
12. Huang, G.-B., Chen, L.: Convex incremental extreme learning machine. Neurocomputing **70**, 3056–3062 (2007)
13. Huang, G.B., Chen, L.: Enhanced random search based incremental extreme learning machine. Neurocomputing **71**(1618), 3460–3468 (2008)
14. Huang, G.-B., Chen, L., Siew, C.-K.: Universal approximation using incremental constructive feedforward networks with random hidden nodes. IEEE Trans. Neural Netw. **17**(4), 879–892 (2006)
15. Huang, G.-B., Zhu, Q.-Y., Siew, C.-K.: Extreme learning machine: theory and applications. Neurocomputing **70**(1–3), 489–501 (2006)
16. Jo, S.H., Chang, T., Ebong, I., Bhadviya, B.B., Mazumder, P., Lu, W.: Nanoscale memristor device as synapse in neuromorphic systems. Nano Lett. **10**(4), 1297–1301 (2010)
17. Kuzum, D., Jeyasingh, R., Lee, B.C., Wong, H.: Nanoelectronic programmable synapses based on phase change materials for brain-inspired computing. Nano Lett. **12**(5), 2179–2186 (2012)
18. Lieberman, M.D., Sankaranarayanan, J., Samet, H.: A fast similarity join algorithm using graphics processing units. In: IEEE 24th International Conference on Data Engineering, ICDE 2008, pp. 1111–1120. IEEE (2008)
19. Mead, C.: Neuromorphic electronic systems. Proc. IEEE **78**(10), 1629–1636 (1990)
20. Mosegaard, J., Sørensen, T.S.: Real-time deformation of detailed geometry based on mappings to a less detailed physical simulation on the GPU. In: IPT/EGVE, pp. 105–111 (2005)
21. Nickolls, J., Dally, W.J.: The GPU computing era. IEEE micro, **30**(2) (2010)

22. Pan, X., Shuai, Y., Wu, C., Luo, W., Sun, X., Zeng, H., Zhou, S., Böttger, R., Ou, X., Mikolajick, T.: Rectifying filamentary resistive switching in ion-exfoliated linbo3 thin films. Appl. Phys. Lett. **108**(3), 032904 (2016)
23. Pershin, Y.V., Di Ventra, M.: Experimental demonstration of associative memory with memristive neural networks. Neural Netw. **23**(7), 881–886 (2010)
24. Prezioso, M., Merrikh-Bayat, F., Hoskins, B., Adam, G., Likharev, K.K., Strukov, D.B.: Training and operation of an integrated neuromorphic network based on metal-oxide memristors. Nature **521**(7550), 61–64 (2015)
25. Strukov, D.B., Snider, G.S., Stewart, D.R., Williams, R.S.: The missing memristor found. Nature **453**(7191), 80–83 (2008)
26. Trancoso, P., Othonos, D., Artemiou, A.: Data parallel acceleration of decision support queries using cell/be and GPUs. In: Proceedings of the 6th ACM Conference on Computing Frontiers, pp. 117–126. ACM (2009)
27. Trimberger, S., Carberry, D., Johnson, A., Wong, J.: A time-multiplexed FPGA. In: Proceedings the 5th Annual IEEE Symposium on Field-Programmable Custom Computing Machines, pp. 22–28. IEEE (1997)
28. Volk, P.B., Habich, D., Lehner, W.: GPU-based speculative query processing for database operations. In: ADMS@ VLDB, 2010, pp. 51–60 (2010)
29. Yang, J.J., Strukov, D.B., Stewart, D.R.: Memristive devices for computing. Nat. Nanotechnol. **8**(1), 13–24 (2013)

Model Research on CFBB's Boiler Efficiency Based on an Improved Online Learning Neural Network

Guoqiang Li, Bin Chen$^{(\boxtimes)}$, Xiaobin Qi, and Lu Zhang

Key Lab of Industrial Computer Control Engineering,
Yanshan University, Qinhuangdao 066004, China
18234077996@163.com

Abstract. Aiming at the accuracy prediction of combustion efficiency for a 300 MW circulating fluidized bed boiler (CFBB), a online sequential circular convolution parallel extreme learning machine (OCCPELM) is proposed and applied to build the models of boiler efficiency. In OCCPELM, the circular convolution theory is introduced to map the hidden layer information into higher-dimension information; in addition, the input layer information is directly transmitted to its output layer, which makes the whole network into a double parallel construction. Four UCI regression problems and Mackey-Glass chaotic time series are employed to verify the effectiveness of OCCPELM. Finally, this paper establishes a model of combustion efficiency for a CFBB. Some comparative simulation results show that OCCPELM with less hidden units owns better generalization performance and repeatability.

Keywords: Circular convolution · ELM · Online sequential learning
Boiler efficiency prediction model

1 Introduction

Along with increasingly growth in electricity demand and coal consumption, the prediction of boiler efficiency play a role in combustion process of a power plant. In the practice, the boiler efficiency can not be directly measured. Therefore, system modeling is an important method and foundation to predict the boiler efficiency, and researchers have built linear system's model to establish high reliability models [1]. However, owing to existed complex nonlinear systems in real world, it is difficult for classical model to forecast precisely the target [2]. So the prediction of boiler efficiency require a accuracy model.

In the face of above problem, artificial neural network (ANN) has very strong nonlinear mapping ability, robustness and generalization ability [3, 4]. For traditional algorithms like BP, all the parameters of network are tuned iteratively, whose process costs a lot of training time [5]. Aiming at the problem, Guang-Bin Huang has proposed Extreme Learning Machine (ELM) which is a simple and fast learning algorithm for ANN [6, 7]. ELM translates a training problem of single hidden layer feed-forward

© Springer Nature Switzerland AG 2019
J. Cao et al. (Eds.): ELM 2017, PALO 10, pp. 135–149, 2019.
https://doi.org/10.1007/978-3-030-01520-6_12

neural networks into a linear least squares problem, and obtains the smallest norm of weights by means of Moore-Penrose generalized inverse.

In face of increasingly expanding data with an very high speed, online sequential extreme learning machine (OSELM) provides a method to analyze incremental data in this era of big data [8]. Compared with popular sequential learning algorithms like stochastic gradient descent BP (SGBP) [9] and resource allocation network (RAN) [10], OSELM can achieve better generalization ability with very fast learning speed [11]. Based on the advantage of OSELM, it has been applied to solve practical problems [12, 13].

In order to further develop OSELM, this paper proposes a online sequential circular convolution parallel extreme learning machine (OCCPELM). It first introduces the circular convolution theory [14] to further extract the information of its hidden layer, forming a circular convolution layer which can acquire high-dimension mapping from hidden layer and double parallel construction [15] is also introduced to enhance relation between input layer and output layer. The proposed algorithm is applied to build a prediction model of boiler efficiency, simulation results show the built model can achieve great generalization ability and stability with less hidden units.

2   Online Sequential Circular Convolution Parallel Extreme Learning Machine (OCCPELM)

In aspect of processing big data, OSELM has a lot great advantages on acquiring good generalization ability, saving the memory space and shorting training time [11]. In order to further improve the performance of OSELM, this section proposes a novel network called online sequential circular convolution parallel extreme learning machine (OCCPELM) and gives its basic principle.

2.1   The Theory of Circular Convolution

In digital signal processing, circular convolution occupies very important position and is the basic theory of achieving discrete fourier transformation (DFT). Suppose, there are two sequences $h(n)$ and $x(n)$, whose length are N and M respectively. The L points circular convolution can be defined as follows:

$$y_c(n) = \left[\sum_{m=0}^{L-1} h(m)x((n-m))_L \right] R_L(n) \tag{1}$$

where L means the length of the circular convolution interval which must satisfy the following relation: $L \geq \max[N, M]$, $x((n-m))_L$ is the periodic signal with a circle of L points, n and m are integer in the internal $[0, L]$.

Considering convenience of calculation, matrix multiplication method is adopted to transform Eq. (1), and the process is as follows:

$$
\begin{bmatrix} y_c(0) \\ y_c(1) \\ \vdots \\ y_c(L-1) \end{bmatrix} = \begin{bmatrix} x(0) & x(L-1) & x(L-2) & \cdots & x(1) \\ x(1) & x(0) & x(L-1) & \cdots & x(2) \\ \vdots & \vdots & \vdots & \ddots & \vdots \\ x(L-1) & x(L-2) & x(L-3) & \cdots & x(0) \end{bmatrix} \begin{bmatrix} h(0) \\ h(1) \\ \vdots \\ h(L-1) \end{bmatrix} \tag{2}
$$

where $\begin{bmatrix} x(0) & x(L-1) & x(L-2) & \cdots & x(1) \\ x(1) & x(0) & x(L-1) & \cdots & x(2) \\ \vdots & \vdots & \vdots & & \vdots \\ x(L-1) & x(L-2) & x(L-3) & \cdots & x(0) \end{bmatrix}$ is the called L points circular

convolution matrix of $x(n)$.

Note: if length of $h(n)$ and $x(n)$ satisfy the following relation: $N < L$ or $M < L$, the end of both of $h(n)$ and $x(n)$ need to fill $L - N$ zero or $L - M$ zero.

2.2 The Description of OCCPELM

The subsection is the main part of the proposed algorithm. The architecture of OCC-PELM is given in Fig. 1. As seen from Fig. 1, OCCPELM is a double parallel feed-forward neural network. And its output nodes are directly and linearly connected with input nodes, simultaneously connected with the nodes of circular convolution layer. And each node of circular convolution layer replaces relevant neuron of hidden layer in

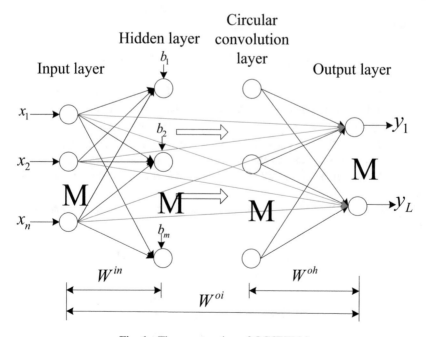

Fig. 1. The construction of OCCPELM

OCCPELM. The specific learning process of OCCPELM is described in detail as follows.

Firstly, given a chunk of N_0 training samples $S_0 = \{(x_i, y_i)\}_{i=1}^{N_0}$, the output of ith hidden neuron for jth sample is as follows:

$$G_0^{ji} = g\left(b_i + \sum_{t=1}^{n} W_{it}^{in} x_{jt}\right) \tag{3}$$

where W^{in} is a weight matrix between hidden layer and input layer.

A case study of the $j-th$ sample, all hidden layer output is described as follows:

$$G_0^j = \begin{bmatrix} G_0^{j1} & G_0^{j2} & \cdots & G_0^{jm} \end{bmatrix}^T \tag{4}$$

According to circular convolution of Sect. 2.1, let $L = m$, then the output of circular convolution layer is described as follows:

$$\tilde{G}_0^j = \begin{bmatrix} \tilde{G}_0^{j1} & \tilde{G}_0^{j2} & \cdots & \tilde{G}_0^{jm} \end{bmatrix}^T = \begin{bmatrix} G_0^{j1} & G_0^{jm} & G_0^{j(m-1)} & \cdots & G_0^{j2} \\ G_0^{j2} & G_0^{j1} & G_0^{jm} & \cdots & G_0^{j3} \\ \vdots & \vdots & \vdots & \ddots & \vdots \\ G_0^{jm} & G_0^{j(m-1)} & G_0^{j(m-2)} & \cdots & G_0^{j1} \end{bmatrix} \begin{bmatrix} G_0^{j1} \\ G_0^{j2} \\ \vdots \\ G_0^{jm} \end{bmatrix} \tag{5}$$

Among outputs of the above Eq. (5), the first output value of circular convolution layer can be in detail described as follows:

$$\tilde{G}_0^{j1} = G_0^{j1} * G_0^{j1} + G_0^{jm} * G_0^{j2} + \cdots + G_0^{j2} * G_0^{jm} \tag{6}$$

From the Eq. (6), $\tilde{G}_0^{j1}$ and all attributes of hidden layer keep a non-linear relationship. And the first node output of circular convolution layer can further replace the first node of hidden layer. Importantly, along with the transformation from G_0^{j1} to $\tilde{G}_0^{j1}$, the dimension of circular convolution layer node is also increased from 1 to m. Therefore, each node of its circular convolution layer can be influenced by the total nodes of its hidden layer, which can intensify the communication among nodes of hidden layer.

Therefore, for all of samples, output matrix $\tilde{G}_0$ of circular convolution layer is as follows:

$$\tilde{G}_0 = \begin{bmatrix} \tilde{G}_0^{11} & \tilde{G}_0^{21} & \cdots & \tilde{G}_0^{j1} & \cdots & \tilde{G}_0^{N_01} \\ \tilde{G}_0^{12} & \tilde{G}_0^{22} & \cdots & \tilde{G}_0^{j2} & \cdots & \tilde{G}_0^{N_02} \\ \vdots & \vdots & \ddots & \vdots & \ddots & \vdots \\ \tilde{G}_0^{1m} & \tilde{G}_0^{2m} & \cdots & \tilde{G}_0^{jm} & \cdots & \tilde{G}_0^{N_0m} \end{bmatrix} \tag{7}$$

Then, output of OCCPELM can be expressed by following matrix format:

$$Y_0 = W_0^{oi} X_0 + W_0^{oh} \tilde{G}_0 = \begin{bmatrix} W_0^{oh} & W_0^{oi} \end{bmatrix} \begin{bmatrix} \tilde{G}_0 \\ X_0 \end{bmatrix} = W_0 H_0 \qquad (8)$$

where $H_0 = [\tilde{G}_0^T \; X_0^T]^T$, W_0^{oh} is weight matrix between the output layer and the circular convolution layer, and W_0^{oi} is the weight matrix between the output layer and the input layer, and $W_0 = [W_0^{oh} \; W_0^{oi}]_{L \times (m+n)}$ is called the output weight matrix, and Y_0 is the target output of samples S_0.

Then, according to least square theory, W_0 could be estimated analytically:

$$W_0 = Y_0 H_0^+ = Y_0 H_0^T (H_0 H_0^T)^{-1} = Y_0 H_0^T K_0^{-1} \qquad (9)$$

where $K_0 = H_0 H_0^T$.

Therefore, according to the W_0, W_0 can be divided into two parts:

$$\begin{cases} W_0^{oh} = W_0(1:L, \; 1:m) \\ W_0^{oi} = W_0(1:L, \; m+1:(m+n)) \end{cases} \qquad (10)$$

Secondly, given a chunk of N_1 training samples $S_1 = \{(x_i, y_i)\}_{i=N_0+1}^{N_0+N_1}$, the output of circular convolution layer $\tilde{G}_1$ can be calculated by the above similar way.

For the combination of S_1 and S_0, the output of OCCPELM can be described by following matrix format:

$$[Y_0 \; Y_1] = \begin{bmatrix} W_1^{oh} & W_1^{oi} \end{bmatrix} \begin{bmatrix} \tilde{G}_0 & \tilde{G}_1 \\ X_0 X_1 \end{bmatrix} = W_1 [H_0 \; H_1] \qquad (11)$$

where $H_1 = [\tilde{G}_1^T \; X_1^T]^T$.

According to least square theory, W_1 of formula (11) can be computed by as follows:

$$W_1 = [Y_0 \; Y_1][H_0 \; H_1]^T ([H_0 \; H_1][H_0 \; H]^T)^{-1} = (Y_0 H_0^T + Y_1 H_1^T)(H_0 H_0^T + H_1 H_1^T)^{-1} \qquad (12)$$

Let $K_1 = H_0 H_0^T + H_1 H_1^T$, then $K_1 = K_0 + H_1 H_1^T$. So formula (12) can be transformed into the following format:

$$W_1 = W_0 + (Y_1 - W_0 H_1) H_1^T K_1^{-1} \qquad (13)$$

Finally, if training samples are extended into $(k+1)th$ chunk, namely $S_{k+1} = \{(x_i, y_i)\}_{i=1+\sum_{j=0}^{k} N_j}^{\sum_{j=0}^{k+1} N_j}$, the output weights can be updated as follows:

$$K_{k+1} = K_k + H_{k+1}H_{k+1}^T \tag{14}$$

$$W_{k+1} = W_k + (Y_{k+1} - W_k H_{k+1})H_{k+1}^T K_{k+1}^{-1} \tag{15}$$

where $H_{k+1} = \begin{bmatrix} \tilde{G}_{k+1}^T & X_{k+1}^T \end{bmatrix}^T$.

Let $P_{k+1} = K_{k+1}^{-1}$, then P_{k+1} can be expressed as follows according to the woodbury formula (11):

$$P_{k+1} = P_k - P_k H_{k+1}\left(I + H_{k+1}^T P_k H_{k+1}\right)^{-1} H_{k+1}^T P_k \tag{16}$$

where $P_k = K_k^{-1}$.

Therefore, the recursive formula of output weight matrix W_{k+1} can be got by the following formula:

$$W_{k+1} = W_k + (Y_{k+1} - W_k H_{k+1})H_{k+1}^T P_{k+1} \tag{17}$$

According to the W_{k+1}, W_{k+1} can be divided into two parts:

$$\begin{cases} W_{k+1}^{oh} = W_{k+1}(1:L,\ 1:m) \\ W_{k+1}^{oi} = W_{k+1}(1:L,\ m+1:(m+n)) \end{cases} \tag{18}$$

Now, the learning procedure of OCCPELM can be summarized as follows. And OCCPELM has also two phases like OSELM, an initialization phase and a sequential learning phase.

(1) Given the initial training samples S_0, the process of initialization is as follows:

(1) Set randomly input weights W^{in} and hidden layer bias b;
(2) Compute the initial hidden layer output matrix G_0 and the initial circular convolution layer output matrix $\tilde{G}_0$;
(3) $\tilde{G}_0$ and X_0 compose a combination H_0, then compute the initial output weights W_0;
(4) W_0 can be divided into W_0^{oi} and W_0^{oh};
(5) Set $k = 0$.

(2) Given $(k+1)th$ chunk of new samples S_{k+1}, the sequential learning phase is as follows:

(1) Compute hidden layer output matrix G_{k+1} and circular convolution layer output matrix $\tilde{G}_{k+1}$ of the $(k+1)th$ chunk;
(2) $\tilde{G}_{k+1}$ and X_{k+1} compose a new combination H_{k+1}, then update the output weight W_{k+1};
(3) W_{k+1} can be divided into W_{k+1}^{oi} and W_{k+1}^{oh};
(4) Set $k = k+1$. Return the sequential learning phase (2) unless all of the sequential learning chunk can be trained.

On the basis of OSELM, this section proposes OCCPELM under the condition of not increasing random parameters of the network, acquiring an aim of mapping lower dimension hidden layer information into higher dimension information. In addition, double parallel construction is adopted to further increase the non-linear mapping ability of the whole network. Do it like this, and its output layer not only receives circular convolution layer information, but also obtains its input layer information, which is good for non-linear mapping ability of whole network.

3 Performance Evaluation

In order to verify performance of the proposed algorithm OCCPELM, four regression data sets, from UCI data sets (http://archive.ics.uci.edu/ml/datasets), and Mackey-Glass chaotic time series are introduced to conduct some experiments compared with ELM and OSELM. All of the experiments in this work have been carried out in MATLAB 7. 14 environment running in a desktop PC with windows 7(64 bit), 3.30 GHz CPU and 4 GB RAM. In addition, learning modes chunk-by-chunk choose 1-by-1, 20-by-20 and [10, 30] which mean a randomly varying chunk size between 10–30. And the initial training samples N_0 adopts $N_0 = L + 50$.

3.1 UCI Regression Problems

The four benchmark problems (Boston housing, Abalone, Auto MPG, and stocks) are selected and information specification of those problems are listed in Table 1.

Table 1. Specification of regression problems

Data sets	Attributes	Observations	Training	Testing
Boston housing	13	506	250	256
Abalone	8	4177	2000	2177
Auto MPG	7	398	200	198
Stocks	9	950	500	450

For each problem, Table 1 shows the number of training samples and testing samples, randomly generated from its whole data set before each trial of simulation. We introduce the cross-validation method to select the nearly optimal number of hidden nodes of OSELM and OCCPELM listed in Table 2. For ELM and OSELM, they are set as the same hidden neurons units. Under the condition different learning mode, Table 2 shows the comparison of testing Root Mean Square Error (RMSE) and testing R-Square (averaged 30 trials) for algorithms.

As observed from the Table 2, compared with ELM and OSELM for different modes, OCCPELM can acquire the higher R-Square value, which is closer to 1 for the model prediction, the better the prediction ability is. And for RMSE, OCCPELM has the lowest error in different modes. For each case, OCCPELM use less hidden units than other algorithms.

Table 2. Comparison of testing RMSE and R-Square

Data sets	Model	Learning mode	RMSE	R-square	#nodes
Boston housing	ELM	Batch	0.1073	0.8509	26
	OSELM	1-by-1	0.1067	0.8528	26
		20-by-20	0.1103	0.8419	
		[10,30]	0.1081	0.8517	
	OCCPELM	1-by-1	0.1049	0.8588	8
		20-by-20	0.1076	0.8501	
		[10,30]	0.1063	0.8571	
Abalone	ELM	Batch	0.0768	0.7439	38
	OSELM	1-by-1	0.0777	0.7395	38
		20-by-20	0.0768	0.7460	
		[10,30]	0.0774	0.7401	
	OCCPELM	1-by-1	0.0771	0.7433	14
		20-by-20	0.0769	0.7450	
		[10,30]	0.0766	0.7445	
Auto MPG	ELM	Batch	0.0811	0.9216	40
	OSELM	1-by-1	0.0799	0.9239	40
		20-by-20	0.0776	0.9285	
		[10,30]	0.0796	0.9252	
	OCCPELM	1-by-1	0.0783	0.9263	13
		20-by-20	0.0764	0.9305	
		[10,30]	0.0777	0.9284	
Stocks	ELM	Batch	0.0361	0.9880	110
	OSELM	1-by-1	0.0356	0.9884	110
		20-by-20	0.0360	0.9881	
		[10,30]	0.0368	0.9876	
	OCCPELM	1-by-1	0.0361	0.9880	90
		20-by-20	0.0360	0.9881	
		[10,30]	0.0366	0.9877	

3.2   Time-Seris Prediction Problem

In order to further verify the effectiveness of the proposed algorithm, OCCPELM is applied in a typical online prediction of Mackey-Glass chaotic time series. For the chaotic series, Mackey and Glass gave the nonlinear differential delay equation:

$$\frac{dx(t)}{dt} = \frac{\alpha x(t-\tau)}{1 + x^{\gamma}(t-\tau)} - \beta x(t) \tag{19}$$

where $\alpha = 0.2, \beta = 0.1, \gamma = 10$ and τ is a adjustable parameter. When $x(0) = 1.2$ and $\tau = 17$, we can acquire a non-periodic and non-convergent time series that is sensitive to initial conditions (let $x(t) = 0$ when $t < 0$). And the time series is generated by the four order Runge-kutta method.

Now, we want to build an model that can can predict $x(t+6)$ from the past values of this time series, $x(t-18), x(t-12), x(t-6), x(t)$. Therefore, data format is $[x(t-18), x(t-12), x(t-6), x(t); x(t+6)]$.

The following description is the data processing: from $t = 119$ to 1118, we collect 1000 data pairs of the above format. The first 500 are used for training while the others are used for checking. The Fig. 2 shows the segment of the time series where data pairs were extracted from. The first 100 data points are ignored to avoid the transient portion of the data.

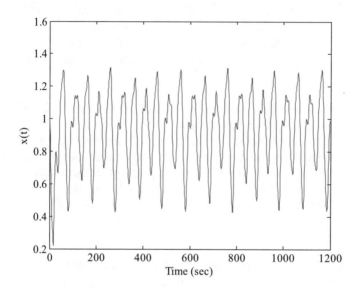

Fig. 2. Mackey-Glass chaotic time series

In the section, the hidden number of three algorithms are assigned as 10 and Sigmoid function is adopted as their activation function. For each learning mode, testing performance comparisons of algorithms are given in Table 3 under the condition averaged 30 trials. As observed from the Table 3, OCCPELM can fulfill the sequence learning target the same as the OSELM. Compared with OSELM, OCC-PELM obtain the best R-Square and the lowest MAPE and RMSE under the same learning mode, meaning that the expected valves of OCCPELM are very close to the actual values. In addition, for the S.D., we can see that OCCPELM has the better repeatability.

4 The Prediction of Boiler Efficiency for CFBB

In a power plant, the combustion efficiency of CFBB's combustion process is vulnerable to some complicated and changeable factors, such as coal quantity, primary air flow of burner inlet, primary air temperature of burner inlet and so on. However, once

Table 3. The comparison of testing performance index

Algorithms	Learning mode	MAPE(%)	R-Square	RMSE	
				Mean	S.D.
ELM	Batch	3.8596	0.9836	0.0395	0.00901
OSELM	1-by-1	3.6752	0.9855	0.0373	0.00818
	10-by-10	3.8946	0.9825	0.0401	0.01215
	20-by-20	4.2236	0.9796	0.0437	0.01187
	[10,30]	4.0501	0.9817	0.0416	0.01053
OCCPELM	1-by-1	2.8282	0.9906	0.0296	0.00795
	10-by-10	2.6801	0.9920	0.0276	0.00664
	20-by-20	2.9141	0.9907	0.0300	0.00574
	[10, 30]	2.8368	0.9910	0.0293	0.00681

the fuel and environmental conditions are specified, combustion efficiency are mainly vulnerable to various operating parameters.

For the prediction of combustion efficiency, it is necessary to set up the function relation between various operational parameters and combustion efficiency. Therefore, this paper adopts the proposed OCCPELM to set up a prediction model of combustion efficiency. This section presents the simulation results of the proposed OCCPELM algorithm applied in combustion thermal efficiency model for a 300 MW CFBB.

4.1   Data Samples

In this section, there are 2880 samples, collected from CFBB in a 300WM under operational conditions of a power plant, and then those samples are applied to set up a prediction model of combustion efficiency for the CFBB. The data includes 27 variables, and those specific variables are listed in Table 4.

In this study, the total 2880 samples are divided into two parts: 1920 samples (above account for two-thirds of the total data) as the training samples and the rest of total samples as testing samples. The training samples are used to build the functional relation between boiler efficiency and other 26 variables (input attributes) based on neural network algorithms. The testing samples are mainly used to test the model's generalization ability.

4.2   Analysis of the Simulation Experimental Results

Before the experiment, in order to avoid the influence of orders of magnitude among attributes, the data samples need employ normalization processing: input attributes have been normalized into $[-1,1]$ while target attributes have been normalized into $[0,1]$. After getting a trained model of boiler efficiency, the expected values need be disposed by processing of reverse normalization.

In term of the proposed algorithm, the number of hidden layer neurons is the main influence factors for prediction ability. In order to illustrate OCCPELM's behavior with the number change of hidden layer neurons, under the condition that OCCPELM

Table 4. The specification of attributes information

No.	Variable	Unit
	Input	
1	Load	%
2	Coal quantity of coal feeder (#1, #2, #3, #4)	t/h
3	Primary air flow of burner inlet (A, B)	KNm3/h
4	Primary air temperature of burner inlet (A, B)	°C
5	Primary air fan inlet temperature (#1, #2)	°C
6	Distribution flow of secondary air (A, B)	KNm3/h
7	Air preheater secondary air outlet air temperature (A, B)	°C
8	Total flow of secondary air	KNm3/h
9	Limestone powder conveying motor current (#1,#2)	AMP
10	Unburned carbon in flue dust (A, B)	%
11	Flue gas oxygen content	%
12	Flue gas temperature	°C
13	Outlet temperature of slag cooler (#1, #2, #3, #4)	°C
	Output	
1	Boiler efficiency	–

applied the sigmoidal function and the learning mode 1-by-1, an experimental study is conducted by varying the number of hidden neurons from 1 to 100 in steps of 1. RMSE (averaged 30 trials)'s variations in combustion efficiency of testing are given in Fig. 3. In addition, the number of initial data is assigned as $N_0 = L + 50$ for boiler efficiency modeling according to Sect. 3.

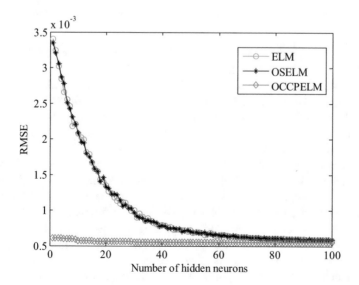

Fig. 3. The generalization performance of OCCELM with a range of number of hidden nodes

As shown from the Fig. 3, OSELM and ELM keep a decreasing trend with the increasing number of hidden neurons. However, compared with OSELM and ELM, RMSE of OCCELM is very stable on a wide range number of of hidden nodes. And Obviously, the bottom curve is for the OCCPELM, meaning that OCCPELM obtains the lowest root-mean-square error.

Figure 4 details the regression results (1-by-1) of the proposed algorithm OCC-PELM with 10 hidden units for combustion thermal efficiency under the condition a single trial.

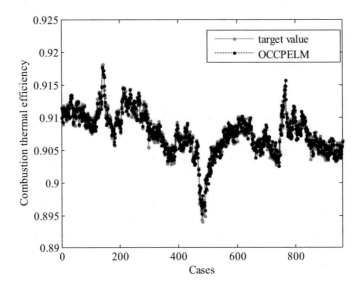

Fig. 4. The regression results for combustion thermal efficiency

In order to further show the forecast results of different learning mode, OCCPELM model chooses the hidden layer neuron number for 10, OSELM for 100 and ELM for 100. Do it like this, those model can reach the their nearly best prediction effect. For sake of verifying that OCCPELM can be applied stably in boiler efficiency modeling, there are 30 trials for the prediction model of boiler efficiency under the condition different learning mode. Table 5 lists the simulation results comparison in terms of some performance indexes.

As observed from Table 5, OCCPELM with different learning mode can obtain best R-Square and lower MAPE and mean of RMSE than ELM, further illustrating the superiority of the proposed online sequential model. Compared with OSELM under the same learning mode, OCCPELM obtains the better performance on various indexes (MAPE, R-Square and mean of RMSE), showing that the expected values of OCC-PELM is closer to the actual values than OSELM. For the training time, 1-by-1 sequential learning mode takes usually longest time than other learning modes including batch mode. However, for the best prediction performance, OCCPELM runs the same order of magnitudes of training time as OSELM. Synthesizing kinds of above

Table 5. Simulation results comparison

Algorithms	Learning mode	MAPE(%)	R-Square	RMSE (Mean)	Training time(s)
ELM	Batch	0.0394	0.9850	0.000581	0.0712
OSELM	1-by-1	0.0396	0.9847	0.000586	0.5803
	10-by-10	0.0398	0.9847	0.000586	0.0894
	50-by-50	0.0391	0.9849	0.000582	0.0567
	[10,50]	0.0395	0.9848	0.000584	0.0671
OCCPELM	1-by-1	0.0374	0.9854	0.000573	0.2439
	10-by-10	0.0374	0.9854	0.000572	0.0879
	50-by-50	0.0374	0.9854	0.000574	0.0759
	[10, 50]	0.0375	0.9853	0.000575	0.0811

simulation results, OCCPELM with fast learning speed can achieve better generalization ability with 10 hidden units to build an accuracy prediction model of combustion efficiency.

For a algorithm, repeatability is also a mainly performance index in industrial field, so combustion thermal efficiency modeling is also no exception in a power plant. Under a condition that OSELM and OCCPELM use a random chunk size between 10 and 50, Fig. 5 presents RMSE in each run of ELM, OSELM and OCCPELM for the combustion efficiency modeling problem. It can be seen that the curve of OCCPELM is affected sightly and below the curves of OSELM and ELM. So OCCPELM obtains the great stability and meets the online testing task of industrial field.

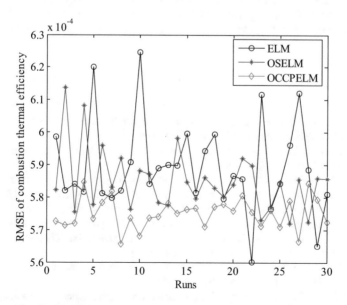

Fig. 5. RMSE in each run of algorithms for combustion efficiency

As observed from Fig. 5, it can be seen that the curve of OCCPELM is affected sightly and below the curves of OSELM and ELM. So OCCPELM obtains the great stability and meets the online testing task of industrial field.

5 Conclusion

In a power plant, boiler efficiency is an important index of various operating parameters for assessing energy utilization. Forecasting boiler efficiency is the basic key for saving and taking full advantage of non-regenerated energy (coal). In this study, a novel variant of OSELM, OCCPELM, was proposed and applied to build a higher accuracy model for forecasting combustion efficiency. In this paper, we first adopt four UCI regression problems and Mackey-Glass chaotic time series to verify the effectiveness of OCCPELM, and simulation results show that OCCPELM can acquire the better generalization ability than OSELM and ELM. At last, we use the verified OCCPELM to build the prediction model for the combustion efficiency in combustion process of a CFBB, in order to further verify its availability and practicability. The experiment results show that OCCPELM obtains more stability from simulation results analysis. And more importantly, compared with ELM and OSELM, OCCPELM can use less hidden units to achieve a extremely high accuracy forecasting and learning speed of its built model can also meet the online forecast requirement in the industrial field.

Acknowledgements. This work was supported by the National Natural Science Foundation of China [Grant No. 61403331, 61573306], the Natural Science Foundation of Hebei Province [Grant No. F2016203427], the China Postdoctoral Science Foundation [Grant No. 2015M571280], the Doctorial Foundation of Yanshan Univerity [Grant No. B847], the natural science foundation for young scientist of Hebei province [Grant No. F2014203099] and the independent research program for young teachers of yanshan university [Grant No. 13LG006].

References

1. Schoukens, J., Pintelon, R.: Identification of Linear Systems: A Practical Guideline to Accurate Modeling. Pergamon Press, Oxford (1991)
2. Antoulas, A.C.: An overview of approximation methods for large-scale dynamical systems. Annu. Rev. Control. 2, 181–190 (2005)
3. Green, M., Ekelund, U., Edenbrandt, L., et al.: Exploring new possibilities for case-based explanation of artificial neural network ensembles. Neural Netw. 22, 75–81 (2009)
4. Iliya, S., Goodyer, E., Gow, J., et al.: Application of Artificial Neural Network and Support Vector Regression in cognitive radio networks for RF power prediction using compact differential evolution algorithm. In: Federated Conference on Computer Science and Information Systems, pp. 55–66. IEEE press (2015)
5. Wang, S., Chung, F.L., Wang, J., Wu, J.: A fast learning method for feedforward neural networks. Neurocomputing 179, 295–307 (2014)
6. Huang, G.-B., Zhu, Q.-Y., Siew, C.-K.: Extreme learning machine: theory and applications. Neurocomputing. 70, 489–501 (2006)

7. Huang, G.B., Zhu, Q.Y., Siew, C.K.: Extreme learning machine: a new learning scheme of feed forward neural networks. In: 2004 IEEE International Joint Conference on Neural Networks, pp. 25–29. IEEE press, New York (2004)

8. Huang, S., Wang, B., Qiu, J., et al.: Parallel ensemble of online sequential extreme learning machine based on MapReduce. Neurocomputing. **174**, 352–367 (2016)

9. LeCun, Y., Bottou, L., Orr, G.B., Müller, K.P.: Efficient backprop. LNCS, vol. 1524, pp. 9–50 (2002)

10. Platt, J.: A resource-allocating network for function interpolation. Neural Comput. **3**, 213–225 (1991)

11. Liang, N.Y., Huang, G.B., Saratchandran, P., et al.: A Fast and Accurate Online Sequential Learning Algorithm for Feedforward Networks. IEEE Trans. Neural Networks **17**, 1411–1423 (2006)

12. Lima, A.R., Cannon, A.J., Hsieh, W.W.: Forecasting daily streamflow using online sequential extreme learning machines. J. Hydrol. **537**, 431–443 (2016)

13. Zhang, J., Feng, L., Yu, L.: A novel target tracking method based on OSELM. Multidimens. Syst. Signal Process. 1–18 (2016)

14. Gao, X.Q., Ding, Y.M.: Digital Signal Processing. Xidian University Press, Xi'an (2009)

15. Wang, J., Wu, W., Li, Z., et al.: Convergence of gradient method for double parallel feed forward neural network. Int. J. Numer. Anal. & Model. **8**, 484–495 (2012)

Robust Multi-feature Extreme Learning Machine

Zhang Jing[✉] and Ren Yonggong

Liaoning Normal University, NO. 1, Liushu South Road, Dalian, Liaonig, China
zhangjing_0412@163.com

Abstract. Extreme learning machine as efficiently and effectively single-hidden layer feedforward neural network is widely applied in pattern classification, feature extraction, data prediction and so on. However, it is difficult that ELM deal with multi-feature data directly, because each feature of multi-feature data has a specific statistical property. Therefore, we proposed a robust multi-feature model based on ELM, namely, multi-feature ELM (MFELM). In particular, MFELM builds models to each single feature, and optimize the combinatorial coefficient of multi-feature by iterating. Obtain the robust output weight and the most reasonable combinatorial coefficient that achieve the minimum of training errors sum in different features. In order to avoid the model degrade into single feature and to enhance the robustness of multi-feature, we induce the higher order combinatorial coefficient. Moreover, MFELM is developed to kernel method, and propose robust kernel multi-feature ELM (MFKELM), which solve the problem that different dimensions lead to operate difficultly. MFELM and MFKELM fully explore the complementary property of multi-feature to improve the recognition accuracy, and retain the robustness to multi-feature data. We demonstrate the performance of proposed in image, text, manual multi-feature datasets.

Keywords: Multi-feature learning · ELM · Kernal method

1 Introduction

Neural network, one of important discriminant classification method, establishes network structure of parallel interconnection by using a number of adaptive basic nodes. Representative network structures include multi-layer perceptron network, self-organizing mapping network, radial basis network, BP neural network, deep learning (DL) network that achieves greater breakthroughs in the current image feature extraction. However, in order to avoid overfitting, DL utilizes a large number of training samples, which induce a slow training speed in training process. Moreover, popular BP has a slowly solving speed, and its solution is easy to fall into local optimal. Aiming to above problems, Huang et al. proposed the extreme learning machine (ELM) method [1,2] that transforms the iterative solving into the linear equation set solving in the learning

© Springer Nature Switzerland AG 2019
J. Cao et al. (Eds.): ELM 2017, PALO 10, pp. 150–161, 2019.
https://doi.org/10.1007/978-3-030-01520-6_13

process. Therefore, ELM can get the display expression of the output weight and it avoids iterative computation. A large number of scholars have carried out in-depth research and improvement. Hung et al. utilize the item of regularization to enhance the generalization of ELM [3]. Frnay et al. introduce the kernel theory that trains data mapping into the high dimensional space for enhancing the classification performance [4]. In the past decade, ELM has been widely used in various fields. Zong et al. induce the weight matrix in the ELM optimization equation to solve imbalance dataset problem [5]. Liu et al. utilize ELM to solve image retrieval problem [6]. Huang et al. proposed semi-ELM algorithm that establishes the relationship between unlabeled samples and label samples based on graph theory [7]. Liu et al. further improved semi-ELM, and proposed semi-KELM [8]. At the same time, Tang et al. proposed HELM algorithm for feature extraction with helping of AE theory and solving process of ELM [9]. Above methods pay focus on learning single feature information of data and fail to take advantage of the complementary information that is provided by multi-feature data. However, this information can help to enhance the performance usually, as shown in the examples below (Shown in Fig. 1). webKB-washington dataset is a subset of webKB text dataset. We record the classification results for two single feature(content, cites) of webKB-washington separately in this graph. We can see that indexes of incorrect samples are different in two single feature by recording (indexes of samples in the feature "content" are: 7, 15, 39, 42, 53. and indexes of samples in the feature "cites" are: 7, 29, 61, 71, 79). This experiment shows that the two different features are complementary in the process of the classification. If two features can be used simultaneously, it will has an enhance for classification performance. Some scholars put forward a kind of multi-view method based on active learning theory. The classification accuracy is improved by modifying the classifier from different view. However, the method is only adjusting the classifier in the single-feature, which is difficult to apply the information of multi-feature simultaneously.

Aiming to the above problem, we develop a multi-feature classification method based on ELM (multi-feature extreme learning machine, MFELM). Firstly, obtain the linear combination of different features of input samples. Secondly, take the combined features as new input of ELM, and solve output weight β. Thirdly, use β to retrain each single feature as well as get training

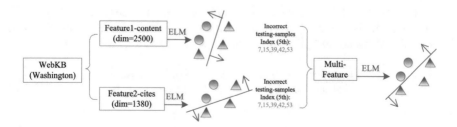

Fig. 1. Classification performance between the single-feature and multi-feature based on ELM.

error of single feature σ. According to σ adjust the coefficient of combination of multi-feature. At the same time, the iteration method is used to operate the firstly step until the sum of training errors tends the minimum value. Finally, the optimum combination coefficient of multi-feature and the output weight β are obtained. The advantages of the proposed are as follows:

1. Realizing the multi-feature combining automatically according to training results, multiple features provide the complementary information for classification.
2. The high order coefficents is leaded into avoid the phenomenon that only single-feature is selected in classification process, and improve the robustness of multi-feature method.
3. MFKELM is proposed to avoid the problem that different dimensions of different features are difficult to calculate.

This paper is organized as follows: we reviews the related work in Sect. 2. Section 3 analyzes problems of the original ELM and propose MFELM and MFKELM model. Section 4 presents and examines the experimental results, and we draw our conclusion in Sect. 5.

2   Related Work

The proposed multi-feature classification method is based on ELM. In order to facilitate the understanding of MFELM and MFKELM, this section briefly reviews the related concepts and theories of ELM and developed KELM.

Extreme learning machine is improved by single hidden layer neural network (SLFNs): assume given N samples (X, T), where $X = [x_1, x_2, ..., x_N]^T \in \mathbb{R}^{d \times N}$, $T = [t_1, t_2, ..., t_N]^T \in \mathbb{R}^{\tilde{N} \times N}$, and $t_i = [t_{i1}, t_{i2}, ..., t_{im}]^T \in \mathbb{R}^m$. The method is used to solve multi-classification problems, and thereby the number of network output nodes is $m(m \geq 2)$. There are $\tilde{N}$ hidden layer nodes in networks, and activation function $h(\cdot)$ can be Sigmoid or RBF: $\sum_{i=1}^{\tilde{N}} \beta_i h(a_i x_j + b_j) = o_j$ where $j = 1, \cdots, \tilde{N}$, $a_j = [a_{j1}, a_{j2}, \cdots, a_{jd}]^T$ is the input weight vector, and $\beta_j = [\beta_{j1}, \beta_{j2}, ..., \beta_{jm}]^T$ is the output weight vector. Moreover, a_j, b_j can be generated randomly, which is known by [2]. Written in matrix form: $H\beta = T$, where $H_i = [h_1(a_1 x_1 + b_1), \cdots, h_N(a_{\tilde{N}} x_1 + b_{\tilde{N}})]$. Moreover, the solution form of $H\beta = T$ can be written as: $\hat{\beta} = H^\dagger T$, where $H^\dagger$ is the generalized inverse matrix of H. ELM minimize both the training errors and the output weights. The expression can be formulated based on optimization of ELM:

$$\text{Minimize} : L_{ELM} = \frac{1}{2}\|\beta\|_2^2 + C\frac{1}{2}\sum_{i=1}^{N}\|\xi_i\|_2^2$$
$$\text{Subject to} : t_i\beta \cdot h(x_i) \geq 1 - \xi_i, i = 1, ..., N \tag{1}$$
$$\xi_i \geq 0, i = 1, ..., N$$

where $\xi_i = (\xi_{i,1} \cdots \xi_{i,m})$ is the vector of the training errors. We can solve the above equation based on KKT theory by Lagrange multiplier, and can obtain the analytical expression of the output weight: $\hat{\beta} = H^T(\frac{I}{C} + HH^T)^{-1}T$. The output function of ELM is: $f(x) = h(x)\hat{\beta} = h(x)H^T(\frac{I}{C} + HH^T)^{-1}T$. The matrix expression of ELM kernel function can be defined as: $\Omega_{ELM} = HH^T : \Omega ELM_{i,j} = h(x_i) \cdot h(y_j) = K(x_i, x_j)$. ELM output function can be obtained, where can be showed as kernel form: $f(x) = [K(x, x_1), \cdots, K(x, x_N)]^T(\frac{I}{C} + \Omega ELM)^{-1}T$. Here, the proper kernel function can be selected, such as RBF kernel function $K(u, v) = exp(-\gamma \|u - v\|^2)$. And $K(x_i, x_j) = h(x_i) \cdot h(x_j)$ is called as kernel ELM algorithm (KELM).

3 Proposed Algorithms

3.1 Multi-feature ELM and Multi-feature Kernel ELM

From ELM model, we can obtains mapping by activation function from ELM decision function, and then the samples that meet the conditions are activated by the activation function. However input samples contains multiple features, this decision function is difficult to find the solution. Therefore, single feature ELM cannot utilize the information of provided multi-feature samples for classification. In this paper, a multi-feature model that is called MFELM is proposed for above problem.

Given N the different samples, each sample contains two kinds of features by collecting in different ways. The feature 1 corresponds to the samples are: $(X_i^{(1)}, t_i^{(1)})$, where $X^{(1)} = \left[x_1^{(1)}, \ldots, x_N^{(1)}\right]^T \in \mathbb{R}^{d_1 \times N}$, $t_i^{(1)} = \left[t_{i1}^{(1)}, \ldots, t_{im}^{(1)}\right]^T \in \mathbb{R}^m$. The feature 2 corresponds to the samples for: $(X_i^{(2)}, t_i^{(2)})$, where $X^{(2)} = \left[x_1^{(2)}, \ldots, x_N^{(2)}\right]^T \in \mathbb{R}^{d_2 \times N}$, $t_i^{(2)} = \left[t_{i1}^{(2)}, \ldots, t_{im}^{(2)}\right]^T \in \mathbb{R}^m$. Meanwhile, the same sample of different features corresponds to the same class, then there is: $t_i^{(1)} = t_i^{(2)} = t_i$. The optimization equation of multi-feature ELM is as follows:

$$\text{Minimize} : L_{MFELM} = 1/2 \|\beta\|_2^2 + C/2 \sum_{i=1}^N \left\|\xi_i^{(1)}\right\|_2^2 + C/2 \sum_{i=1}^N \left\|\xi_i^{(2)}\right\|_2^2$$
$$\text{Subject to} : k_1 H^{(1)}\beta = t_i^T - C/2(\xi_i^{(1)})^T,$$
$$k_2 H^{(1)}\beta = t_i^T - C/2\left(\xi_i^{(2)}\right)^T, i = 1, \ldots, N \qquad (2)$$
$$\sum_{j=1}^Q k_q = 1, k > 0, q = 1, 2.$$

where k_q is the combined parameter that corresponds to the single-features. $\xi_i^{(1)} = \left(\xi_{i1}^{(1)}, \ldots, \xi_{im}^{(1)}\right)$ is training error vector that corresponds to feature 1. $\xi_i^{(2)} = \left(\xi_{i1}^{(2)}, \ldots, \xi_{im}^{(2)}\right)$ is training error vector that corresponds to feature 2. β is the output weight vector for different feature spaces in the equation (2). According to the equation (2), the target obtains the minimum value of the training error that combine the different feature spaces. However, from the equation (2), the solution of $[k_1, k_2]$ may be $(0,1)/(1,0)$. In this situation, only the single-feature is selected in solving processing, and the other feature is failed. In order to avoid this problem and to enhance the robustness of multi-feature, the higher order coefficient is leaded into equation (2). We define r and $r \geq 2$, the optimization equation of MFELM as follows:

$$\text{Minimize} : L_{MFELM} = 1/2\|\beta\|_2^2 + C/2\sum_{i=1}^{N}\left\|\xi_i^{(1)}\right\|_2^2 + C/2\sum_{i=1}^{N}\left\|\xi_i^{(2)}\right\|_2^2$$
$$\text{Subject to} : k_1^r H^{(1)}\beta = t_i^T - C/2(\xi_i^{(1)})^T,$$
$$k_2^r H^{(1)}\beta = t_i^T - C/2(\xi_i^{(2)})^T, i = 1,\ldots,N \tag{3}$$
$$\sum_{j=1}^{Q} k_q = 1, k > 0, j = 1, 2.$$

According to the KKT condition, we can solve above MFELM model:

$$L_{MFELM} = \frac{1}{2}\|\beta\|_2^2 + \frac{C}{2}\sum_{i=1}^{N}\left\|\xi_i^{(1)}\right\|_2^2 + \frac{C}{2}\sum_{i=1}^{N}\left\|\xi_i^{(2)}\right\|_2^2$$
$$-\sigma\left[\left(k_1^r H^{(1)}\right)\beta - t_i^T + \frac{C}{2}(\xi_i^{(1)})^T\right] - \omega\left[\left(k_2^r H^{(1)}\right)\beta - t_i^T + \frac{C}{2}(\xi_i^T)^{(2)}\right]$$
$$-\lambda\left(\sum_{j=1}^{C} k_j - 1\right)$$

where $(\sigma_i, \omega_i, \lambda)$ are Lagrange multipliers. By the KKT condition, obtain the partial derivative of $\beta, \xi_i, \sigma_i, \omega_i, \lambda$, and set the value is 0. Have the following results:

$$\frac{\partial L}{\partial \beta} = 0 \rightarrow \beta = k_1^r (H^{(1)})^T \sigma_i + k_2^r (H^{(1)})^T \omega_i \tag{4}$$

$$\frac{\partial L}{\partial \xi_i^{(1)}} = 0 \rightarrow= \xi_i^{(1)} = \sigma_i, \frac{\partial L}{\partial \xi_i^{(2)}} = 0 \rightarrow= \xi_i^{(2)} = \omega_i \tag{5}$$

$$\frac{\partial L}{\partial \sigma_i} = 0 \rightarrow \left(\xi_i^{(1)}\right)^T = t_i^T - k_1^r H^{(1)}, \frac{\partial L}{\partial \omega_i} = 0 \rightarrow \left(\xi_i^{(2)}\right)^T = t_i^T - k_1^r H^{(2)} \tag{6}$$

$$\frac{\partial L}{\partial \lambda_j} = 0 \rightarrow k_j = \frac{\left(1/\sum_{i=1}^{N}\beta^T h(x_i)\right)^{1/(r-1)}}{\sum_{j=1}^{C}\left(1/\sum_{i=1}^{N}\beta^T h(x_i)\right)^{1/(r-1)}} \tag{7}$$

Let the Eqs. (5–6) into the Eq. (4), we can obtain the displayed expressions of the output weights $\hat{\beta}$ and the multi-feature coefficient of combination k^r:

$$\hat{\beta} = \left[(k_1^r H^{(1)})^T + (k_2^r H^{(2)})^T\right] \cdot$$
$$\left[I/C + k_1^r H^{(1)}(k_1^r H^{(1)})^T + k_2^r H^{(2)}(k_2^r H^{(2)})^T\right]^{-1} \cdot T \tag{8}$$

And,

$$k_j = \frac{\left(1/\sum_{i=1}^{N}\beta^T h(x_i)\right)^{1/(r-1)}}{\sum_{j=1}^{C}\left(1/\sum_{i=1}^{N}\beta^T h(x_i)\right)^{1/(r-1)}} \tag{9}$$

When the testing sample x is obtained, the discriminant function is:

$$f(x) = \left[\left(k_1^r h(x)^{(1)}\right)^T + \left(k_2^r h(x)^{(2)}\right)^T\right] \cdot \left[\left(k_1^r H^{(1)}\right)^T + \left(k_2^r H^{(2)}\right)^T\right] \cdot$$
$$\left[I/C + k_1^r H^{(1)}\left(k_1^r H^{(1)}\right)^T + k_2^r H^{(2)}\left(k_2^r H^{(2)}\right)^T\right]^{-1} \cdot T \tag{10}$$

However, the different features on the same sample may be have different dimensions $d_1 \neq d_2$. This phenomenon induces that the Eqs. (8) and (10) are failure. Aiming to this problem, we map the samples to the kernel space based on the kernel theory, so that the Eq. (8) can be transformed into the following kernel form:

$$\hat{\beta}_{\text{ker nel}} = \left[k_1^r \phi(x^{(1)})^T + k_2^r \phi(x^{(2)})^T\right] \cdot$$
$$\left[I/C + k_1^r \phi(x^{(1)})\left(k_1^r \phi(x^{(1)})\right)^T + k_2^r \phi(x^{(2)})\left(k_2^r \phi(x^{(2)})\right)^T\right]^{-1} \cdot T \tag{11}$$

Therefore, the output discriminant function expression can be expressed as the form of the multi-feature kernel ELM (MFKELM):

$$f(x) = \begin{bmatrix} K(z, z_1) \\ \vdots \\ K(z, z_N) \end{bmatrix} \cdot \left[I/C + k_1^{2r}\Omega_1 + k_2^{2r}\Omega_2 + \ldots + k_f^{2r}\Omega_f\right]^{-1} \cdot T \tag{12}$$

At the same time, the input samples are mapped to the kernel space, then the algorithm can be used to classify for the multi-feature samples of different dimensions. Moreover, the algorithm can be extended to more than two kinds of features.

MFELM and MFKELM can be summarized as follows: firstly, the combined parameter k_q^r is set to the mean value of the number of features, and utilize combined multi-feature samples and the Eqs. (8) or (11) to obtain the output weight β. Then, using β train each single-feature and get the training error. Thirdly, revising k_q^r by the training error and the Eq. (9), and execute step 1 to calculate new β iteratively. Finally, the output weight and the most reasonable combination way of multi-feature for the current samples is obtained.

3.2   Discussion

According to above describe and derivation of MFELM and MFKELM, we can obtain display expressions of the output weight and the coefficient of combination, and β, r would be able to compute by iterating. Therefore, we note that MFELM and MFKELM have one obvious advantage is efficiency. In addition, proposed algorithms achieve better robustness, we can discussion as follow: in order to ensure that proposed algorithms do not degrade to single-feature models and use the information of multi-feature fully, we induce the high order coefficient of combination r. We know that the solution of the model does not exist this expression: $[0, \cdots, 0, 1, 0, \cdots, 0]$ when $r > 2$ and only one feature is available from the solving processing. Therefore, r can avoid above situation efficiently, and enhance the robustness for multi-feature data. Moreover, we note that the solving processing of the output weight depends on both of input data and the combination approach of multi-feature from expressions (8) and (11). Therefore, we can achieve optimal discriminant models (10), (11) for the current distribution when our models can combine multi-feature reasonably.

4   Experiments

In this paper, sigmod function is used to activation function, and set $C = 10^0$, $\tilde{N} = 300$. In order to further verify the classification performance of multi-feature data, we compare with single-feature methods including classical and effective ELM, KELM and CELM. Moreover, we compare with ELM, KELM, CELM when all features of each sample are connected. We utilize three group datasets including text WebKB datasets, image datasets, manual dataset to comprehensive evaluate performances for algorithms.

4.1   Datasets Description

Image are usually represented by multi-feature. Our experiments include humane face and Getgure datasets. AR face dataset is created by Martinez Robert and Benavente in the Aleix computer visual center. This dataset is consists of 4000 individuals (70 males and 56 females) face images. Subjects were collected under different illumination conditions, facial expression and occlusion condition (sunglasses and scarf). The size of each image sample is 40*50. It is the most popular human face dataset [10]. YaleB face dataset is established by Georghiades et al. from [11], which contains 16128 face images. This dataset is collected by 28 subjects in 9 postures and 64 case of illuminations. The size of each image sample is 32 * 32. ORL face dataset contains 40 different shooting conditions, shooting time and different illumination condition. The subjects include different expressions (eyes, smile, serious) with different props (glasses etc.). Each case contains 10 pictures, a total of more than 400 pieces of face image [12]. The size of each image sample is 32 * 32. PIE face dataset was created by the Carnegie Mellon University in 2000 12–10th, including 41368 images of 68 volunteers with different pose, illumination and facial expression. In this dataset include a various of pose and illumination. The size of each image sample is 32 * 32. At present, this dataset gradually becomes an important testing set in the field of face recognition [13]. Getgure dataset has more samples than the face dataset, and features of the samples are more similar. The dataset is proposed by KIM et al. from [14]. The dataset contains 9 types of gestures, and each types contains 900 gesture images (this dataset can be download from: http://www.iis.ee.ic.ac.uk/icvl/ges_db.htm.). The size of each image sample is 240 * 320 * 3. The authors have compiled a collection of 5 subsets of samples under different illumination conditions. In this paper, we select 2 subsets (Getgure1, Getgure3) to verify that algorithms. According to the characters of images, experiments extract three kinds of features including color feature (extracting method: histogram, dimension: 256), texture feature (extracting method: LBP, dimension: 256) and key points (extracting method: SIFT, dimension: 256) respectively.

In addition to the image datasets, we utilize webKB multi-feature text dataset to verified the proposed algorithm. The dataset contains 4 subsets which has six classes: cornell, texas, washington, wisconsin of where wisconsin includes two kinds of feature for easy to compare results between single-feature and double features (this dataset can be download from: http://lig-membres.imag.fr/grimal/data.html).

The manual dataset comes from [15], which contains three kinds of features. Each feature is generated by two Gauss hybrid model. In this paper, we choose two features to verify the algorithm. The two Gauss distribution centers: in the feature1, centers are $\mu_1^{(1)} = (1,1)$, $\mu_2^{(1)} = (3,4)$, and variance are $\Sigma_1^{(1)} = \begin{pmatrix} 1 & 0.5 \\ 0.5 & 1.5 \end{pmatrix}$, $\Sigma_2^{(1)} = \begin{pmatrix} 0.3 & 0.2 \\ 0.2 & 0.6 \end{pmatrix}$. In the feature2, centers of

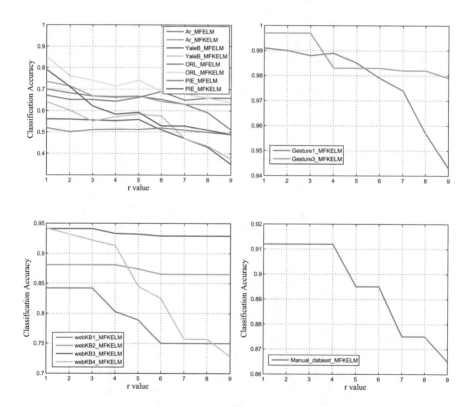

Fig. 2. The classification accuracy with changing r value in MFELM and MFKELM algorithm. (a) Human face datasets (b) Gesture datasets (c) webKB datasets (d) Manual dataset

the two Gauss distribution are $\mu_1^{(2)} = (1,2)$, $\mu_2^{(2)} = (2,2)$, and variance are $\Sigma_1^{(2)} = \begin{pmatrix} 1 & -0.2 \\ -0.2 & 1 \end{pmatrix}$, $\Sigma_2^{(2)} = \begin{pmatrix} 0.6 & 0.1 \\ 0.1 & 0.5 \end{pmatrix}$

4.2   Experimental Results and Analysis

The variable r in the Eqs. (8) and (11) decides how to combine different feature, and impact on the classification performance directly. If the value of r is relatively small, then the combined feature more tend to single-feature in all features. In contrast, the value of r is large, then the combined feature more tend to average feature. In Fig. 2 we record the effect on the classification accuracy with changing of r value. Moreover, only human face datasets obtain MFELM results due to same dimension. Fig. 2 shows that the classification accuracy achieves maximum value when $r = 2$ in all datasets. The r value ensure that MFELM and MFKELM do not degenerate single-feature model, and provided information of difference feature will complement each other for enhancing the performance. Meanwhile, we set $r = 2$ in the follow-up experiments.

We collect statistics the mean value and variance of 50 times experiments, and summarize in Tables 1, 2, 3 and 4, of which SELMF1, SELMF2, SELMF3 respectively correspond the feature 1, 2, 3 of ELM. SKELMF1, SKELMF2, SKELMF3 respectively correspond the feature 1, 2, 3 of KELM. SCELMF1, SCELMF2, SCELMF3 respectively correspond the feature 1, 2, 3 of CELM. ELMCF, KELMCF, CELMCF are results of connecting three kinds of features of ELM, KELM, CELM. Tables 1 and 3 show that proposed MFELM and

Table 1. Mean values of classification results in image datasets are statistical in table

Algorithm	MEAN					
	Ar	YaleB	ORL	PIE	Gesture1	Gesture3
SELMF1	0.586	0.475	0.373	0.180	0.873	0.765
SELMF2	0.609	0.558	0.602	0.504	0.733	0.889
SELMF3	0.518	0.257	0.446	0.150	0.729	0.853
ELMCF	0.429	0.208	0.524	0.570	0.615	0.837
SKELMF1	0.688	0.560	0.224	0.188	0.981	0.983
SKELMF2	0.680	0.540	0.475	0.356	0.986	0.941
SKELMF3	0.516	0.304	0.447	0.152	0.931	0.952
KELMCF	0.400	0.513	0.396	0.324	0.786	0.950
SCELMF1	0.681	0.316	0.484	**0.789**	0.980	0.978
SCELMF2	0.692	0.521	0.788	0.788	0.981	0.979
SCELMF3	0.586	0.556	0.774	0.687	0.969	0.969
CELMCF	0.599	0.512	0.621	0.654	0.850	0.893
MFELM	0.698	0.564	0.790	0.517	-	-
MFKELM	**0.731**	**0.637**	**0.851**	0.730	**0.999**	**0.996**

Table 2. Variance values of classification results in image datasets are statistical in table

Algorithm	VAR					
	Ar	YaleB	ORL	PIE	Gesture1	Gesture3
SELMF1	1.8e−4	8.4e−5	0	1.2e-6	9.2e−3	7.4e−3
SELMF2	1.6e−4	5.3e−5	7.5e−4	5.2e-5	3.8e-3	4.0e−3
SELMF3	3.2e−5	5.3e−5	3.2e−4	0	3.8e−3	9.1e−3
ELMCF	1.2e−4	9.2e−5	4.4e−4	3.8e−3	1.3e-3	8.4e-3
SKELMF1	**0**	**0**	**0**	**0**	**0**	**0**
SKELMF2	**0**	**0**	**0**	**0**	**0**	**0**
SKELMF3	**0**	**0**	**0**	**0**	**0**	**0**
KELMCF	**0**	**0**	**0**	**0**	**0**	**0**
SCELMF1	1.9e−3	1.6e−3	**0**	**0**	**0**	7.2e-4
SCELMF2	1.1e−3	0.48	1.6e−3	1.6e−3	1.2e−4	9.0e−4
SCELMF3	7,6e−4	3.4e−4	4.6e−3	2.5e−3	5.4e−4	6.3e−4
CELMCF	5.1e−4	6.3e−4	5.1e−3	5.2e−4	1.9e−3	1.2e−3
MFELM	1.9e−4	1.9e−4	8.9e−4	8.8e−5	-	-
MFKELM	**0**	**0**	**0**	**0**	1.3e−4	**0**

Table 3. Mean values of classification results in webKB and Manual datasets are statistical in table

Algorithm	MEAN				
	webKB1	webKB2	webKB3	webKB4	Manual
SELMF1	0.501	0.668	0.745	0.880	0.809
SELMF2	0.418	0.717	0.736	0.920	0.824
SELMF3	0.544	0.800	0.829	-	-
ELMCF	0.219	0.490	0.709	0.714	0.742
SKELMF1	0.355	0.684	0.798	0.824	0.738
SKELMF2	0.350	0.803	0.548	0.918	0.903
SKELMF3	0.581	0.789	0.857	-	-
KELMCF	0.311	0.656	0.686	0.818	0.733
SCELMF1	0.534	0.778	0.881	0.871	0.779
SCELMF2	0.531	0.789	0.787	0.887	0.887
SCELMF3	0.611	0.805	0.836	-	-
CELMCF	0.544	0.800	0.829	-	-
MFKELM	**0.622**	**0.869**	**0.893**	**0.941**	**0.942**

MFKELM model achieve better performance than other algorithm. MFELM and MFKELM solve the output weight by iterating, and optimize the coefficient of combination of different features. The proposed models adaptive choices the feature that is suitable for the current data. Moreover, utilize r value ensure robustness of multi-feature method. Let the different feature provide complementary information for classification. From Tables 1 and 3, we observe that the results of connecting feature are lower than results of single-feature in most of human face datasets. This situation demonstrate that the way of connecting briefly adds noise information in the original distribution in image datasets. Same problem appears in webKB and Manual datasets. We can see that simple feature concatenation does not help much. In fact, it reduce the performance due to increase redundance information. However, our method fully utilize complementary information that are provided by different feature, and achieve better results. We can note that the variance of MFKELM is 0 in all datasets from Tables 2 and 4. Therefore, the proposed MFKELM obtain higher roust and reliability than others causing by inducing kernel method.

Table 4. Variance values of classification results in webKB and Manual datasets are statistical in table

Algorithm	VAR				
	webKB1	webKB2	webKB3	webKB4	Manual
SELMF1	1.1e−3	5.1e−3	0.0005	8.6e−3	3.8e-3
SELMF2	1.7e−3	1.3e−3	0.0009	1.8e−3	8.6e-3
SELMF3	2.0e−3	1.4e−3	0.0011	_	_
ELMCF	1.7e−3	5.6e−3	2.3e−3	4.09e−03	3.8e−3
SKELMF1	**0**	**0**	**0**	**0**	**0**
SKELMF2	**0**	**0**	**0**	**0**	**0**
SKELMF3	**0**	**0**	**0**	_	_
KELMCF	**0**	**0**	**0**	**0**	**0**
SCELMF1	3.4e−4	3.5e−4	1.9e−3	1.5e−4	8.9e−3
SCELMF2	6.47e−03	2.2e−4	1.1e−3	2.1e−5	4.4e−5
SCELMF3	8.6e−4	1.2e−4	7,6e−4	_	_
CELMCF	7.2e−3	1.4e−3	3.5e−3	8.8e-3	5.7e−4
MFKELM	**0**	**0**	**0**	**0**	**0**

5   Conclusion

With the advent of the big data era, more and more samples can be collected in every time, and each sample can be described by multiple features. It is becoming a worthy problem to utilize multi-feature of samples for data analysis. In this paper, we propose the MFELM model based on extreme learning machine theory. This algorithm reasonable combines all features on same sample, and then use the combined feature to solve the output weight of the model. In order to avoid the problem that is difficult to calculate the different dimension of different features, we propose the MFKELM model. Moreover, aiming to enhance multi-feature robustness, we induce the higher order combinatorial coefficients. However, samples arrive on real-time in some applications. How to obtain an effective incremental learning model to dynamic multi-feature samples become the focus on the next research.

References

1. Huang, G.B.: An insight into extreme learning machines: random neurons, random features and kernels. Cogn. Comput. **6**(3), 376–390 (2014)
2. Huang, G., Huang, G.B., Song, S.: Trends in extreme learning machines: a review. Neural Netw. **61**, 32–48 (2015)
3. Huang, G.B., Zhou, H., Ding, X.: Extreme learning machine for regression and multiclass classification. IEEE Trans. Syst. Man Cybern. Part B Cybern. **42**(2), 513–529 (2012)
4. Frnay, B., Verleysen, M.: Parameter-insensitive kernel in extreme learning for non-linear support vector regression. Neurocomputing, **74**(16), 2526–2531 (2011)
5. Zong, W., Huang, G.B., Chen, Y.: Weighted extreme learning machine for imbalance learning. Neurocomputing **101**, 229–242 (2013)
6. Liu, S., Feng, L., Liu, Y., Wu, J., Sun, M., Wang, W.: Robust discriminative extreme learning machine for relevance feedback in image retrieval. Multidimens. Syst. Signal Process., 1–19 (2016). https://doi.org/10.1007/s11045-016-0386-3
7. Huang, G., Song, S., Gupta, J.N.D.: Semi-supervised and unsupervised extreme learning machines. IEEE Trans. Cybern. **44**(12), 2405–2417 (2014)
8. Shenglan, L., Lin, F., Yao, X., Huibing, W.: Robust activation function and its application: semi-supervised kernel extreme learning method. Neurocomputing **144**, 318–328 (2014)
9. Kasun, L.L.C., Yang, Y., Huang, G.B.: Dimension reduction with extreme learning machine. IEEE Trans. Image Process. **25**(8), 3906–3918 (2016)
10. Martinez, A.M.: The AR face database. CVC Technical report, 24 (1998)
11. Georghiades, A.S., Belhumeur, P.N., Kriegman, D.J.: From few to many: illumination cone models for face recognition under variable lighting and pose. IEEE Trans. Pattern Anal. Mach. Intell. **23**(6), 643–660 (2001)
12. Guo, G., Li, S.Z., Chan, K.: Face recognition by support vector machines. In: Proceedings, Fourth IEEE International Conference on Automatic Face and Gesture Recognition, pp. 196–201. IEEE (2000)
13. Sim, T., Baker, S., The, Bsat M., CMU pose, illumination, and expression (PIE) database. In: Proceedings, Fifth IEEE International Conference on Automatic Face and Gesture Recognition, pp. 46–51. IEEE (2002)
14. Kim, T.K., Wong, S.F., Cipolla, R., Tensor canonical correlation analysis for action classification. In: IEEE Conference on Computer Vision and Pattern Recognition, CVPR 2007, pp. 1–8. IEEE (2007)
15. Kumar, A., Rai, P., Daume, H.: Co-regularized multi-view spectral clustering. In: Advances in Neural Information Processing Systems, pp. 1413–1421 (2011)

Person Recognition via Facial Expression Using ELM Classifier Based CNN Feature Maps

Ulas Baran Baloglu[1(✉)], Ozal Yildirim[1], and Ayşegül Uçar[2]

[1] Department of Computer Engineering, Munzur University, Tunceli, Turkey
ulasbaloglu@gmail.com, yildirimozal@hotmail.com
[2] Department of Mechatronics Engineering, Firat University, Elazig, Turkey

Abstract. Extreme learning machine (ELM) and deep learning methods are well-known with their efficiency, accuracy, and speed. In this study, we focus on the application of ELM to a deep learning structure for person recognition with facial expressions. For this purpose, a new convolutional neural network (CNN) model containing Kernel ELM classifiers was constructed. In this model, ELM was not used only as a fully connected layer replacement and energy function was employed to generate feature maps for the ELM. There are two advantages of the proposed model. First, it is fast and successful in face recognition studies. Second, it can drastically improve the performance of a partially-trained CNN model. Consequently, the proposed model is very suitable for CNN models, where the learning process requires a lot of time and computational power. The model is tested with the Grimace data set and experimental results are presented in details.

Keywords: Convolutional neural networks · Deep ELM
Extreme learning machine (ELM) · Kernel ELM

1 Introduction

Recently, extreme learning machine (ELM) and deep learning studies have gained a lot of popularity in computer vision and machine learning studies. Both methods are renowned for their efficiency and speed. ELM method is developed by Huang et al. [1,2] because of several performance problems related to weight and bias tuning of single-hidden layer feed-forward networks (SLFNs). In an ELM weights from the input layer to hidden layer are randomly chosen. The idea comes from the reality that there should not be tuning or training steps in brain cells so that weight tuning in SLFNs is an unnecessary process. The application of this idea drastically increases the performance of neural networks.

Deep learning is another human brain-inspired method. In this method, abstract concepts are used to develop improved learning algorithms, which do not comprise too many feature engineering tasks [3]. High-level features of complex

© Springer Nature Switzerland AG 2019
J. Cao et al. (Eds.): ELM 2017, PALO 10, pp. 162–171, 2019.
https://doi.org/10.1007/978-3-030-01520-6_14

data sets are usually extracted through a multi-layer neural network structure, such as convolutional neural networks (CNNs) [4]. Different than ELM, deep learning structures usually have multiple hidden layers instead of a single one. Different methodologies are applied on these layers to create various forms of deep learning structures. The motivation of this study is to investigate the ELM as a part of a deep learning structure and to analyze its benefits for particularly face recognition studies.

There are several approaches in the literature which use multi-layered ELM networks to form deep neural network architectures. Additional layers in a hierarchical architecture are promising to have a better abstract representation of the model and to achieve a faster classification performance [5]. Deep extreme learning machine (Deep ELM) naming usually refers to multi-layered or hierarchical ELM structures. It is possible to define the model as a hybrid architecture [6]. Although naming is done in different ways, the common finding is the fact that ELMs can be used as a stack or as a deep structure to improve classification performance [5–9]. While traditional ELM method is used to solve single task learning problems, Deep ELM structures can be utilized for cross-task learning problems as well [10–12].

Hierarchical ELM architectures may increase a model's learning performance radically when they are used in other convolutional architectures [13]. Several ELMs can be combined to form a deep structure, or an ELM can be used within a CNN structure. Zhang et al. [9] used the ELM as a replacement for fully connected layer in a model adopted from AlexNet [14]. Likewise, ELM used as a classifier in the fully connected layer [15]. In this study, a new CNN model containing ELM was formed, but ELM was not used only as a fully connected layer replacement. Energy function was also used to extract features and generate feature maps for the ELM. Furthermore, the proposed model is tested with the Grimace data set to show its success in face recognition with facial expressions. After using Kernel ELM, the recognition speed is significantly improved in the model. We need to consider how to take advantage of various ELM architectures in such circumstance.

Up to now, a lot of methods have been developed to enhance the performance of face recognition by employing different models and techniques. In recent years, the number of studies emphasizing the success of ELM has increased not only in face recognition studies but also in other research areas [16–20]. Support vector machine (SVM) and ELM classifiers showed a much better performance than the nearest neighbor (NN) classifier in face recognition studies when there is a limited number of images for the training phase [16]. Similarly, curvelet transform with bidirectional two-dimensional principal component analysis (B2DPCA) was combined with ELM to outperform kNN classifier in recognition accuracy and speed [17]. ELM structure can be modified and enhanced to improve the classification performance as in the discriminative graph regularized ELM (GELM) study [18].

In this study, we focus on the application of ELM to a deep learning structure for face recognition with facial expressions. The main advantages of the proposed method can be summarized as its speed and accuracy in face recognition and its success for increasing the performance of a partially-trained CNN model. Due to this advantages, this paper proposes a technique which is very suitable for CNN models, where the learning process requires a lot of time and computational power.

The rest of this paper is organized as follows. Section 2 presents the ELM classifier based CNN model and preliminaries. Section 3 presents the training and testing results of the proposed model for face recognition. Finally, Sect. 4 concludes this paper.

2 ELM Classifier Based CNN Feature Maps

In this section, we present the ELM classifier based CNN model for face recognition. In Sect. 2.1 a brief introduction to CNNs is given for a better understanding of the proposed model. In Sect. 2.2 and Sect. 2.3 mathematical foundations of ELMs and kernel ELMs are provided. Finally, we introduce the proposed model in Sect. 2.4.

2.1 Convolutional Neural Networks

Fukhushima proposed a computational model called as Neocognitron, which was using a local connection method between hierarchically organized neurons [21]. This model opened a path for the development of neural networks. A convolutional neural network (CNN) is characteristically a hierarchical neural network incorporating successive layers, such as convolution layers, activation layers, pooling layers and fully-connected layers. Convolution is a linear mathematical operation, and it gives the name of the structure. CNN structures can be trained forwardly and backwardly to accomplish the learning process. Using essential CNN layers in different combinations help researchers to develop various solutions for many problem types of machine learning and computer vision. Development of CNN models became very widespread because they do not require hand designed features for particular tasks and CNN structures have the ability to learn their features. There are accepted basic CNN models, such as AlexNet [14], GoogleNet [22], Clarifai [23], and VGG [24]. These models can be used for deriving new CNN models to construct sophisticated recognition and learning systems. Additionally, mathematical operations such as convolution can be threaded on a GPU to increase the performance of those systems.

2.2 Extreme Learning Machines

The structure of basic ELMs is similar to that of feedforward neural networks (FNNs) having a single hidden layer. The back-propagation methods used for training FNNs require a very long time and high computational cost. Moreover, they may result in low generalization accuracy. Basic ELMs were proposed to obtain extremely fast training and better generalization ability than FNNs at both regression and classification problems using only matrix calculations [1,2]. ELMs were later extended in for hierarchical form.

Given the training data $\{x^i, y^i\}_1^L$, $x \in \mathfrak{R}^n$, $y \in \{-1, 1\}$, the initial ELMs are initialized by weight w_j and bias b_j of random values. A linear equation system is constructed to calculate the output parameters v_j of the ELM with P hidden layers as follows:

$$\sum_{i=1}^{L} \left\| y^i - f\left(x^i\right) \right\|^2 = 0 \tag{1}$$

$$f(x) = \sum_{j}^{P} v_j s\left(w_j x^i + b_j\right) = s(x) V, \quad i \in \{1, 2, \ldots, L\}. \tag{2}$$

where $s(.)$ is a piecewise continuous function [1,2].

Multi-class classification problem, ELMs uses the organization of one against all. Prepared output matrix $Y = \left[y^1, y^2, \ldots, y^L\right] \in \mathfrak{R}^{m \times L}$, the output of ELM is calculated as follows:

$$\hat{Y} = SV \tag{3}$$

$$V = [v_1, \ v_2, \ldots, v_m]^T_{P \times m} \tag{4}$$

$$S = \begin{bmatrix} s\left(w_1 x^1 + b_1\right) & \ldots & s\left(w_P x^1 + b_P\right) \\ \vdots & \ldots & \vdots \\ s\left(w_P x^L + b_1\right) & \ldots & s\left(w_P x^L + b_P\right) \end{bmatrix}_{L \times P} \tag{5}$$

where V is output weight matrix and S is the hidden layer output matrix [1,2].

On the other hand, the optimization problem of regularized ELM aims to minimize both the norm of output weights $\|V\|^2$ and training error ξ as follows:

$$\min_{V} \ J(V) = \frac{1}{2} \|V\|^2 + \frac{1}{2} \sum_{i=1}^{L} \left\| \xi^i \right\|^2 \tag{6a}$$

s.t.

$$\forall i : s\left(x^i\right) V = y^i - \xi^i \tag{6b}$$

where μ is regularization constant achieving trade off between training error and generalization error. The dual form of (6a) is generated to solve the primal optimization problem by using Lagrange multiple method

$$J(V, \xi, \beta) = \frac{1}{2} \|V\|^2 + \frac{1}{2} \sum_{i=1}^{L} \left\| \xi^i \right\|^2 - \frac{1}{2} \sum_{i=1}^{L} \beta^i \left(s\left(x^i\right) V - y^i + \xi^i \right) \tag{7}$$

$$\frac{\partial J}{\partial V} = 0 \to V = \sum_{i=1}^{L} \beta^i s\left(x^i\right)^T = S^T \beta, \tag{8}$$

$$\frac{\partial J}{\partial \xi^i} = 0 \to \beta^i = \xi^i, \ i = 1, \ldots, L, \tag{9}$$

$$\frac{\partial J}{\partial \beta^i} = 0 \rightarrow s\left(x^i\right)\ \beta - y^i + \xi^i = 0, \quad i = 1, \ldots, L. \tag{10}$$

In [25], it is proposed two different solutions to the Karush-Kuhn-Tucker (KKT) conditions taking into consideration the size of training samples. In case of L≥P, the closed form of V is calculated as

$$V = S^{\dagger}Y = \left(S^T S + \frac{I_{PxP}}{\mu}\right)^{-1} S^T Y \tag{11}$$

where $S^{\dagger}$ is the Moore-Penrose generalized inverse of S and I is identify matrix. In that case, the output function is computed as

$$f\left(x\right) = s(x)V = s(x)\left(S^T S + \frac{I_{PxP}}{\mu}\right)^{-1} S^T Y. \tag{12}$$

In case of L<P, an alternative form of V and the output function are written, respectively as

$$V = S^{+}Y = S^T \left(S^T S + \frac{I_{LxL}}{\mu}\right)^{-1} Y, \tag{13}$$

$$f\left(x\right) = s(x)V = s(x)S^T \left(S^T S + \frac{I_{LxL}}{\mu}\right)^{-1} Y. \tag{14}$$

The output labels of ELM are predicted as

$$label\left(x\right) = arg_{j=1,\ldots,m}\max f_j\left(x\right). \tag{15}$$

In this equation, for each argument $f_j\left(x\right)$ expresses the function defining jth output node, $f\left(x\right) = [f_1\left(x\right), \ldots, f_m\left(x\right)]^T$.

2.3   Kernel Extreme Learning Machines

Like SVMs, KELMs use kernel function and Mercer condition for an unknown feature mapping $s\left(x\right)$ [26]. For input vectors x^j and x^i, a kernel function is defined as

$$K\left(x^i, x^j\right) = s(x^i).s(x^j). \tag{16}$$

The kernel functions such as Wavelet, Polynomial, and RBF are explicitly given. The feature space dimension P and the feature mapping $s(x)$ are implicit.

By using the output weights in (13), the output $f\left(x\right)$ of KELM is calculated as

$$f\left(x\right) = s\left(x\right) V = s\left(x\right) S^T \left(S^T S + \frac{I_{LxL}}{\mu}\right)^{-1} Y = \begin{bmatrix} K\left(x, x^i\right) \\ \vdots \\ K\left(x, x^L\right) \end{bmatrix} \left(\Omega + \frac{I_{LxL}}{\mu}\right)^{-1} Y \tag{17}$$

where

$$\Omega = S^T S = \begin{bmatrix} K\left(x^1, x^1\right) & \cdots & K\left(x^1, x^L\right) \\ \vdots & \ddots & \vdots \\ K\left(x^L, x^1\right) & \cdots & K\left(x^L, x^L\right) \end{bmatrix}. \tag{18}$$

2.4 The Proposed Model

In this study, a novel model for face recognition was developed. ELM and CNN were used together to construct the proposed model, where feature vectors obtained from the outputs of the CNN layers are used as an input for the ELM classifier. In addition, a new CNN model was constructed for the face data set, and a new method was developed for processing the input data set. The block representation of the proposed model for face recognition is given in Fig. 1.

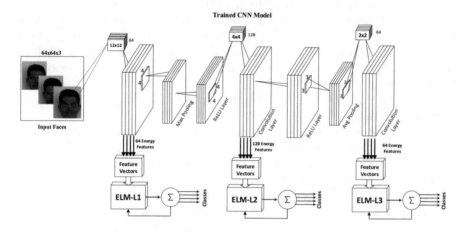

Fig. 1. Block diagram of the ELM based CNN model for face recognition.

One of the motivations of this study is face recognition of someone with different facial expressions. For this reason, different facial expressions are used in the gray-scale format as an input, instead of using the color depth of RGB images. Consequently, the visual input data of the CNN includes three different facial expressions in 64×64 size. An 8-layer CNN model with three convolution layers was designed for this purpose. There are also Pooling and ReLU layers as it is shown in Fig. 1. An ELM classifier is connected to each convolution layer. These classifiers are named as ELM-L1, ELM-L2, and ELM-L3 according to the hierarchical ranking of the convolution layer which they are connected. In the training stage, generated feature maps of these layers have been transformed into feature vectors for the ELM classifiers.

The energy function is applied on the feature maps of the convolution layer's output to construct ELM feature vectors. 64 different 19×19 feature matrices are constructed from the output of the first convolution layer. The energy of each matrix is calculated first, and then 64 features are applied to the input of the ELM-L1. Similarly, there are 128 features for the input of the ELM-L2 and 64 features for the input of the ELM-L3.

3   Experiments

Grimace [27] facial expression data set was used in the experimental study. This database consists of 360 images of 18 subjects. There is a sequence of 20 images, which are taken by a fixed camera in successive frames of 0.5 seconds, per individual. The position of the head and facial expression of subject alters at each frame. When the sequence goes forward, the variation between the images excessively increases. All images in the data set have a resolution of 180×200 pixels. Some sample images from the Grimace dataset is shown in Fig. 2.

Twenty different image data per person were produced from the dataset for the training of the proposed model. When these images are being created, three face expressions belonging to the related person are selected randomly every time. Ten out of twenty from different facial expression images are used for the training. The remaining ten images are used to construct the test data. Therefore, there is a total of 360 images (18-person × 20-images) in the training and test data sets.

Kernel ELM was used in the model. Parameters were set as 1 for the kernel parameter and 50 for the regularization coefficient. Kernel type was RBF. Performance comparison of the different classifiers are given in are given in Table 1 for 22 epochs. At this epoch, all classifiers reached.

In addition to the accuracy evaluation, it is also important to evaluate classifiers in terms of speed for the same data. For this purpose, recognition speed of the models was calculated, and performance comparisons are given in Table 2.

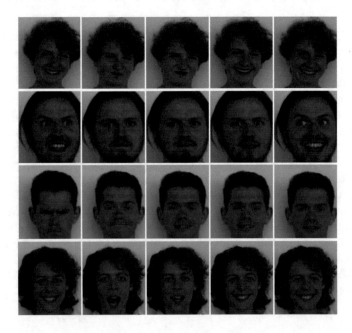

Fig. 2. Facial expression examples from the Grimace dataset

Table 1. The performance of the classifiers for the Grimace dataset.

Accuracy(%)										
Epoch										
	2.ep	4.ep	6.ep	8.ep	10.ep	12.ep	14.ep	16.ep	18.ep	20.ep	22.ep
CNN	16.66	16.66	16.66	16.66	16.66	22.22	27.77	61.11	77.77	94.44	**100**
ELM-L1	77.78	94.44	94.44	**100**	100	100	100	100	100	100	100
ELM-L2	44.44	**100**	100	100	100	100	100	100	100	100	100
ELM-L3	11.11	83.33	**100**	100	100	100	100	100	100	100	100

Table 2. Performance comparisons (in seconds) of classifier models.

Epochs	CNN		ELM-L1		ELM-L2		ELM-L3	
	Train	Val	Train	Test	Train	Test	Train	Test
2.ep	0.480	13.907	0.0024	0.0011	0.0021	0.0011	0.0023	0.0013
4.ep	0.605	13.885	0.0023	0.0014	0.0023	0.0014	0.0027	0.0012
6.ep	0.526	14.281	0.0024	0.0014	0.0025	0.0014	0.0028	0.0014
8.ep	0.559	14.481	0.0025	0.0014	0.0024	0.0013	0.0023	0.0015
10.ep	0.580	15.512	0.0025	0.0012	0.0023	0.0011	0.0023	0.0014
12.ep	0.605	13.736	0.0022	0.0014	0.0025	0.0011	0.0023	0.0012
14.ep	0.565	14.707	0.0025	0.0012	0.0024	0.0014	0.0023	0.0014
16.ep	0.576	14.748	0.0023	0.0014	0.0023	0.0011	0.0025	0.0014
18.ep	0.554	13.067	0.0023	0.0012	0.0023	0.0014	0.0023	0.0015
20.ep	0.538	14.564	0.0024	0.0014	0.0020	0.0011	0.0027	0.0014
22.ep	0.589	14.325	0.0024	0.0012	0.0025	0.0012	0.0023	0.0014

We can observe that the time spent in each epoch of the training stage for the CNN is considerably higher than the training of other ELM classifiers. This finding suggests that the recognition goal can be achieved more quickly with the help of ELM classifiers and in this model, there is no need to wait for the completion of the CNN's training.

4   Conclusion

In this study, a new model is proposed in which ELM and CNN structures are used together. Feature maps, which were obtained from the convolution layers of the trained CNN model, were used as an input for the ELM classifiers. The most interesting finding of this study is the observed increase in efficiency and recognition accuracy. The recognition speed is significantly improved after using the Kernel ELM. This result is promising for future research on the subject.

ELM is a very efficient structure especially when the input is simplified. A convoluted CNN output becomes a preferred input type for the ELM classifier.

If ELM was not used and the face recognition process was completed by using the CNN only, the computational cost would be higher, and the recognition accuracy would be poorer. In the opposite case, if the ELM was used alone, the difficulty of the input data would prevent us from achieving an anticipated performance. The critical point here is the development of unsupervised methods that can accurately predict at which epoch convoluted data will be passed to the ELM. We will investigate optimization techniques to achieve this goal in our future studies.

References

1. Huang, G.B., Zhu, Q.Y., Siew, C.K.: Extreme learning machine: a new learning scheme of feedforward neural networks. In: Proceedings of the International Joint Conference on Neural Networks (IJCNN 2004), Budapest, pp. 985–990 (2004)
2. Huang, G.B., Zhu, Q.Y., Siew, C.K.: Extreme learning machine: theory and applications. Neurocomputing **70**, 489–501 (2006)
3. Bengio, Y., Lee, H.: Editorial introduction to the neural networks special issue on deep learning of representations. Neural Netw. **64**, 1–3 (2015)
4. Kim, S., Choi, Y., Lee, M.: Deep learning with support vector data description. Neurocomputing **165**, 111–117 (2015)
5. Lv, Q., Niu, X., Dou, Y., Xu, J., Lei, Y.: Classification of hyperspectral remote sensing image using hierarchical Local-Receptive-Field-Based extreme learning machine. IEEE Geosci. Remote Sens. Lett. **13**(3), 434–438 (2016)
6. Yu, J.S., Chen, J., Xiang Z.Q., Zou, Y.X: A hybrid convolutional neural networks with extreme learning machine for WCE image classification. In: 2015 IEEE International Conference on Robotics and Biomimetics (ROBIO), Zhuhai, pp. 1822-1827 (2015)
7. Tissera, M.D., McDonnell, M.D.: Deep extreme learning machines: supervised autoencoding architecture for classification. Neurocomputing **174**(A), 42-49 (2016)
8. Tissera, M.D., McDonnell, M.D.: Enhancing deep extreme learning machines by error backpropagation. In: 2016 International Joint Conference on Neural Networks (IJCNN), Vancouver, BC, pp. 735–739 (2016)
9. Ibrahim, W., Abadeh, M.S.: Extracting features from protein sequences to improve deep extreme learning machine for protein fold recognition. J. Theor. Biol. **421**, 1–15 (2017)
10. Zhang, L., He, Z., Liu, Y.: Deep object recognition across domains based on adaptive extreme learning machine. Neurocomputing **239**, 194–203 (2017)
11. Zhou, H., Huang, G.B., Lin, Z., Wang, H., Soh, Y.C.: Stacked extreme learning machines. IEEE Trans. Cybern. **45**(9), 2013–2025 (2015)
12. Sun, K., Zhang, J., Zhang, C., Hu, J.: Generalized extreme learning machine autoencoder and a new deep neural network. Neurocomputing **230**, 374–381 (2017)
13. Zhu, W., Miao, J., Qing, L., Huang, G.B.: Hierarchical extreme learning machine for unsupervised representation learning. In: 2015 International Joint Conference on Neural Networks (IJCNN), Killarney, pp. 1–8 (2015)
14. Krizhevsky, A., Sutskever, I., Hinton, G.E.: Imagenet classification with deep convolutional neural networks. In: Proceedings of the NIPS (2012)
15. Weng, Q., Mao, Z., Lin, J., Guo, W.: Land-Use classification via extreme learning classifier based on deep convolutional features. IEEE Geosci. Remote Sens. Lett. **14**(5), 704–708 (2017)

16. Zong, W., Huang, G.B.: Face recognition based on extreme learning machine. Neurocomputing **74**, 2541–2551 (2011)
17. Mohammed, A.A., Minhas, R., Jonathan Wu, Q.M., Sid-Ahmed, M.A.: Human face recognition based on multidimensional PCA and extreme learning machine. Pattern Recognit. **44**, 2588–2597 (2011)
18. Peng, Y., Wang, S., Long, X., Lu, B.L.: Discriminative graph regularized extreme learning machine and its application to face recognition. Neurocomputing **149**, 340–353 (2015)
19. Uçar, A., Özalp, R.: Efficient android electronic nose design for recognition and perception of fruit odors using kernel extreme learning machines. Chemometr. Intell. Lab. Syst. **166**, 69–80 (2017)
20. Uçar, A., Demir, Y., Güzeliş, C.: A new facial expression recognition based on curvelet transform and online sequential extreme learning machine initialized with spherical clustering. Neural Comput. Appl. **27**(1), 131–142 (2016)
21. Fukushima, K.: Neocognitron: a self-organizing neural network model for a mechanism of pattern recognition unaffected by shift in position. Biol. Cybern. **36**(4), 193–202 (1980)
22. Szegedy, C., Liu, W., Jia, Y., Sermanet, P., Reed, S., Anguelov, D., Erhan, D., Vanhoucke, V., Rabinovich, A.: Going deeper with convolutions. In: Proceedings of the CVPR (2015)
23. Zeiler, M.D., Fergus, R.: Visualizing and understanding convolutional neural networks. In: Proceedings of the ECCV (2014)
24. Simonyan, K., Zisserman, A.: Very deep convolutional networks for large-scale image recognition. In: Proceedings of the ICLR (2015)
25. Zhang, L., Zhang, D.: Evolutionary cost-sensitive extreme learning machine. IEEE Trans. Neural Netw. Learn. Syst. **99**, 1–16 (2016)
26. Huang, G.B., Zhou, H., Ding, X., Zhang, R.: Extreme learning machine for regression and multiclass classification. IEEE Trans. Syst. Man. Cybern. B Cybern. **42**, 513–529 (2012)
27. Spacek, L.: Face Directories. http://cswww.essex.ac.uk/mv/allfaces/grimace.html

A New Asynchronous Architecture for Tabular Reinforcement Learning Algorithms

Xingyu Zhao, Shifei Ding$^{(\boxtimes)}$, and Yuexuan An

School of Computer Science and Technology,
China University of Mining and Technology, Xuzhou 221116, China
dingsf@cumt.edu.cn

Abstract. In recent years, people have combined deep learning with reinforcement learning to solve practical problems. However, due to the characteristics of neural networks, it is very easy to fall into local minima when facing
small scale discrete space path planning problems. Traditional reinforcement
learning uses continuous updating of a single agent when algorithm executes,
which leads to a slow convergence speed. In order to solve the above problems,
we combine asynchronous methods with existing tabular reinforcement learning
algorithms, propose a parallel architecture to solve the discrete space path
planning problems, and present four new variants of asynchronous reinforcement learning algorithms. We apply these algorithms on the standard reinforcement learning environment FrozenLake problem, and the experimental
results show that these methods can solve discrete space path planning problems
efficiently. One of these algorithms, which is called Asynchronous Dyna-Q,
surpasses existing asynchronous reinforcement learning algorithms, can well
balance the exploration and exploitation.

Keywords: Reinforcement learning · Path planning · Dyna architecture
Asynchronous methods · Discrete space

1 Introduction

As an important machine learning method, reinforcement learning (RL) enables agent
to learn the mapping relationship from states to actions by trial and error in the
continuous interaction with the environment, so as to maximize the long-term cumulative reward [1]. At present, reinforcement learning methods have been widely used in
intelligent control, robotics and other fields [2–4].

In traditional reinforcement learning, the continuous update of a single agent has
been used for the updating of parameters. This may lead to slow convergence speed
and convergence to local minimum. To solve these problems, Mnih et al. proposed an
asynchronous architecture for deep reinforcement learning in 2016 [5]. Asynchronous
reinforcement learning adopts several parallel actor-learners to explore the environment, and each actor-learner online updates the global parameters. Using this approach,
the training time could be greatly shortened. However, when using deep reinforcement
learning to solve small discrete space problems, there usually exists problems like local

© Springer Nature Switzerland AG 2019
J. Cao et al. (Eds.): ELM 2017, PALO 10, pp. 172–180, 2019.
https://doi.org/10.1007/978-3-030-01520-6_15

minimum and slow convergence speed. Therefore, this method has not been extended to small scale discrete space problems.

How to make full use of existing knowledge is another major issue in reinforcement learning. Sutton proposed a reinforcement learning architecture, called Dyna learning, through establishing environment model and planning to aid the learning process, made reinforcement learning out of simple trial and error learning and with the ability to "know" [6]. In the past few years, some scholars have improved the Dyna-Q algorithm and achieved some results. But these research works are carried out on single agent, the improvements of algorithm efficiency are not obvious.

To solve the above problems, we extend the asynchronous method which applied in deep reinforcement learning algorithms to traditional reinforcement learning, and propose an asynchronous reinforcement learning algorithm architect that can effectively solve small scale discrete space problems. We also present our asynchronous variants of standard reinforcement learning algorithms, and propose an asynchronous Dyna-Q algorithm. Experiments show that our asynchronous architecture can significantly improve the learning efficiency and learning performance of the algorithm.

2 Basic Theory

We consider tasks in which an agent interacts with an environment ε over a number of discrete time steps. At each time-step, the agent receives a state s and selects a legal action a according to its policy π, where $\pi : S \rightarrow A$ is a mapping from states to actions. The agent executes the selected action a, then the state turns into s' and the environment returns a reward r to the agent. The agent then selects the next action based on the reward signal (Fig. 1).

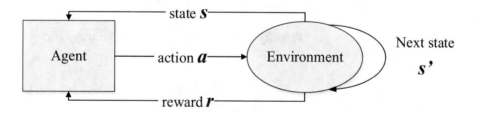

Fig. 1. The framework of reinforcement learning

We define $V^\pi(s)$ as the expected total return of the learning system from the state s according to the policy π, and $Q^\pi(s,a)$ as the expected total return of the learning system from the state-action pair (s,a) according to the policy π. Then there are:

$$V^\pi(s) = E^\pi[\sum_{k=0}^{\infty} \gamma^k r_{t+k+1} | s_{t=s}] \tag{1}$$

$$Q^{\pi}(s) = E^{\pi} \left[\sum_{k=0}^{\infty} \gamma^k r_{t+k+1} | s_t = s, a_t = a \right] \tag{2}$$

Where $\gamma \in (0, 1)$ denotes the discount factor. Since the best policy π^* is the policy of maximizing the value function, there are:

$$\pi^* = \arg\max_{a \in A} V^{\pi}(s) \tag{3}$$

$$\pi^* = \arg\max_{a \in A} Q^{\pi}(s, a) \tag{4}$$

We usually use standard model-free reinforcement learning methods such as Q-learning and Sarsa learning to solve problems [7–12]. In Q-learning, the Q value is updated as follows:

$$Q(s_t, a_t) \leftarrow Q(s_t, a_t) + \alpha[r_{t+1} + \gamma \max_a Q(s_{t+1}, a) - Q(s_t, a_t)] \tag{5}$$

Where α denotes the learning rate, $\gamma \in (0, 1)$ denotes the discount factor.

In Sarsa learning, we use the actual Q value for iteration. So Sarsa learning is called the on-policy method. In Sarsa learning, the Q value is updated as follows:

$$Q(s_t, a_t) \leftarrow Q(s_t, a_t) + \alpha[r_{t+1} + \gamma Q(s_{t+1}, a_{t+1}) - Q(s_t, a_t)] \tag{6}$$

In order to solve the temporal credit assignment problems, We apply the eligibility traces to the calculation of Q value which can form a kind of calculating methods such as TD(λ) and Sarsa(λ). The calculating formula of Sarsa (λ) an be written as follows:

$$V(s_t) \leftarrow V(s_t) + \alpha[r_{t+1} + \gamma V(s_{t+1}) - V(s_t)]e(s_t) \tag{7}$$

Where $e(s_t)$ denotes the eligibility trace of state s_t. The eligibility trace can be updated as follows:

$$e_t(s) = \begin{cases} 1, s = s_t \\ \gamma \lambda e_t(s), s \neq s_t \end{cases} \tag{8}$$

In addition to the model-free methods mentioned above, we can also use the model-based methods such as Dyna to implement the reinforcement learning algorithms. Dyna architecture integrate learning and planning, which makes agent can use the experience to build environment model and use the environment model to generate hypothesis experience as learning resource, can effectively improve the convergence speed of the value function (Fig. 2).

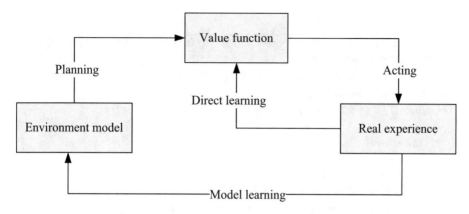

Fig. 2. Information flow in the Dyna architecture

3 Asynchronous Framework for Tabular Reinforcement Learning Algorithms

3.1 Model-Free Asynchronous Reinforcement Learning Algorithms

We now present our multi-threaded asynchronous variants of model-free reinforcement learning algorithms. We use asynchronous actor-learners, similarly to the Asynchronous RL Framework [5]. But for the shortcomings of neural networks in small scale problems, we use tabular methods instead of deep neural networks. We make parallel actor-learners run in its own thread and update parameters asynchronously.

Asynchronous one-step tabular Q-learning: Pseudocode for our asynchronous one-step tabular Q-learning is shown in Algorithm 1. Since it is a tabular algorithm, we cannot use the neural network as the error optimization method. So we define a new parameter β as the asynchronous updating rate to make parallel actor-learners can update global parameters. When an actor-learner runs a certain number of steps, it will update the global parameters and get new global parameters as its own parameters. The equation of updating can be written as follows:

$$\mathbb{Q}^-(s,a) \leftarrow \mathbb{Q}^-(s,a) + \beta \left[Q(s,a) - \mathbb{Q}^-(s,a)\right] \tag{9}$$

Where $\mathbb{Q}^-(s,a)$ denotes the value of global Q-table. As for exploration policy, in order to facilitate the comparison of different algorithms, we use $\varepsilon - $ greedy exploration uniformly.

Algorithm 1 The proposed asynchronous one-step tabular Q-learning - pseudocode for each actor-learner thread.

// *Assume global shared* $Q^-(s,a)$, β *and counter* $T = 0$

Initialize thread step counter $t \leftarrow 0$

Initialize $Q^*(s,a)$ for all $s \in \mathcal{S}$ and $a \in \mathcal{A}$

repeat

    Initialize S

    repeat

        With probability ε select a random action A

        Otherwise $A \leftarrow \max_a Q^*(S,a)$

        Execute action A, observe reward R and the next state S'

        $Q^*(S,A) \leftarrow Q^*(S,A) + \alpha[R + \gamma \max_a Q^*(S',a) - Q^*(S,A)]$

        $S \leftarrow S'$, $T \leftarrow T+1$ *and* $t \leftarrow t+1$

        if $t \bmod t \bmod I_{AsynUpdate} == 0$ or S is terminal **then**

            $Q^-(S,A) \leftarrow Q^-(S,A) + \beta\,[Q^*(S,A) - Q^-(S,A)]$

            $Q^*(S,A) \leftarrow Q^-(S,A)$

        end if

    until S is terminal

until $T > T_{max}$

Asynchronous one-step tabular Sarsa: The asynchronous one-step tabular Sarsa algorithm is the same as asynchronous one-step tabular Q-learning as given in Algorithm 1 except that it uses a different target value for $Q(S,A)$. In one-step tabular Sarsa, the equation of calculating $Q(S,A)$ is given by formula (6). We again use the asynchronous updating rate β to update the global parameters.

Asynchronous tabular Sarsa(λ): In this algorithm, we introduce eligibility traces into our asynchronous framework. Each asynchronous actor-learner has its own eligibility traces table. In this way, we can maintain the diversity of different actor-learners. The equations of calculating $Q(S,A)$ and updating eligibility traces are given by formula (7) and formula (8).

3.2 Asynchronous Tabular Dyna-Q Algorithm

We then introduce Dyna-Q into the asynchronous framework. In our variant of Dyna-Q algorithm, all actor-learners use the same environment model. We divide actor-learners into two categories: actors and learners. The responsibilities of actors are exploring the environment and storing experience in the model while that of learners is only learning from the model, not exploring the environment. In this way, we can not only reduce the time required for the algorithm, but also conveniently control the status of each actor-learner.

We define a parameter η to indicate the ratio of actor-learner in different states. It can be changed according to the state of environment. Note that although our actors perform exploratory tasks, they still have their own Q-tables and can also update global Q-table after act some steps. We do this because when experience is insufficient, learners cannot learn well based on the model, and that may lead to a wrong way to learn Q-table. Pseudocode for our asynchronous tabular Dyna-Q is shown in Algorithm 2.

Algorithm 2 The proposed asynchronous tabular Dyna-Q - pseudocode for each actor-learner thread.

// *Assume global shared* $Q^-(s,a)$, β, *counter* $T = 0$ *and* $Model(s,a)$

Initialize thread step counter $t \leftarrow 0$

Initialize $Q^*(s,a)$ for all $s \in \mathcal{S}$ and $a \in \mathcal{A}$

repeat

    if actor thread **then**

        With probability ε select a random action A

        Otherwise $A \leftarrow \max_a Q^*(S,a)$

        Execute action A, observe reward R and the next state S'

        $Q^*(S,A) \leftarrow Q^*(S,A) + \alpha[R + \gamma \max_a Q^*(S',a) - Q^*(S,A)]$

        $S \leftarrow S'$, $T \leftarrow T+1$ *and* $t \leftarrow t+1$

        if $t \bmod I_{AsynUpdate} == 0$ or S is terminal **then**

            $Q^-(S,A) \leftarrow Q^-(S,A) + \beta\,[Q^*(S,A) - Q^-(S,A)]$

            $Q^*(S,A) \leftarrow Q^-(S,A)$

        end if

    end if

    if learner thread **then**

        $Q^*(S,A) \leftarrow Q^-(S,A)$

        repeat n times:

            $S \leftarrow$ random state from $Model(s,a)$

            $A \leftarrow$ random action previously taken in S from $Model(s,a)$

            $R, S' \leftarrow Model(S,A)$

            $Q^*(S,A) \leftarrow Q^*(S,A) + \alpha[R + \gamma \max_a Q^*(S',a) - Q^*(S,A)]$

        $Q^-(S,A) \leftarrow Q^*(S,A)$, $T \leftarrow T+1$

    end if

until $T > T_{max}$

---

4 Experiments

4.1 Experimental Environment

Our experiments ran under the software environment of Windows 10 and Python 3.5.2. The CPU we used is Inter i5-3317U with 4 cores, and the memory is 10 GB. Our Experimental environment is the the standard reinforcement learning environment

FrozenLake problem. Like the interface provided by OpenAI Gym platform, we implemented an interface to the problem. The only difference from the environment of OpenAI Gym is the punishment mechanism for failure that we added.

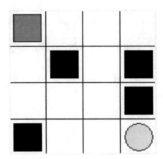

Fig. 3. Experimental environment of "FrozenLake"

Figure 3 shows the environment named "FrozenLake". There are sixteen grids in the environment. The agent start with the upper left corner, and its goal is successfully reached the lower right corner. There exist four holes where the ice has melted. If stepping into one of those holes, our exploration will fail immediately. Our experiments setting is that when we achieve the goal, we receive a reward of 10, and when we fall into the ice hole, we get a reward of –10. In addition, in order to compare the results of each algorithm, we set that when the agent executes an action, it get a reward of –1.

4.2   Experiments of Tabular RL Method and Deep RL Method

We first compared the performances of Q-learning, Sarsa learning and Deep Q-network algorithms in the environment of "FrozenLake". The results are shown in Fig. 4. The cumulative reward has been smoothed. In the experiment, we set the learning rate of

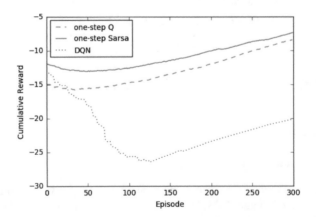

Fig. 4. Performance of Q-Learning, Sarsa and DQN in "FrozenLake"

DQN is 0.01, the replace iteration number is 300, and the memory size is 500. All algorithms used $\varepsilon -$ greedy policy with $\varepsilon = 0.1$. In this experiment, we found that deep reinforcement learning algorithms had poor performance when facing small scale discrete space problems. So we decided not to use deep reinforcement learning algorithms to carry out our experiments.

4.3   Experiments of Asynchronous Framework

Then we compared our four algorithms proposed in Sect. 3. The results are shown in Fig. 5. All algorithms used $\varepsilon -$ greedy policy with $\varepsilon = 0.1$. We set the value of β to 0.25 for all algorithms. This experiment shows that our asynchronous methods for tabular reinforcement learning algorithms are effective. The application of eligibility traces can efficaciously improve the performance of tabular algorithms, and our asynchronous Dyna-Q architecture is more efficient than other methods.

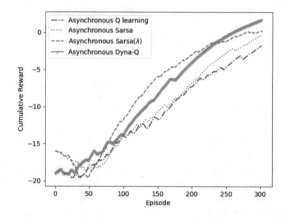

Fig. 5.   Performance of asynchronous tabular algorithms in "FrozenLake"

5   Conclusions

We have proposed an asynchronous architecture for tabular reinforcement learning methods in this paper. Then we combined it with existing model-free algorithms like Q-learning, Sarsa and Sarsa(λ), and proposed three new reinforcement learning algorithms. After that, we introduced Dyna framework into our architecture, which can efficaciously improve the performance of tabular algorithms, and proposed a new algorithm name Asynchronous Dyna-Q. Finally, we proved the effectiveness of the above algorithms through experiments. In the future work, we will introduce the phased method into our architecture to improve the performance of our asynchronous Dyna-Q algorithm. Moreover, we will combine asynchronous methods with more model-based algorithms, to make this architecture more general.

Acknowledgments. This work is supported by the National Natural Science Foundation of China (Nos. 61672522, 61379101).

References

1. Sutton, R.S., Barto, A.G.: Reinforcement Learning: An Introduction. MIT press, Cambridge (1998)
2. Mnih, V., Kavukcuoglu, K., Silver, D., et al.: Playing atari with deep reinforcement learning. In: Proceedings of Workshops at the 26th Neural Information Processing Systems Lake Tahoe, USA, pp. 201–220 (2013)
3. Mnih, V., Kavukcuoglu, K., Silver, D., et al.: Human-level control through deep reinforcement learning. Nature **518**(7540), 529–533 (2015)
4. Silver, D., Huang, A., Maddison, C.J., et al.: Mastering the game of Go with deep neural networks and tree search. Nature **529**(7587), 484–489 (2016)
5. Mnih, V., Badia, A.P., Mirza, M., et al.: Asynchronous methods for deep reinforcement learning. In: International Conference on Machine Learning, pp. 1928–1937 (2016)
6. Sutton, R.S.: Dyna, an integrated architecture for learning, planning, and reacting. ACM SIGART Bull. **2**(4), 160–163 (1991)
7. Watkins, C.J.C.H.: Learning from delayed rewards. Robot. Auton. Syst. **15**(4), 233–235 (1989)
8. Rummery, G.A., Niranjan, M.: On-Line Q-Learning Using Connectionist Systems. University of Cambridge, Department of Engineering (1994)
9. Sutton, R.S.: Learning to predict by the methods of temporal differences. Mach. Learn. **3**(1), 9–44 (1988)
10. Singh, S.P., Sutton, R.S.: Reinforcement learning with replacing eligibility traces. Mach. Learn. **22**, 123–158 (1996)
11. Silver, D., Lever, G., Heess, N., et al.: Deterministic policy gradient algorithms. In: Proceedings of the 31st International Conference on Machine Learning, pp. 387–395 (2014)
12. Schulman, J., Levine, S., Abbeel, P., et al.: Trust region policy optimization. In: Proceedings of the 32nd International Conference on Machine Learning, pp. 1889–1897 (2015)

Extreme Learning Tree

Anton Akusok[1]([✉]), Emil Eirola[1], Kaj-Mikael Björk[2,5], and Amaury Lendasse[3,4]

[1] Arcada University of Applied Sciences, Helsinki, Finland
anton.akusok@arcada.fi
[2] Arcada University of Applied Sciences, Helsinki, Finland
[3] Department of Mechanical and Industrial Engineering,
The University of Iowa, Iowa City, USA
[4] The Iowa Informatics Initiative, The University of Iowa, Iowa City, USA
[5] Hanken School of Economics, Helsinki, Finland

Abstract. The paper proposes a new variant of a decision tree, called an Extreme Learning Tree. It consists of an extremely random tree with non-linear data transformation, and a linear observer that provides predictions based on the leaf index where the data samples fall. The proposed method outperforms linear models on a benchmark dataset, and may be a building block for a future variant of Random Forest.

Keywords: ELM · Decision tree · Randomized methods

1 Introduction

Randomized methods are a recent trend in practical machine learning [1]. They enable the high performance of complex non-linear methods without the high computational cost of their optimization. Current most prominent examples are randomized neural networks, in both feed-forward [2] and recurrent [3] forms. For the latter, the randomized approach provided an efficient training method for the first time, and enabled achieving state-of-the-art performance in multiple areas [4].

Random forest [5] is one of the best methods for Big Data processing due to its adaptive nearest neighbour behavior [6]. The forest predicts an output based only on local data samples. Such an approach works the better the more training data is available, thus making for a perfect supervised method for Big Data. K-nearest neighbors algorithm benefits from more data as the data itself is the model, but Random Forest avoids the quadratic scaling of k-Nearest neighbors in terms of the data samples, that makes it prohibitively slow for large-scale problems.

Decision tree [7] is a building block of Random Forest. A deep decision tree has high variance but low bias. An ensemble of multiple such trees reduces variance, and improves the prediction performance. Additional measures are taken to make the trees in an ensemble as different as possible, including random subsets of features and boosting [8].

© Springer Nature Switzerland AG 2019
J. Cao et al. (Eds.): ELM 2017, PALO 10, pp. 181–185, 2019.
https://doi.org/10.1007/978-3-030-01520-6_16

The paper proposes a merge between random methods and a decision tree, called an Extreme Learning Tree (ELT). The method builds a tree using expanded data features from an Extreme Learning Machine [9], by splitting nodes on a random feature at a random point. The result is an Extremely Randomized Tree [10]. Then a linear observer is added to the leaves of the tree, that learns a linear projection from the leaves to the target outputs. Each tree leaf is represented by its index, in the one-hot encoding format.

2   Methodology

Extreme Learning Tree consists of three parts. First, it generates random data features using an Extreme Learning Machine (ELM) [11]. Second, it builds a random tree from these features, similar to Extremely Randomized Trees [10]. Each data sample is then represented by the index of its leaf from the tree, in one-hot encoding. Third, a linear regression is learned from the dataset in that one-hot encoding to the target outputs.

ELT follows the random methods paradigm as it has an untrained random part (the tree), and a learned linear observer (a linear regression model from leaves of the tree to the target outputs).

An ELT tree has two hyper parameters: the minimum node size, and the maximum thee depth. A node data is split by a random feature using a random split point. Split points that generates nodes under the minimum size are rejected. Nodes that reach the maximum depth or under twice the minimum size become leafs. Node splitting continues until there are non-leaf terminal nodes.

3   Experimental Results

The Extreme Learning Tree is tested on well-known Iris flower dataset [12], in comparison with a Decision Tree, an L2 regularized ELM [13], and Ridge regression. Decision Tree implementation is from the Scikit-Learn library[1].

The random tree in the ELT method splits data samples into groups of similar ones. The resulting structure in the original data space is shown on Fig. 1. The tree works as a adaptive nearest neighbour, combining together similar samples. Then the target variable information from these samples is used by a linear observer to make predictions.

A formal performance comparison is done on Iris dataset. The data is randomly split into 70% training and 30% test sets, and the test accuracy is calculated for all the methods. The whole experiment is repeated 100 times. Mean accuracy and its standard deviation are presented in Table 1.

---

[1] http://scikit-learn.org/stable/auto_examples/tree/plot_iris.html.

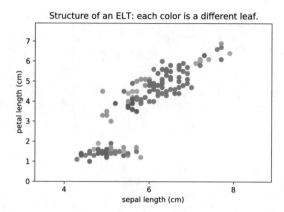

Fig. 1. Leaf structure of an ELT, each color represents a different leaf. The random tree works as an approximated nearest neighbour method, joining together similar data samples.

Table 1. Average accuracy and its standard deviation on a test subset of Iris dataset.

Method	Accuracy ±std, %
Ridge regression	82.7 ± 5.1
Extreme Learning Tree	87.2 ± 6.1
ELM	90.9 ± 5.0
Decision Tree	94.1 ± 3.2

In this experiment, an Extreme Learning Tree performs under ELM and Decision Tree methods. However, it outperforms a linear model (in the form of Ridge regression) by a significant margin. Outperforming a linear model is an achievement for a single ELT, as it represents each data sample by a single number – an index of its leaf in the tree.

Decision surface of ELT is visualized on Fig. 2. The boundaries between classes have complex shape, but the classes are unbroken. Class boundaries of the original Decision Tree (shown on Fig. 3) break into each other creating false predictions. They are always parallel to an axis, while ELT learns class boundaries of an arbitrary shape.

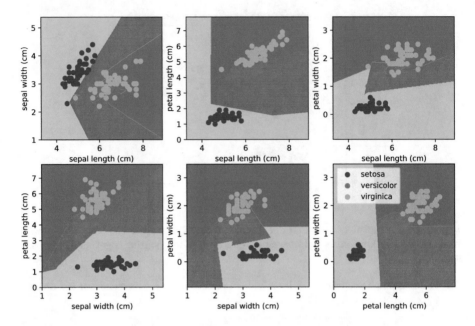

Fig. 2. Decision surface of an ELT on Iris dataset, using different pairs of features. Different colors correspond to the three different classes of Iris flowers.

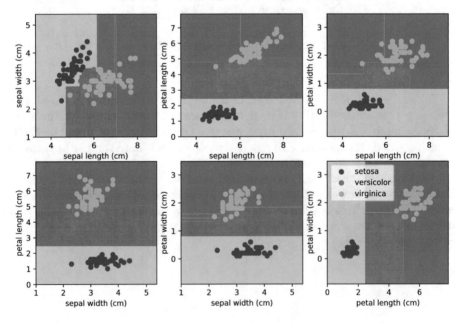

Fig. 3. Decision surface of a Decision Tree on Iris dataset, using different pairs of features. Note that all decision boundaries are parallel to axes.

4 Conclusions

The paper proposes a new version of decision tree, that follows the random methods paradigm. It consists of an untrained random non-linear tree, and a learned linear observer. The method provides decision boundaries of a complex shape and with less noise than an original decision tree. It outperforms a purely linear model in accuracy despite representing the data samples only by a corresponding tree leaf index.

Future works will examine an application of Extreme Learning Tree to an ensemble method similar to Random Forest.

References

1. Gallicchio, C., Martin-Guerrero, J.D., Micheli, A., Soria-Olivas, E.: Randomized machine learning approaches: Recent developments and challenges. In: ESANN 2017 Proceedings, European Symposium on Artificial Neural Networks, Computational Intelligence and Machine Learning, pp. 77–86. d-side publi., Bruges, Belgium, 26–28 April 2017
2. Huang, G.B.: What are extreme learning machines? Filling the gap between Frank Rosenblatt's Dream and John von Neumann's Puzzle. Cogn. Comput. **7**(3), 263–278 (2015)
3. Lukoševičius, M., Jaeger, H.: Reservoir computing approaches to recurrent neural network training. Comput. Sci. Rev. **3**(3), 127–149 (2009)
4. Jaeger, H., Haas, H.: Harnessing nonlinearity: predicting chaotic systems and saving energy in wireless communication. Science **304**(5667), 78 (2004)
5. Ho, T.K.: The random subspace method for constructing decision forests. IEEE Trans. Pattern Anal. Mach. Intell. **20**(8), 832–844 (1998)
6. Lin, Y., Jeon, Y.: Random forests and adaptive nearest neighbors. J. Am. Stat. Assoc. **101**(474), 578–590 (2006)
7. Breiman, L., Friedman, J., Stone, C.J., Olshen, R.A.: Classification and Regression Trees. CRC press (1984)
8. Breiman, L.: Random Forests. Mach. Learn. **45**(1), 5–32 (2001)
9. Huang, G.B., Zhou, H., Ding, X., Zhang, R.: Extreme learning machine for regression and multiclass classification. IEEE Trans. Syst. Man Cybern. Part B Cybern. **42**(2), 513–529 (2012)
10. Geurts, P., Ernst, D., Wehenkel, L.: Extremely randomized trees. Mach. Learn. **63**(1), 3–42 (2006)
11. Huang, G.B., Zhu, Q.Y., Siew, C.K.: Extreme learning machine: Theory and applications. In: Neural Networks Selected Papers from the 7th Brazilian Symposium on Neural Networks (SBRN 2004) 7th Brazilian Symposium on Neural Networks, vol. 70(1-3), pp. 489–501, December 2006
12. Fisher, R.A.: The use of multiple measurements in taxonomic problems. Ann. Eugen. **7**(2), 179–188 (1936)
13. Miche, Y., van Heeswijk, M., Bas, P., Simula, O., Lendasse, A.: TROP-ELM: a double-regularized ELM using LARS and Tikhonov regularization. In: Advances in Extreme Learning Machine: Theory and Applications Biological Inspired Systems. Computational and Ambient Intelligence Selected papers of the 10th International Work-Conference on Artificial Neural Networks (IWANN2009), vol. 74(16), 2413–2421, September 2011

Forecasting Solar Power Using Wavelet Transform Framework Based on ELM

Dandan Zhang, Yuanlong Yu$^{(\boxtimes)}$, and Zhiyong Huang

College of Mathematics and Computer Science, Fuzhou University,
Fuzhou 350116, Fujian, China
yu.yuanlong@fzu.edu.cn

Abstract. Forecasting solar power with good precision is necessary for ensuring the reliable and economic operation of electricity grid. In this paper, we consider the task of predicting a given day photovoltaic power (PV power) outputs in 30 min intervals from previous solar power and weather data. We proposed a method combining extreme learning machine and wavelet transform (WT-ELM), which build a separation prediction model for every moment using weather data and corresponding PV power, the weather characteristics are treated as input features and the PV power data are the corresponding ground truth. In addition, we also compared our method with K Nearest Neighbour (K-NN) and support vector machine (SVM) based on clustering using the same data. Then we evaluated the performance of our approach for all data with different time interval. The results show that our result performs much more better than KNN based on clustering.

Keywords: Wavelet transform · Extreme learning machine
Solar power forecasting · Time series prediction

1 Introduction

A photovoltaic (PV) system, is a power system used to supply solar power. Largescale PV systems have been built to connect to the electricity grid by many countries. Forecasting solar power more precisely will result in an optimized production due to the adjustments of output from other power production units. However, predicting solar power is a challenge task because of the instability of the data.

The research focuses on machine learning methods such as K nearest neighbor (K-NN), Neural networks (NNs) and support vector regression (SVR) [1]. These approaches treat solar power as high-dimensional data without considering temporal information in time series signals. Wavelet was widely used since the real wavelet basis and multi-scaling analysis were constructed by mathematicians. Wavelet transform (WT) was originally introduced in signal analysis

---

Y. Yu—This work is supported by National Natural Science Foundation of China (NSFC) under grant 61473089.

J. Cao et al. (Eds.): ELM 2017, PALO 10, pp. 186–202, 2019.
https://doi.org/10.1007/978-3-030-01520-6_17

which was compared with short-time Fourier transform [2]. As we all know that wavelet analysis has a big advantage on time series signal, what's more, wavelet decomposition [3–6] emerges as a power tool for approximation and has already been used successfully in image processing, signal denoising, signal compression. Then more and more people began to consider using wavelet transform in regression. There is no doubt that wavelet transform can be used in machine learning including classification and regression. In this paper, we present a new type of neural network inspired by both the WT and extreme learning machine (ELM).

Pati [7] has been demonstrated that it is possible to construct a theoretical formulation of a feedforward neural network in terms of wavelet decomposition. Zhang [8] proposed a wavelet network which combine wavelet transform and artificial neural networks. The parameters in the wavelet network were trained in the way of back propagation (BP) strategy. [9] algorithm is a method of training multi-hidden layers neural networks, it minimizes the training error and the parameters including weights and biases were trained in the method of gradient descent. However, BP algorithm has some weak points. Firstly, the training process takes a lot of time inevitably. Secondly, since it used gradient descent method, it may be trapped into local optimal point. In 1992, based on wavelet transform theory, Zhang and Benveniste proposed a new notion– wavelet network, which could be used to approximate arbitrary non-liner functions. Unlike the wavelet network, we integrated Fast Wavelet Transform (FWT) into designing the structure of the neural network in light of lower computational complexity and efficient data processing of the algorithm.

Huang [10] proposed a single hidden layer feedforward neural network called ELM whose weights connecting input layer and hidden layer were generated randomly. What's more, ELM not only used in classification, but also can be used in regression [11]. ELM spend less time in training stage for the reason that the weights and biases were random generation. In addition, it also has a better generalization ability than other neural networks. But ELM also has a big weak point concerning its advantage, that is: ELM could not get the detail feature of input data. Based the above mentioned theories, we thought that it is possible to approximate a time-series data by ELM combining WT.

The majority of our work focuses on predicting the daily solar power given the weather characteristics at near time. Different from other complicated methods require many and hard to obtainable meteorological measurements, we use only temperature and sunlight that are easily available. We treated sunlight and temperature value at multiple steps as input of model to predict the solar power value at one step. The data at half-hourly intervals was used to build separate prediction models for each half-hourly moment according to given weather vector in experiments. The data in a whole year was applied to train the model, and we tested it by evaluating the data of another year. The motivation is that the data is with the characteristics of seasonal, a whole year of data is enough to build a general prediction model. In general, wavelet does well in dealing with time series signals and fitting continuous outputs.

The main contributions of this paper are:

1. Combining ELM and WT, we proposed a new method use wavelet feature replace the non-liner feature mapping between input layer and hidden layer of ELM.
2. All the wavelet filters we used were generated randomly, we can get the corresponding scaling filters according to the relationship between scaling filter and wavelet filter. Random filters have two important purposes: on the one hand, we could overcome the limitation of existing wavelet types; on the other hand, we could acquire much more multiply features.
3. We analysis problems from the perspective of machine learning, weather characteristics were treated as the input feature, and PV power in the same day were regarded as ground truth.

This paper is organized as follows. Section 2 provides an overview of related work. The theory of our method was described in Sect. 3, the detail was presented in Appendix. Section 4 presented the data we used in the experiments, as well as experimental results.

2   Related Work

Based on the same dataset, we also compared with other techniques. Wang [1] proposed several clustering based methods, including K nearest neighbor (K-NN), Neural networks (NNs) and support vector regression (SVR). They group the days based on the weather characteristics and then build a separate prediction model for each cluster using the solar power data. They evaluated their results with Mean absolute error (MAE) and root mean squared error (RMSE).

Mashud Rana's method [12] is the same as Zheng Wang that used clustering in the first place, the different is that Rana predicted the next day photovoltaic power(PV power) from the previous values without using any exogenous data. They used two standard performance measures to assess the predictive accuracy, Mean Absolute Error (MAE) and Mean Relative Error (MRE).

Shi [13] proposed an algorithm for forecasting power output of PV systems based on weather classification and support vector machine (SVM). In this paper, they firstly classified the history data into four groups according to weather condition: sunny day, foggy day, rainy day and cloudy day. Then normalized the input data which means the historical PV power, which has a 15-min interval. At last, four SVM PV power forecasting models are setup to deal with different weather condition. MRE and RMSE were used to evaluated the forecasting accuracy.

Huang [14] presents an application of a new hybrid model in forecasting daily global solar radiation. The hybrid model incorporates the WT and Gaussian process regression (GPR). The WT was used to extract meaningful time-frequency information via decomposing the clearness index time series into a set of well-designed subseries. Then they forecast the future clearness index by means of welled trained GPR model using those subseries. They use MAE, RMSE and normalized root mean square error (nRMSE) to evaluate the forecasting accuracy.

Ji [15] proposed a new approach that contains two phases to predict the hourly solar radiation series. The first phrase was called the detrending phase which was used to remove the non-stationary trend lying in the solar radiation series. The second phrase was called prediction phrase combining Autoregressive and Moving Average (ARMA) model and Time Delay Neural Network (TDNN) to do the prediction. Still, RMSE and nRMSE were used to measure the goodness of prediction.

Hocao [16] proposed a two-dimensional (2D) representation model of the hourly solar radiation data. The model uses image processing method to forecast hourly solar radiation. To use the image model, they scan the rows and columns corresponding to days and hours respectively. In addition, to test the forecasting efficiency of the model, they tested nine different liner filters. At last, they compared the model with feed-forward NN based on the same data in the sense of RMSE.

Ravinesh [17] proposed a wavelet-coupled support vector machine (W-SVM) model to forecasting global incident solar radiation based on sunshine hours, minimum temperature, maximum temperature and windspeed. They use discrete wavelet transform to decompose the input data into their sub-series and summed up to create new series with one approximation and four levels of details with Daubechies-2 wavelet. Then the sub-series were used to train SVM. They use MAE, RMSE, mean absolute percentage (MAPE) and RMSE to evaluate the prediction errors.

3   The Proposed WT-ELM Algorithm

Our proposed approaches that predicting the PV power combine ELM and wavelet transform. The key point is to use the wavelet feature replace the non-liner mapping feature in ELM.

Figure 1 shows the architecture of our approaches in predicting the PV power. Every hidden neuron consists of a wavelet filter and a scaling filter. And every wavelet filter was generated randomly and then we could get its corresponding scaling filter.

3.1   Wavelet Transform

In wavelet analysis [2,18], there are two important functions: Scaling function φ and wavelet function ψ. These two function can generate a serious of functions that can be used to decompose and reconstruct signals. What's more, sometimes φ and ψ are called father wavelet and mother wavelet respectively.

According to the relationship between scaling and wavelet subspace and Eqs. 9 and 16, we have that any wavelet function can be expressed as a weighted sum of shifted, double-resolution scaling functions. i.e.

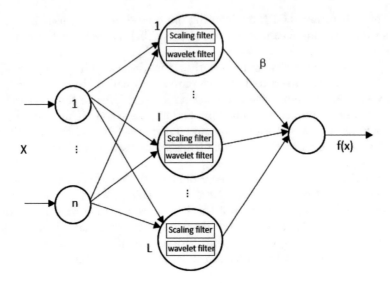

Fig. 1. The architecture of our proposed approach

$$\psi(x) = \sum_n h_\psi(x)\sqrt{2}\varphi(2x - n) \qquad (1)$$

Similarly, $h_\psi(n)$ are called the wavelet function coefficients and h_ψ is the wavelet vector. Using the condition that shown in Fig. 10 and that integer wavelet translates are orthogonal, the relationship between h_φ and h_ψ can be write

$$h_\varphi(n) = (-1)^n h_\psi(k - 1 - n) \qquad (2)$$

where k is the number of coefficients.

According to the theory of fast wavelet transform (FWT), we have the structure of FWT shown in Fig. 2. The detail process was presented in Sect. 5.

According to Eq. 2, we know that as long as we have wavelet filter, then we could get the corresponding scaling filter. Considering the reason that random input weight could speed the learning process of neuron networks like ELM,

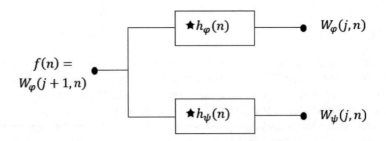

Fig. 2. The FWT structure

we generated all the wavelet filters randomly generated from -1 to 1, so the expression $h\varphi(-n)$ in Eq. 28 can be transformed as $h_\varphi(n)$ in Fig. 2.

Figure 2 shows the structure of Fast Wavelet Transform (FWT). The detail of FAT theory was deduced in Appendix.

3.2 Output Weights β

Given an input sample (x,t), the output of our network is

$$y = \sum_{i=1}^{l}(\beta_i h_i(x)) = \mathbf{h(x)}\beta \tag{3}$$

where β is the output weight connecting the hidden layer and the output layer. $\mathbf{h(x)}$ means the result of wavelet decomposition(output of hidden layer). Our algorithm aims to minimize the training error and the norm of output weight, which is proved that the smaller norm it is, the better generalization it has. Then the target of our method is

$$Minimize : \|H\beta - T\| + C\|\beta\| \tag{4}$$

where H is the hidden-layer output matrix.

$$H = \begin{bmatrix} h_1(x_1) & \cdots & h_L(x_1) \\ \vdots & \vdots & \vdots \\ h_1(x_N) & \cdots & h_L(x_N) \end{bmatrix} \tag{5}$$

The calculation of β can be expressed as

$$\beta = H^T(\frac{I}{C} + HH^T)^{-1}T^T \tag{6}$$

4   Experiments

4.1   DataSet

In this paper, our purpose is to predict the PV power. And the continuous data we used is collected form the Australia's largest PV system located at the University of Queensland in Brisbane. The original data measured at 1 min interval, which means that we could have a PV power value every minute, as long as its corresponding input feature(Sunlight and temperature).

We collected two complete years of data from 1 January 2013 to 31 December 2014. The original data of a day is from 5:00 to 17:00. But when we predict the PV power only ranging from 7am to 17pm, since most of the PV power outside the time window was either close to zero or not available [12]. We predict 20 data points [1] of PV power output for each day and there are 14,160 data points in total ($2 \times 354 \times 20$, except unavailable days).

In addition, we also collected temperature and sunlight data from the PV
site, sunlight and temperature are also measured at 1 min interval. All the data
we used in this paper are available online at [19]. Figure 3 shows three different
components: temperature, sunlight and PV power in the year of 2013. The red
line stands for the PV power of the whole year, below the red line there are two
lines stands for input attributes, the blue one is sunlight and the green one is
temperature.

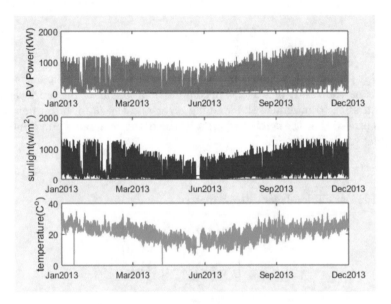

Fig. 3. Three different components of data

4.2   Data Preprocessing

And then using the same method as dealing with the PV power (divide them
into three different time span).

In this paper, we divided the data into two non-overlapping subsets: training
data and testing data. Based on the fact that the weather is of the character of
regularity and seasonal, so both the training and testing data we used were each
whole year. In detail, the data in 2013 was used to train the predict model and
the data of 2014 was used to evaluated the accuracy of the model.

To find the most appropriate input vector, we constructed four different
vector with different time span.

$$I_1^d = [T_{t-5}^d, T_{t-4}^d, T_{t-3}^d, T_{t-2}^d, T_{t-1}^d, T_t^d, T_{t+1}^d, T_{t+2}^d, T_{t+3}^d, T_{t+4}^d, T_{t+5}^d, S_{t-5}^d,$$
$$S_{t-4}^d, S_{t-3}^d, S_{t-2}^d, S_{t-1}^d, S_t^d, S_{t+1}^d, S_{t+2}^d, S_{t+3}^d, S_{t+4}^d, S_{t+5}^d].$$

And the expression of I_2^d, I_3^d, I_4^d are the same as I_1^d, but they present different
time scale. I_1^d means day d weather vector whose time span is 5-min, i.e. the
time interval between T_{t-5}^d and T_{t-4}^d is 5-min and other time span is the same

as mentioned before. I_2^d represents the time interval is 10-min, similarly, I_3^d, I_4^d respectively represents 15-min time interval and 20-min time interval. As we have get four different weather vector with different time span, we can construct five different input weather vectors and get the best compound mode via evaluating their performances.

$W_1^d = [I_1^d, I_2^d, I_3^d, I_4^d]^T$, the combination of four different time intervals.

$W_2^d = [I_1^d, I_2^d, I_3^d]^T$, the combination of 5-min interval, 10-min interval and 15-min interval.

$W_3^d = [I_1^d, I_2^d, I_4^d]^T$, the combination of 5-min interval, 10-min interval and 20-min interval.

$W_4^d = [I_1^d, I_3^d, I_4^d]^T$, the combination of 5-min interval, 15-min interval and 20-min interval.

$W_5^d = [I_2^d, I_3^d, I_4^d]^T$, the combination of 10-min interval, 15-min interval and 20-min interval.

4.3   Results and Discussion

In order to assess the accuracy of the prediction model, we use standard performance measure: Mean Absolute Error (MAE) as follows:

$$MAE = \frac{1}{D * n} \sum_{i=1}^{N} | P^i - \hat{P}^i | \tag{7}$$

where P^i and $\hat{P}^i$ are the vector of actual and predicted power output for every moment, D is the number of days in the testing set and n is the number of predicted points output for a day ($n = 20$). When we compute every moment's MAE, n equals to 1. Figure 4 is the prediction result of moment 9:00 moment, 12:00 and moment 16:00 seperately.

Table 1 presents the MAE of training and testing set at different moments with different input time span. Figure 5 show the MAE of different input feature, from the result above, we obviously know: The best input feature is W_3 and the most accurate prediction model, in terms of MAE measure, is the model predict moment 16:30. It achieved MAE achieves 25.3891 KW which is smaller than KNN based clustering's best prediction model of C2 whose MAE is 60.4391 KW at moment 16:30. We also compare our result with KNN, the result is shown in Fig. 6. Table 3 is the result using ELM to predict the PV power. We also compare the MAE of ELM with ours, the result is shown in Fig. 7. Apparently, our approach performs much better than KNN based on clustering and ELM at any moment. Table 4 presents the result of support vector regression (SVR) based on clustering, which is the same as in KNN. We also compared the SVR prediction result with our method, the result is shown in Fig. 8. In this figure, we know that there may have one cluster error smaller than ours at some moments. The most important thing is that there is fewer samples in that cluster. As long as there exists more samples, the predicted error becomes much bigger (Table 2).

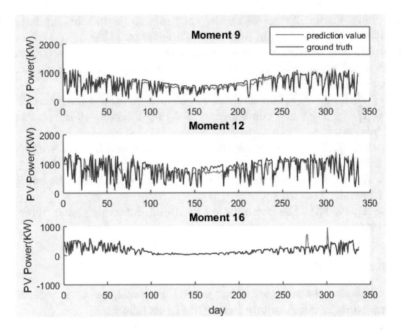

Fig. 4. Three different moments' result

Table 1. The MAE at different moments with different input feature

Moment	W1(KW)	W2(KW)	W3(KW)	W4(KW)	W5(KW)
7:00	33.0005	33.8411	33.1098	32.9963	33.294
7:30	31.0239	32.2287	30.8642	30.8823	32.2696
8:00	46.7317	43.4603	46.7881	47.5631	46.3471
8:30	57.9534	57.6777	57.271	58.1627	64.7974
9:00	68.9019	69.6363	69.8861	67.2568	66.8304
9:30	88.3363	89.3093	85.8097	85.0523	92.0719
10:00	100.4888	102.8835	99.5678	99.0452	93.5539
10:30	110.0812	101.9193	98.8839	96.5477	107.2683
11:00	120.8059	110.6003	97.5327	113.2896	113.6289
11:30	104.3205	105.1657	108.1814	104.3613	109.051
12:00	109.5554	105.9821	105.4075	105.8312	119.9618
12:30	104.1089	102.7174	103.0446	104.3222	111.3581
13:00	89.7254	94.7009	87.8382	85.9668	105.5415
13:30	90.7403	88.3078	88.1757	88.5917	100.0605
14:00	73.8710	77.3602	71.1952	74.5801	79.6301
14:30	73.4642	75.2076	70.5194	71.336	79.2136
15:00	45.7651	44.4861	44.744	44.5364	53.4195
15:30	45.0558	44.4038	43.0508	44.5403	51.6328
16:00	26.9345	26.2594	27.0902	26.8905	28.2154
16:30	24.5548	26.2847	25.3891	24.6789	25.457

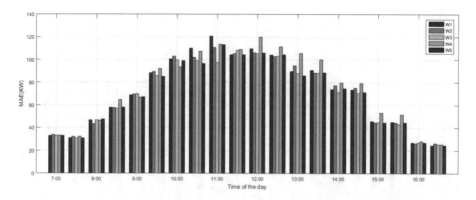

Fig. 5. The result of different input with different input feature

Table 2. The result of KNN based on clustering

Moment	KNN-C1(KW)	KNN-C2(KW)	KNN-C3(KW)	KNN-C4(KW)
7:00	432.1460	234.9540	69.3919	507.1680
7:30	552.5229	342.5104	636.4860	133.8042
8:00	577.0757	400.1100	726.3960	226.7015
8:30	807.0176	656,1070	205.4119	442.1655
9:00	912.3470	563.3171	282.8065	736.3266
9:30	664.8951	985.4599	802.5837	292.3232
10:00	925.8147	713.1261	1049.6	345.5542
10:30	1082.0	304.3973	735.5722	875.5141
11:00	1122.3	726.1614	970.8103	350.0654
11:30	395.8394	1139.4	960.8237	792.1939
12:00	1156.2	385.2765	805.0329	938.7061
12:30	1130.3	954.0328	744.8258	359.9103
13:00	1037.3	883.4772	634.9487	291.6483
13:30	1030.0	862.5054	275.6405	598.6466
14:00	531.7492	8893.9222	256.6194	714.1504
14:30	250.7884	896.9904	731.1124	545.0920
15:00	686.3402	315.9741	519.5322	175.8139
15:30	166.2463	532.8698	696.5332	330.5320
16:00	363.7657	186.1867	369.3800	61.0832
16:30	456.5389	60.4301	175.1482	331.9228

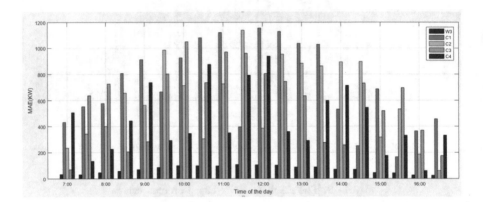

Fig. 6. The result of wavelet transform and KNN based on clustering

Table 3. The result using prediction model of ELM

Moment	Testing error (KW)	Moment	Testing error(KW)
7:00	42.6094	12:00	161.0336
7:30	57.1309	12:30	159.8838
8:00	82.4146	13:00	152.6769
8:30	92.8792	13:30	159.7345
9:00	108.5371	14:00	126.8373
9:30	131.7588	14:30	120.9033
10:00	140.6327	15:00	81.7031
10:30	149.3832	15:30	78.4309
11:00	157.6794	16:00	40.4451
11:30	163.6189	16:30	41.9569

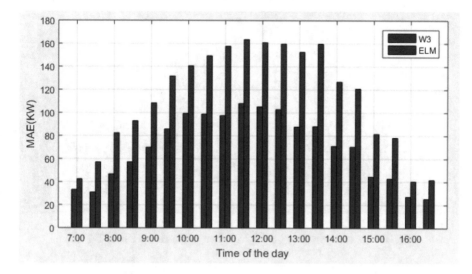

Fig. 7. The comparation of ELM and our method

Table 4. The predicted result of SVR based on clustering

Moment	SVR-C1(KW)	SVR-C2(KW)	SVR-C3(KW)	SVR-C4(KW)
7:00	62.679	81.4499	59.6203	30.1666
7:30	77.39	52.7871	42.7755	76.3762
8:00	85.6031	74.8124	82.5004	116.5491
8:30	88.6965	114.6286	82.2725	75.6495
9:00	159.0239	118.0262	77.3163	76.8351
9:30	175.8084	87.1247	79.5021	158.2512
10:00	155.8326	70.0816	110.5258	178.6433
10:30	81.7949	89.9383	191.4011	197.7948
11:00	75.4059	216.4801	206.9218	132.8159
11:30	72.7239	120.934	188.9833	208.8542
12:00	196.2383	163.1083	221.6833	63.0466
12:30	205.9302	138.1297	77.5116	201.333
13:00	70.5929	180.9001	102.1509	187.9714
13:30	69.3277	164.7661	192.2414	107.0902
14:00	146.871	163.6777	106.5339	72.645
14:30	135.7363	133.5488	115.3778	105.0714
15:00	73.8262	110.0823	88.117	64.7573
15:30	76.3939	97.6916	76.8379	72.051
16:00	80.3573	70.3244	27.6503	44.5928
16:30	85.5627	88.9782	25.1799	57.8541

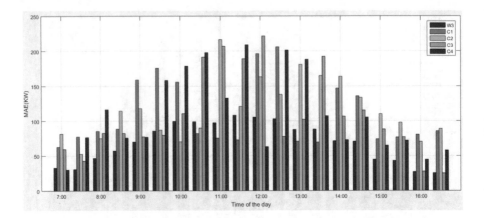

Fig. 8. The comparison of SVR based on clustering with WT-ELM

5   Conclusion

This paper proposed a new method for learning single hidden layer feedforward neural network. It used fast wavelet transform as feature mapping to connect the input layer and hidden layer. In this way, it can get more detail feature of input data and improve the accuracy of our method. Our method not only used to approximate the PV power(regression), but also can be used to classification. Even experiments in this paper we have made proved that our approach performs better than other algorithm, we also have something to improve. In classification, our method performs better in multi-class classification, but not so good as ScELM in dealing with Binary classification problems. In the near future, we will overcome this weak point. What's more, as we know that wavelet filter could capture the detail feature of input data, maybe combine wavelet transform and deep learning such as convolutional neural network (CNN) could have wonderful results. So it is also an challenging for future work.

Appendix: Wavelet Transform

In multi-resolution analysis (MRA), a scaling function is used to create a series of approximations of a function or image [20]. Consider the set of expansion functions composed of integer translation and binary scaling of the real, square-integrable function $\varphi(x)$; this is the set $\varphi_{j,k}(x)$, where

$$\varphi_{j,k}(x) = 2^{j/2}\varphi(2^j x - k) \tag{8}$$

for all j,k∈Z and $\varphi(x) \in L^2(\mathrm{R})$ [6]. Here, k determines the position of $\varphi_{j,k}(x)$ along the x-axis, and j determines the width of $\varphi_{j,k}(x)$. Generally, we denote the subspace spanned over k for any j as

$$V_j = \overline{Span_k\{\varphi_{jk}(x)\}} \tag{9}$$

$$if \ f(x) \in V_j, \ f(x) = \sum_k \alpha_k \varphi_{jk}(x) \tag{10}$$

MRA has four requirements as following.

Requirement 1: the scaling function is orthogonal to its integer translates:

$$\langle \varphi_{j,k}(x), \varphi_{j,k'}(x) \rangle = \int \varphi_{j,k}^*(x) \varphi_{j,k'}(x) dx = 0, k \neq k' \tag{11}$$

Requirement 2: The subspaces spanned by the scaling function at low scales are nested within those spanned at higher scales. As shown in Fig. 9. That is,

$$V_{-\infty} \subset ... \subset V_{-1} \subset V_0 \subset V_1 \subset V_2 \subset ...V_{\infty} \tag{12}$$

Requirement 3: The only function that is common to all V_j is f(x) = 0. i.e. $V_{-\infty} = 0$.

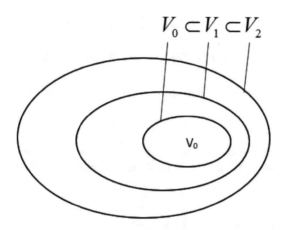

Fig. 9. The nested function spaces spanned by a scaling function

Requirement 4: Any function can be represented with arbitrary precision. That is,

$$V_{\infty} = \{L_2(R)\} \tag{13}$$

Under these conditions, the expansion functions of subspace V_j can be expressed as a weighted sum of the expansion function of subspace V_{j+1}. Using Eq. 10, we let

$$\varphi_{j,k}(x) = \sum_n \alpha_n \varphi_{j+1,n}(x) \tag{14}$$

combine Eqs. 14 and 8 we have

$$\varphi_{j,k}(x) = \sum_n h_\varphi(x) 2^{(j+1)/2} \varphi(2^{j+1}x - n) \tag{15}$$

because $\varphi = \varphi_{0,0}(x)$, Eq. 15 can be expressed as

$$\varphi(x) = \sum_n h_\varphi(n)\sqrt{2}\varphi(2x - n) \tag{16}$$

The $h_\varphi(n)$ coefficients in Eq. 16 are called scaling function coefficients; h_φ is called the scaling vector.

Given a scaling function that meet the MRA requirements, we have a wavelet function $\psi(x)$, together with its integer translates and binary scalings functions $\psi_{j,k}(x)$, which spans the difference between any two adjacent scaling subspaces V_j and V_{j+1}. The set $\{\psi_{j,k}(x)\}$ of wavelet was defined as:

$$\psi_{j,k}(x) = 2^{j/2}\psi(2^j x - k) \tag{17}$$

for all k ∈ Z that span the W_j spaces in the Fig. 10. The same as scaling functions, wavelet subspaces can be write

$$W_j = \overline{Span_k\{\psi_{jk}(x)\}} \tag{18}$$

$$if \ f(x) \in W_j, \ f(x) = \sum_k \alpha_k \psi_{j,k}(x) \tag{19}$$

The relationship between scaling and wavelet function subspaces are as

$$V_{j+1} = V_j \oplus W_j \tag{20}$$

where $\oplus$ denotes the union of spaces. And W_j is the orthogonal complement of V_j in $V_j + 1$. That is all members of V_j are orthogonal to the members of W_j.

$$\langle \varphi_{j,k}(x), \psi_{j,k'}(x)\rangle = 0 \tag{21}$$

We can now express the space of all measurable, square-integrable functions as

$$L^2(\mathbf{R}) = V_0 \oplus W_0 \oplus W_1 \oplus ... \tag{22}$$

or

$$L^2(\mathbf{R}) = V_{j_0} \oplus W_{j_0} \oplus W_{j_0+1} \oplus ... \tag{23}$$

where j_0 is an arbitrary starting scale.

According to the relationship between scaling and wavelet subspace and Eqs. 9 and 16, we have that any wavelet function can be expressed as a weighted sum of shifted, double-resolution scaling functions (Eq. 1).

Similarly, $h_\psi(n)$ are called the wavelet function coefficients and h_ψ is the wavelet vector. Using the condition that shown in Fig. 10 and that integer wavelet translates are orthogonal, the relationship between h_φ and h_ψ can be write as Eq. 2.

Consider Eq. 16 again, scaling x by 2^j, translating it by k, and letting $m = 2k + n$ gives

$$\varphi(2^i - k) = \sum_n h_\varphi(n)\sqrt{2}\varphi(2(2^j x - k) - n)$$
$$= \sum_n h_\varphi(n)\sqrt{2}\varphi(2^{j+1} x - 2k - n) \tag{24}$$
$$= \sum_n h_\varphi(m - 2k)\sqrt{2}\varphi(2^{j+1} x - 2m)$$

Equation 24 can be written as Eq. 25, similarly, we could get Eq. 26.

$$W_\varphi(j, k) = \sum_m h_\varphi(m - 2k)W_\varphi(j + 1, m) \tag{25}$$

$$W_\psi(j, k) = \sum_m h_\psi(m - 2k)W_\varphi(j + 1, m) \tag{26}$$

Therefore, the coefficients expression of FWT can be translated into Eqs. 27 and 28. The structure of FWT are shown as Fig. 2.

$$W_\varphi(j, k) = h_\varphi(-n) * W_\varphi(j + 1, n) \,|_{n=2k, k \geq 0} \tag{27}$$
$$W_\psi(j, k) = h_\psi(-n) * W_\varphi(j + 1, n) \,|_{n=2k, k \geq 0} \tag{28}$$

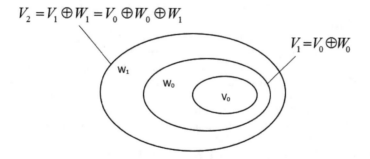

$$V_2 = V_1 \oplus W_1 = V_0 \oplus W_0 \oplus W_1$$
$$V_1 = V_0 \oplus W_0$$

Fig. 10. The relationship between scaling and wavelet function spaces

References

1. Wang, Z., Koprinska, I., Rana, M.: Clustering based methods for solar power forecasting. In: 2016 International Joint Conference on Neural Networks (IJCNN), pp. 1487–1494. IEEE (2016)
2. Daubechies, I.: The wavelet transform, time-frequency localization and signal analysis. IEEE Trans. Inf. Theory **36**(5), 961–1005 (1990)
3. Mallat, S.G.: A theory for multiresolution signal decomposition: the wavelet representation. IEEE Trans. Pattern Anal. Mach. Intell. **11**(7), 674–693 (1989)

4. Mallat, S.G.: Multiresolution approximations and wavelet orthonormal bases of l^2 (r). Trans. Am. Math. Soc. **315**(1), 69–87 (1989)
5. Daubechies, I., Grossmann, A., Meyer, Y.: Painless nonorthogonal expansions. J. Math. Phys. **27**(5), 1271–1283 (1986)
6. Daubechies, I.: Orthonormal bases of compactly supported wavelets ii: variations on a theme. SIAM J. Math. Anal. **24**(2), 499–519 (1993)
7. Pati, Y.C., Krishnaprasad, P.S.: Analysis and synthesis of feedforward neural networks using discrete affine wavelet transformations. IEEE Trans. Neural Netw. **4**(1), 73–85 (1993)
8. Zhang, Q., Benveniste, A.: Wavelet networks. IEEE Trans. Neural Netw. **3**(6), 889–898 (1992)
9. Rumelhart, D.E., Hinton, G.E., Williams, R.J.: Learning internal representations by error propagation. Technical report, DTIC Document (1985)
10. Huang, G.-B., Zhu, Q.-Y., Siew, C.-K.: Extreme learning machine: theory and applications. Neurocomputing **70**(1), 489–501 (2006)
11. Huang, G.-B., Zhou, H., Ding, X., Zhang, R.: Extreme learning machine for regression and multiclass classification. IEEE Trans. Syst. Man Cybern. Part B (Cybern.) **42**(2), 513–529 (2012)
12. Rana, M., Koprinska, I., Agelidis, V.G.: Forecasting solar power generated by grid connected PV systems using ensembles of neural networks. In: 2015 International Joint Conference on Neural Networks (IJCNN), pp. 1–8. IEEE (2015)
13. Shi, J., Lee, W.-J., Liu, Y., Yang, Y., Wang, P.: Forecasting power output of photovoltaic systems based on weather classification and support vector machines. IEEE Trans. Ind. Appl. **48**(3), 1064–1069 (2012)
14. Huang, C., Zhang, Z., Bensoussan, A.: Forecasting of daily global solar radiation using wavelet transform-coupled Gaussian process regression: case study in Spain. In: Innovative Smart Grid Technologies-Asia (ISGT-Asia), pp. 799–804. IEEE (2016)
15. Ji, W., Chee, K.C.: Prediction of hourly solar radiation using a novel hybrid model of ARMA and TDNN. Solar Energy **85**(5), 808–817 (2011)
16. Hocaoğlu, F.O., Gerek, Ö.N., Kurban, M.: Hourly solar radiation forecasting using optimal coefficient 2-D linear filters and feed-forward neural networks. Solar Energy **82**(8), 714–726 (2008)
17. Deo, R.C., Wen, X., Qi, F.: A wavelet-coupled support vector machine model for forecasting global incident solar radiation using limited meteorological dataset. Appl. Energy **168**, 568–593 (2016)
18. Boggess, A., Narcowith, F.J.: A First Course in Wavelet with Fourier Analysis, (2nd edn.)
19. University of Queensland. http://solar-energy.uq.edu.au/
20. Gonzalez, R.V., Woods, R.E.: Digital Image Processing, 3rd edn.

Distance Estimation for Incomplete Data by Extreme Learning Machine

Emil Eirola[1]([✉]), Anton Akusok[1], Kaj-Mikael Björk[3,4],
and Amaury Lendasse[2]

[1] RiskLab at Arcada University of Applied Sciences, Helsinki, Finland
emil.eirola@arcada.fi
[2] Department of Mechanical and Industrial Engineering and The Iowa
Informatics Initiative, The University of Iowa, Iowa City, USA
[3] Arcada University of Applied Sciences, Helsinki, Finland
[4] Hanken School of Economics, Helsinki, Finland

Abstract. Data with missing values are very common in practice, yet
many machine learning models are not designed to handle incomplete
data. As most machine learning approaches can be formulated in terms
of distance between samples, estimating these distances on data with
missing values provides an effective way to use such models. This paper
present a procedure to estimate the distances using the Extreme Learning Machine. Experimental comparison shows that the proposed approach achieves competitive accuracy with other methods on standard
benchmark datasets.

Keywords: Machine learning · Extreme learning machine
Incomplete data · Missing values · Distance estimation

1 Introduction

Recent developments in machine learning have provided valuable solutions to a
large number of important tasks. The research behind these algorithms is based
on statistical methods, which too often ignore some of the practicalities relevant
to the final use case. A particularly significant issue is the problem of missing
values in the data. Discarding incomplete samples from the dataset can lead to
biases in the data distribution, affecting the statistical validity of the method
that is applied. All machine learning procedures should be able to handle data
with missing values, so that this problem can be avoided.

One possibility is to impute missing values by plausible values (e.g., the
mean), but this also distorts the distribution of the data. As the majority of
machine learning methods can be formulated in terms of either distances or sim-
ilarities between data samples, the better solution is to estimate these distances.
Having estimates of all pairwise distances in a dataset enables application of a
large number of standard tools (nearest neighbours, support vector machines,
radial basis function models, etc.).

© Springer Nature Switzerland AG 2019
J. Cao et al. (Eds.): ELM 2017, PALO 10, pp. 203–209, 2019.
https://doi.org/10.1007/978-3-030-01520-6_18

This paper presents a procedure to estimate the distances by using the Extreme Learning Machine (ELM) [1]. ELM is an ideal regression model for this task, as the problem is decidedly nonlinear, and ELM is computationally efficient enough to be able to handle the large number of data; considering the number of pairwise distances amounts to $N(N-1)/2$ instances in a dataset with N samples.

The idea of distance estimation has previously been studied in [2,3]. Other approaches to extend ELM to work with incomplete data have been presented in [4–10]. For an overview of statistical considerations relevant to data with missing values, see [11].

The rest of this paper is structured as follows: after reviewing the standard ELM, the proposed approach is detailed in Sect. 2. Section 3 presents an experimental comparison to other common approaches on two standard datasets. The conclusions are summarised in Sect. 4.

2   Estimating Distances by ELM

2.1   Standard ELM

The proposed approach is based on the Extreme Learning Machine (ELM) [1], which is a single hidden-layer feed-forward neural network where only the output weights β_k are optimised, and all the weights w_{kj} between the input and hidden layer are assigned randomly. With input vectors x_i and the targets collected as a vector y, it can be written as

$$\mathbf{H}\boldsymbol{\beta} = \boldsymbol{y} \quad \text{where} \quad H_{ik} = h\big(\boldsymbol{w}_k^T \boldsymbol{x}_i\big). \tag{1}$$

Here $h(\cdot)$ is a non-linear activation function applied elementwise. Training this model is simple, as the optimal output weights β_k can be calculated by ordinary least squares. The method relies on the idea of random projection: mapping the data randomly into a sufficiently high-dimensional space means that a linear model is likely to be relatively accurate. As such, the number of hidden-layer neurons needed for achieving equivalent accuracy is often much higher than in a multilayer perceptron trained by back-propagation, but the computational burden for training the model is still considerably lower.

The optimal weights $\boldsymbol{\beta}$ can be calculated as the least squares solution to Eq. (1), or formulated by using the Moore–Penrose pseudoinverse as follows:

$$\boldsymbol{\beta} = \mathbf{H}^+ \boldsymbol{y} \tag{2}$$

A high number of neurons in the hidden layer introduces concerns of overfitting, and regularised versions of the ELM have been developed to remedy this issue. These include the *optimally pruned ELM* (OP-ELM) [12], and its Tikhonov-regularised variant TROP-ELM [13]. In the current case, Tikhonov regularisation is applied when solving the least square problem in Eq. (1). The value of the regularisation parameter can be optimised by minimising the leave-one-out error (efficiently calculated via the PRESS statistic [13]).

2.2   Applying ELM to Distance Estimation

Distance estimation is inherently an unsupervised machine learning task. In this case, it is reformulated for a supervised model, by setting the distance between two samples as the target, while having the values from both samples (or transformed versions) as inputs to the regression task. The overall approach is to collect all fully known samples (i.e., without any missing values) to assemble a training set, as the distances between these are known. Then, the real scenario of estimating with missing values is simulated by removing values from the training data.

If the original data X contains samples $x_i \in \mathbb{R}^d$ for $i \in \{1, \ldots, N\}$, the training set is constructed as follows. For all i, j such that x_i has no missing values and $i \neq j$, targets are constructed as $t_{ij} = \|x_i - x_j\|^2$, and the corresponding inputs as

$$z_{ij} = [x_{i,1}, x_{i,2}, \ldots x_{i,d}, x_{j,1}, x_{j,2}, \ldots x_{j,d}] \tag{3}$$

To apply the ELM to the test data (sample pairs where values are missing), the missing values are replaced by the mean of that variable. To simulate this scenario, missing values are introduced to the samples z_{ij} by replacing a random selection of values by the mean of the variable. The values to remove should be chosen randomly in a way that represents the real scenario, i.e., the proportion of missing values for each variable $1, \ldots, d$ in z_{ij} should equal that of the original data X.

For convenience, the samples t_{ij} and z_{ij} can be re-indexed to avoid the double index ij.

It is important to note that the dimensionality of the machine learning task grows from d to $2d$, but the number of samples grows much more: from N to $M(M-1)$ where M is the number of completely known samples. While there are only $M(M-1)/2$ pairs of samples, each pair should be considered twice (switching the order of which sample is first) to ensure that the ELM is able to learn the inherent symmetry in the task. The ELM is an appropriate regression model to choose for this task, as it is computationally efficient even on such large data, and expressive enough to learn the required non-linear function.

An ELM model is trained to learn the regression task of predicting the targets t_{ij} from the inputs z_{ij} where missing values have been inserted. As the number of data points in this task is so large, regularising the model is not necessary, but a small Tikhonov regularisation is recommended to avoid potential issues with numerical instability. To estimate the distances between samples with missing values, data is constructed in accordance with Eq. (3), but the missing values are replaced by the mean. Then the sample is used as input to the ELM, which provides an estimate of the (squared) distance.

To ensure that the distribution of the completely known samples (used for generating the training data) is representative of the full data, it is necessary to assume that samples are missing completely at random (MCAR) [11].

3   Experimental Evaluation

An experimental evaluation is conducted to compare the accuracy of the proposed method to two other approaches:

1. Distance calculation after imputing missing values by the mean
2. Partial Distance Strategy (PDS) [14,15]. Calculate the sum of squared differences of the mutually known components and scale to the missing components:

$$\hat{d}_{ij}^2 = \frac{d}{d - |M_i \cup M_j|} \sum_{l \notin M_i \cup M_j} (x_{i,l} - x_{j,l})^2 \tag{4}$$

where M_i is the set of missing components in $\boldsymbol{x}_i$.

The comparison evaluation is conducted on two standard datasets from the UCI machine learning repository [16]:

1. Stocks, with 950 samples and 9 variables
2. Boston Housing, with 506 samples and 13 variables

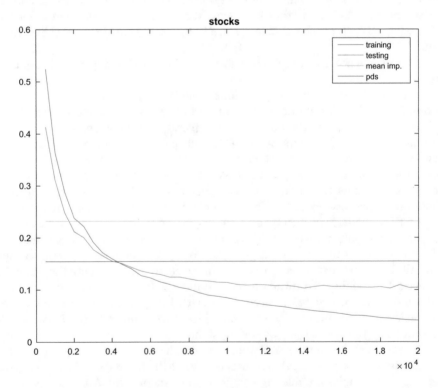

Fig. 1. Plot of the training and test error of the ELM approach against the number of neurons in the hidden layer on the stocks data, compared to the partial distance strategy (PDS) and mean imputation.

The datasets have no missing values, and values can be removed in a controlled manner to study the performance of the methods. As this evaluation only concerns the distances between samples, the output targets of the original regression tasks are ignored. The datasets are standardised to zero mean and unit variance.

Half of the samples of each data set are used for training, and the remaining half for testing. No missing values are introduced to the training set, and as such the number of data points available for the ELM is $N_t^2 - N_t$, with N_t samples in the training set. As this can become intractable for the standard desktop computer used to run the experiment – e.g., 225,150 points in the case of the Stocks data – a random subselection of 50,000 points is chosen to make the computation practical. From the testing set, 10% of the values are randomly removed, and the pairwise distances estimated by each method are compared by the mean squared error (MSE) with the true distances (as calculated before removing values).

Results of the MSE for each method on the two datasets are shown in Figs. 1 and 2. The training and testing accuracy of the ELM is drawn for an increasing number of neurons in the hidden layer, while the MSE of the other methods is shown as horizontal lines.

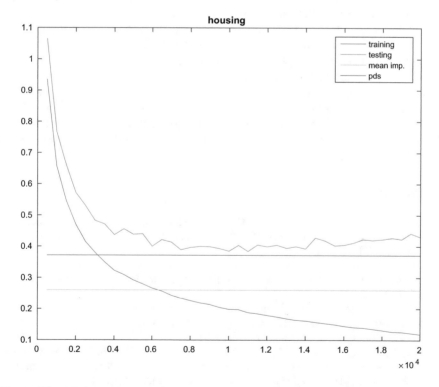

Fig. 2. Plot of the training and test error of the ELM approach against the number of neurons in the hidden layer on the housing data, compared to the partial distance strategy (PDS) and mean imputation.

The accuracy of the ELM increases consistently when adding more neurons to the hidden layer. For the stocks data in Fig. 1, ELM clearly outperforms both methods with a sufficient number of neurons. For the housing data in Fig. 2, the accuracy of the ELM is comparable to the other methods.

Interestingly, PDS outperforms mean imputation for the stocks data, where variables are highly correlated, but the opposite is true for the less correlated housing data. ELM performs even better than PDS on the stocks data, suggesting that the proposed approach is particularly useful in situations where there is considerable correlation between the variables.

4   Conclusions

This paper describes an approach to use ELM to estimate the pairwise distances in a dataset with missing values. Accurate estimates enable the use of a large number of standard machine learning methods, which otherwise could not be applied to incomplete datasets.

The experimental evaluation of the approach reveals that the achieved accuracy is competitive with other methods, and can outperform other methods particularly when variables are highly correlated.

Further research can be conducted to study the problem and improve the method. If the ELM could be extended to structurally recognise the symmetry of the first and second sample in the inputs, it would no longer be necessary to use all sample pairs twice. Currently, missing values are replaced by the mean to indicate that they are missing, and it would be worthwhile to study whether other alternatives could be more appropriate. It is not clear why such a large number of neurons (10,000 or more) is needed for accurate results. In addition, the approach should be evaluated on a more extensive collection of datasets to assess its usefulness.

References

1. Huang, G.B., Zhu, Q.Y., Siew, C.K.: Extreme learning machine: theory and applications. Neurocomputing **70**(13), 489–501 (2006). https://doi.org/10.1016/j.neucom.2005.12.126
2. Eirola, E., Doquire, G., Verleysen, M., Lendasse, A.: Distance estimation in numerical data sets with missing values. Inf. Sci. **240**, 115–128 (2013). https://doi.org/10.1016/j.ins.2013.03.043
3. Eirola, E., Lendasse, A., Vandewalle, V., Biernacki, C.: Mixture of Gaussians for distance estimation with missing data. Neurocomputing **131**, 32–42 (2014). https://doi.org/10.1016/j.neucom.2013.07.050
4. Yu, Q., Miche, Y., Eirola, E., van Heeswijk, M., Sverin, E., Lendasse, A.: Regularized extreme learning machine for regression with missing data. Neurocomputing **102**, 4551 (2013). https://doi.org/10.1016/j.neucom.2012.02.040
5. Sovilj, D., Eirola, E., Miche, Y., Björk, K., Nian, R., Akusok, A., Lendasse, A.: Extreme learning machine for missing data using multiple imputations. Neurocomputing **174**(Part A), 220–231 (2016). https://doi.org/10.1016/j.neucom.2015.03.108.

6. Gao, H., Liu, X.W., Peng, Y.X., Jian, S.L.: Sample-based extreme learning machine with missing data. Math. Probl. Eng. **2015** (2015) https://doi.org/10.1155/2015/145156.
7. Xie, P., Liu, X., Yin, J., Wang, Y.: Absent extreme learning machine algorithm with application to packed executable identification. Neural Comput. Appl. **27**(1), 93–100 (2016). https://doi.org/10.1007/s00521-014-1558-4
8. Yan, Y.T., Zhang, Y.P., Chen, J., Zhang, Y.W.: Incomplete data classification with voting based extreme learning machine. Neurocomputing **193**, 167–175 (2016). https://doi.org/10.1016/j.neucom.2016.01.068
9. Eirola, E., Akusok, A., Björk, K.M., Johnson, H., Lendasse, A.: Predicting huntingtons disease: extreme learning machine with missing values. In: Proceedings of ELM-2016, pp. 195–206. Springer, Cham (2017)
10. Akusok, A., Eirola, E., Björk, K.M., Miche, Y., Johnson, H., Lendasse, A.: Bruteforce missing data extreme learning machine for predicting huntington's disease. In: Proceedings of the 10th International Conference on PErvasive Technologies Related to Assistive Environments, pp. 189–192. ACM (2017)
11. Little, R.J.A., Rubin, D.B.: Statistical Analysis with Missing Data. 2nd edn. Wiley-Interscience (2002). https://doi.org/10.1002/9781119013563
12. Miche, Y., Sorjamaa, A., Bas, P., Simula, O., Jutten, C., Lendasse, A.: OP-ELM: optimally-pruned extreme learning machine. IEEE Trans. Neural Netw. **21**(1), 158–162 (2010). https://doi.org/10.1109/TNN.2009.2036259
13. Miche, Y., van Heeswijk, M., Bas, P., Simula, O., Lendasse, A.: TROP-ELM: a double-regularized ELM using LARS and Tikhonov regularization. Neurocomputing **74**(16), 2413–2421 (2011). https://doi.org/10.1016/j.neucom.2010.12.042
14. Dixon, J.K.: Pattern recognition with partly missing data. IEEE Trans. Syst. Man Cybern. **9**(10), 617–621 (1979)
15. Himmelspach, L., Conrad, S.: Clustering approaches for data with missing values: comparison and evaluation. In: 2010 Fifth International Conference on Digital Information Management (ICDIM), pp. 19–28, July 2010
16. Lichman, M.: UCI machine learning repository (2013)

A Kind of Extreme Learning Machine Based on Memristor Activation Function

Hanman Li[1,2,3], Lidan Wang[1,2,3(✉)], and ShuKai Duan[1,2,3]

[1] The College of Electronic and Information Engineering,
Southwest University, Chongqing 400415, China
ldwang@swu.edu.cn
[2] National and Local Joint Engineering Laboratory of Intelligent Transmission
and Control Technology, Southwest University, Chongqing 400415, China
[3] Brain-Inspired Computing and Intelligent Control of Chongqing Key Lab,
Southwest University, Chongqing 400415, China

Abstract. Extreme learning machine is a new type of algorithm for single hidden layer feedforward neural network. Compared with the traditional algorithms, ELM avoids long time iteration and has the advantages of high speed, small errors. Among them, the activation function plays an important role in the system. Whereas the general ELM usually uses Sigmod function as the activation function, a new kind ELM using memristor's memristance-charge function as activation function is proposed in this article. Experiments show that, compared with the ELM and the traditional neural network algorithms, the extreme learning machine based on memristance-charge activation function can shorten the time and improve the accuracy. In a word, it has better classification and regression performances.

Keywords: Extreme learning machine · Memristance-charge function
Regression and classification performances

1 Introduction

The traditional gradient algorithms (such as BP neural network) has been widely used in multilayer feedforward neural network training [1]. However due to its slow learning speed and easy to fall into local optimal solution, the traditional gradient algorithms' development encounters bottlenecks.

The extreme learning machine(ELM) is a kind of learning algorithm based on single hidden layer feedforward neural network [2], which is proposed by Professor Huang and others in 2004. Because of its rapidity and effectiveness, it has attracted more and more attention in different fields in recent years. In the past few years, ELM has been widely used in pattern recognition, the disease diagnosis of brain wave

---

The work was supported by National Natural Science Foundation of China (Grant Nos. 61571372, 61672436), Fundamental Research Funds for the Central Universities (Grant Nos. XDJK2016A001, XDJK2017A005).

© Springer Nature Switzerland AG 2019
J. Cao et al. (Eds.): ELM 2017, PALO 10, pp. 210–218, 2019.
https://doi.org/10.1007/978-3-030-01520-6_19

prediction based on brain computer interaction, human-computer interaction, image processing, face recognition, gesture recognition of sign language, handwriting recognition, object recognition, image real-time satellite remote sensing, network security, constructing the resolution of the image from the low resolution image and so on. In this algorithm, the input weight and the hidden layer bias are randomly generated, and the output layer weight is calculated, so it avoids long time iterations.

The memristor is a new kind of two port passive devices, which was first proposed by Chua in 1971 [3]. A physical memristor based on titanium dioxide was developed for the first time by the HP Company in 2008 [4].The memristor has attracted extensive attention in academia and industry, which is considered the fourth basic electronic components now, and the other three kinds of electronic components includes resistance, capacitance and inductance [5].

This paper presents a new kind of extreme learning machine based on memristor activation function. It uses memristance-charge function as the activation function, realizing the mapping from the hidden layer to the output layer. The memristance-charge function simplifies the general activation function, improves the performance of extreme learning machine and achieves the regression and classification of data effectively.

2 Extreme Learning Machine

ELM is actually a machine learning algorithm based on a single hidden layer feed-forward neural networks (SLFNs). The structure of the network is simple, consisting of three layers: the input layer, the hidden layer and the output layer. Its structure is shown in Fig. 1. Different from the other algorithms, the parameters in the ELM do not need to be adjusted according to the errors. The weights between the input layer and the hidden layer and the hidden layer bias are randomly generated, the output layer weights are calculated through the minimum norm least-squares solution.

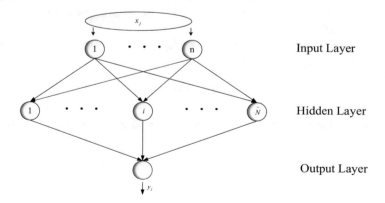

Fig. 1. The structure of the ELM

For N arbitrarily different training sets (x_i, t_i), $x_i = [x_{i1}, x_{i2}, \cdots x_{in}]^T \in R^n$, $t_i = [t_{i1}, t_{i2}, \cdots y_{im}]^T \in R^m$, x_i is the input data vector,t_i is the target vector for each sample. Suppose there are $\tilde{N}$ neuron in the hidden layer, then the output of the ELM is as follows:

$$\sum_{i=1}^{\tilde{N}} \beta_i(w_i \cdot x_j + b_i) = t_j, j = 1, 2 \cdots N \tag{1}$$

Here, $w_i = [w_{i1}, w_{i2}, \cdots w_{in}]^T$ and $\beta_i = [\beta_{i1}, \beta_{i2}, \cdots \beta_{in}]^T$ represent the weights of the input and output layers respectively.b_i and $g(x)$ are bias of the hidden layer and activation function respectively.

Equation (1) can be Simplified as follows:

$$H\beta = T \tag{2}$$

Where, H represents the output matrix of the hidden layer.

$$H = \begin{bmatrix} g(w_1 \cdot x_1 + b_1) & \cdots & g(w_{\tilde{N}} \cdot x_1 + b_{\tilde{N}}) \\ \vdots & \cdots & \vdots \\ g(w_1 \cdot x_N + b_1) & \cdots & g(w_{\tilde{N}} \cdot x_N + b_{\tilde{N}}) \end{bmatrix}_{N \times \tilde{N}} \tag{3}$$

$$\beta = \begin{bmatrix} \beta_1^T, & \beta_2^T, & \cdots, & \beta_{\tilde{N}}^T \end{bmatrix}_{\tilde{N} \times m}^T \tag{4}$$

$$T = \begin{bmatrix} T_1^T, & T_2^T, & \cdots, & T_N^T \end{bmatrix}_{N \times m}^T \tag{5}$$

As mentioned above, the input weights w_i and hidden layer bias b_i are randomly generated and they can be assumed to be known. Then, the network can be considered as a linear system, and the output weights β_i can be obtained by the minimum norm least-squares solution.

The weights of hidden layer and output layer are obtained by solving the least square solution of Eq. (2).

$$\hat{\beta} = H^+ T \tag{6}$$

H^+ is the Moore-Penrose inverse matrix of the hidden layer output matrix.

3   ME-ELM

3.1   Memristor

The memristor is a hypothetical non-linear passive two-terminal electrical component relating electric charge and magnetic flux linkage. The behavior of the memristor can be summarized as follows: its resistance depends on the current following through the device, in other words, it depends on the total charge of the device.

This paper adopts the memristor which is developed by HP laboratory. The charge control model of the HP memristor is described as follows:

$$M(t) = \begin{cases} R_{off}, x(t) < c_1 \\ M(0) + kx(t), c_1 \leq x(t) < c_2 \\ R_{on}, x(t) \geq c_2 \end{cases} \quad (7)$$

Where

$$c_1 = \frac{R_{off} - M(0)}{k} \quad (8)$$

$$c_2 = \frac{R_{on} - M(0)}{k} \quad (9)$$

k is a constant.

$$k = \frac{(R_{on} - R_{off})\mu_v R_{on}}{D^2} \quad (10)$$

D is the thickness of the memristor film device. $M(0)$ is the initial value of the memristance. R_{on} and R_{off} are the limit memeristance when the thickness of the TiO_{2-x} layer is D and 0, respectively. μ_v is the average mobility of oxygen hole. The memristor parameter in this article is set to:

$R_{on} = 100\,\Omega, R_{off} = 20\,k\Omega, \mu_v = 10^{-14} m^2 \cdot s^{-1} \cdot V^{-1}, M(0) = 10\,k\Omega, D = 10\,nm.$

According to the above formula, we can obtained memristor's memristance-charge curve, which is shown as below (Fig. 2):

It is shown that the curve of memristance-charge is a broken line shape. Compared with the commonly used activation function-Sigmod, when we reverse the memristance-charge curve, it very similar with the Sigmod, but because of its characteristics of piecewise linear, the feature mapping is more simple, the speed of algorithm is accelerated.

3.2   The Structure of ME-ELM

Given the training set N, the activation function of a hidden layer $g(\cdot)$, and the numbers of hidden layer's neurons, the training algorithm of the memristor-based extrem learning machine can be summarized as follows:

(1) Randomly assign the input weights w_i and the biases of the hidden layer neurons b_i;

(2) Using the memristance-charge curve as the activation function, calculate the output matrix of hidden layer H.

(3) Calculate the output weight β.

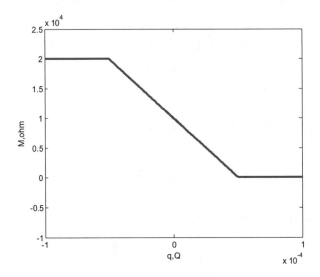

Fig. 2. The memristance-charge curve of memristor

4 Performance Evaluation

4.1 ME-ELM for Regression

To test the regression effect of the ME-ELM, this paper makes a one-dimensional Sinc function regression experiment. The experiment results compared with BP neural network based on LM algorithm (LMBP), ELM and support vector machine (SVM) count in the Table 1.

The expression of Sinc function is

$$f(x) = \begin{cases} \frac{\sin x}{x}, & x \neq 0 \\ 1, & x = 0 \end{cases} \tag{11}$$

Select 5000 data sets $\{x_i, f(x_i)\}$ as training sample sets, $x_i \subset (-10, 10)$. At the same time, select 5000 training sample sets $\{x_i, f(x_i)\}$, x_i is equal interval sampling belonging to $(-10, 10)$.

With the increase of hidden layer nodes, the training time, training error, test error of ME-ELM and ELM are recorded separately on the table. Taking into account the volatility of the test results, each trial was done in 20 groups and then averaged.

As we can see from following table, with the increase of the hidden layer nodes, the training time of the network becomes longer because of the increase of complexity of the system's structure. However, comparing the datas in the table, we can see that when the numbers of hidden layer nodes is less, the training time is very short, which is approximately zero and can be ignored, while the training and test errors of the new ME-ELM are smaller than ELM. When the hidden layer nodes' number is large, the training error and the test error of the two methods are basically remain unchanged, but the ME-ELM's training time is much shorter than ELM. Moreover, ME-ELM achieves stability when the hidden layer node is 15, and the error reaches the minimum.

Table 1. The influence of the number of nodes in the hidden layer on ELM and ME-ELM

Numbers of hidden layer neurons	Training time/s		Training RMS		Testing RMS	
	ELM	ME-ELM	ELM	ME-ELM	ELM	ME-ELM
5	0	0	0.1681	0.1404	0.1218	0.0832
10	0	0	0.1248	0.1170	0.0696	0.0153
15	0.0156	0	0.1162	0.1160	0.0104	0.0073
20	0.0468	0	0.1159	0.1158	0.0073	0.0072
50	0.0312	0.0156	0.1158	0.1158	0.0081	0.0083
100	0.0936	0.0468	0.1158	0.1158	0.0084	0.0083
150	0.1284	0.0780	0.1158	0.1158	0.0085	0.0085
1000	4.0404	3.5100	0.1158	0.1158	0.0085	0.0083

The number of hidden layer nodes is set as 20, and using ME-ELM to do the fitting experiment of Sinc function. The fitting curve is shown in the Fig. 3. The curve predicted by ME-ELM and the real curve are basically consistent, and the error is very small.

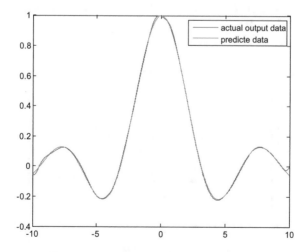

Fig. 3. Sinc function predicted by ME-ELM

The Table 2 gives the performance comparison of the four algorithms: ME-ELM, ELM, SVM and LMBP. The ME-ELM's hidden layer nodes are 15. The ELM's hidden layer nodes are selected as 20. The SVM uses libsvm toolbox. LMBP uses MATLAB's own neural network toolbox function for training. The comparison of the four algorithms is shown in the following table.

Table 2. Comparison of regression performance between four algorithms

Algo	Training time/s	Training RMS	Testing RMS
ME-ELM	0	0.1158	0.0073
ELM	0.0468	0.1159	0.0073
SVM	9.7969	0.1130	0.0064
LMBP	36	0.0011	0.0091

As we can see from the above table, for the one-dimensional Sinc function regression experiment, these four algorithms fit well and the errors are relatively small. The ME-ELM requires only 15 hidden layer nodes to achieve the same effect when the ELM requires 20 hidden layer node. And the ME-ELM is faster.

4.2   ME-ELM for Classification

In order to further verify the effectiveness of the ME-ELM algorithm, we did a simple test of dividing the datas into two categories on MATLAB. The test results of ME-ELM are shown in the following figure. The red and blue dots represent the two class divided by ME-ELM. Similarly, each trial was done in 20 groups and then averaged. Comparing the results of ME-ELM and ELM, the testing accuracy is very high because the data is simple. The error rate of ELM classification is 0.08, while the error rate of ME-ELM is 0.03 (Fig. 4).

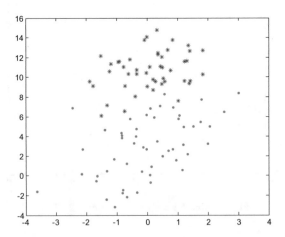

Fig. 4. Classification experiment with ME-ELM

As same as the ELM, ME-ELM can be not only used for the regression and classification of general experimental data, but also for large and complex data in real life. For instance, the prediction of house price, the prediction of forest vegetation types and the classification of diabetes patients. In the classification of diabetes, the learning machine uses various indexes to judge whether the patient is diabetic or not. So it is necessary to improve the judgment accuracy and improve the algorithm time. Next, we take diabetes classification for example, ME-ELM, ELM, LMBP and SVM four algorithms are compared (Table 3).

Table 3. Comparison of classification performance between four algorithms

Algo	Training time/s	Training rate	Testing rate
ME-ELM	0.0156	83.16%	78.65%
ELM	0.0468	81.77%	75.00%
SVM	0.1860	78.45%	77.39%
LMBP	15.281	92.77%	62.81%

From the above table, it can be seen that LMBP has the longest training time and the worst judgment. The accuracy rates of ME-ELM, ELM, and SVM are relatively high, but ME-ELM is better than the other two in terms of accuracy and training time. From this we can conclude that the memristor based extreme learning machine has greatly improved.

5   Conclusion

This paper presents a extreme learning machine based on memristor, making the memristance-charge function as the activation function of the ELM and using the least squares solution to calculate the output weight matrix. When the number of hidden nodes is small, the ME-ELM can reach high accuracy; When the numbers of hidden nodes increases, the experiment time shorten. We give the detailed steps of the algorithm, and simulation experiments prove the good performance of regression and classification.

References

1. Zheng, Y.X.: SVM parameter optimization based on hybrid bionic algorithm. J. Guangxi Norm. Univ. **29**(2), 114–118 (2011)
2. Huang, G.B., Zhu, Q.Y., Siew, C.K.: Extreme learning machine: a new learning scheme of feedforward neural networks. In: IEEE International Joint Conference on Neural Networks, Proceedings IEEE, vol. 2, pp. 985–990 (2005)
3. Chua, L.O.: Memristor-the missing circuit element. IEEE Trans. Circ. Theor. **18**(5), 507–519 (1971)
4. Strukov, D.B., Snider, G.S., Stewart, D.R., Williams, R.S.: The missing memristor found. Nature **453**(7191), 8083 (2008)

5. Pershin, Y.V., Di, V.M.: Spin memristive systems: spin memory effects in semiconductor spintronics. Phys. Rev. B **78**(11), 113309 (2008)
6. Zong, W., Huang, G.B., Chen, Y.: Weighted extreme learning machine for imbalance learning. Neurocomputing **101**(3), 229–242 (2013)
7. Geng, Z.Q., Qin, L., Han, Y.M., et al.: Energy saving and prediction modeling of petrochemical industries: a novel ELM based on FAHP. Energy **122**, 350–362 (2017)
8. Bi, W.C.: A particle swarm optimization based on extreme learning machine. J. Zhengzhou Univ. **45**(1), 100–104 (2013)
9. Zong, W., Huang, G.B.: Face recognition based on extreme learning machine. Neurocomputing **74**(16), 2541–2551 (2011)
10. Huang, G.B., Zhou, H., Ding, X., et al.: Extreme learning machine for regression and multiclass classification. IEEE Trans. Syst. Man Cybern. **42**(2), 513–529 (2012)
11. Liang, N.Y., Huang, G.B., Saratchandran, P., et al.: A fast and accurate online sequential learning algorithm for feedforward networks. IEEE Trans. Neural Netw. **17**(6), 1411–1423 (2006)
12. Samat, A., Du, P., Liu, S., et al.: Ensemble extreme learning machines for hyperspectral image classification. IEEE J. Sel. Top. Appl. Earth Obs. Remote. Sens. **7**(4), 1060–1069 (2014)
13. Bi, H.Y.: The extreme learning machine application in handwritten numeral recognition. Zhengzhou University (2014)
14. Wang, L., Drakakis, E., Duan, S., et al.: Memristor model and its application for chaos generation. Int. J. Bifurc. Chaos **22**(8), 241–252 (2012)
15. Strukov, D.B., Snider, G.S., Stewart, D.R., et al.: The missing memristor found. Nature **453** (7191), 80–83 (2008)
16. Liu, P.: Research on image classification algorithm based on ELM. China Jiliang University (2012)
17. Yaming, Xu, Lidan, Wang, Shukai, Duan: Magnetron TiO2 memristor chaotic system and FPGA hardware implementation. Phys. Sci. J. **65**(12), 62–74 (2016)

Classification of Burden Distribution Matrix Based on ELM

Yanan Liu[1,2], Sen Zhang[1,2(✉)], Yixin Yin[1,2],
Xiaoli Su[1,2], and Jie Dong[1,2]

[1] School of Automation and Electrical Engineering,
University of Science and Technology Beijing, Beijing 100083, China
zhangsen@ustb.edu.cn
[2] Key Laboratory of Knowledge Automation for Industrial Processes of Ministry
of Education, School of Automation and Electrical Engineering,
University of Science and Technology Beijing, Beijing 100083, China

Abstract. The burden distribution matrix is an important means of controlling the burden distribution of blast furnace. In order to provide references to foremen of blast furnaces, a model is established in this paper, which is a relationship model between blast furnace parameters and burden distribution matrix, and is based on the extreme learning machine algorithm (ELM). The model decides if the next burden distribution matrix should be adjusted through a series of blast furnace parameters, such as the gas utilization rate, the blast volume, the blast pressure, the top pressure, the blast velocity, the permeability index, and the utilization coefficient. Finally, compared to other methods used LSSVM and PNN, the method based on ELM is faster and more accurate, so it is more suitable.

Keywords: Blast furnace · Burden distribution matrix
Extreme learning machine · Classification

1 Introduction

During the blast furnace smelting process, when the equipments of blast furnace, the raw materials and the conditions under the blast furnace are determined, the burden distribution matrix is an important means to regulate the distribution of material surface. The burden distribution matrix is usually decided by the experts who make decisions according to the analysis of a series of blast furnace parameters before opening the blast furnace. During the operation of blast furnaces, the change of burden distribution matrix is also determined by furnace foremen and furnace foremen do not change burden distribution matrix regularly according to the blast furnace condition, to control the shape of blast furnace burden surface, improve blast furnace conditions [1]. At present, there are many studies on the relationship between burden distribution matrix and material surface of blast furnace at home and abroad, but there is little research on the relationship between burden distribution matrix and blast furnace parameters and also little research on how to adjust the burden distribution matrix. However, blast furnace operators can not directly obtain the information of blast

© Springer Nature Switzerland AG 2019
J. Cao et al. (Eds.): ELM 2017, PALO 10, pp. 219–229, 2019.
https://doi.org/10.1007/978-3-030-01520-6_20

furnace burden surface, and generally adjust the burden distribution system by learning from the empirical analysis. Because of blast furnace process control system information diversification, uncertainty characteristics and human factors, blast furnace accidents occurred. Therefore, reduce the damage caused by man-made judgments, improve the judgment accuracy of operator and stability of blast furnace operation is a very critical problem.

Burden distribution matrix is both the means to adjust the blast furnace conditions, and equipment work basis of bell-less top blast furnace [2]. Burden distribution matrix is constituted of the angles of rotating chute and rounds numbers of burden distribution. The angle of rotating chute is one of the key factors to form a reasonable surface, which determines the falling track of the materials. Corresponding to different angles, the rounds number of burden distribution is another key parameter. The rounds numbers of burden distribution and weight of materials have directly connection [3]. The following is the expression of burden distribution matrix:

$$\begin{bmatrix} P_1 & \cdots & P_i & \cdots & P_n \\ N_1 & \cdots & N_i & \cdots & N_n \end{bmatrix} \tag{1}$$

$P_1, \cdots P_i \cdots P_n$ are the angles of rotating chute, $N_1, \cdots N_i \cdots N_n$ are the rounds numbers of burden distribution.

Extreme learning machine (ELM) [4] is a simple and effective learning algorithm, which based on the research of single hidden layer feedforward neural network. ELM does not need complicated iteration and optimization in the training process, and it can quickly calculate the global finest solution. ELM has high operation speed, strong generalization ability and will not fall into the local minimum, so it is widely used in various fields [5].

The burden distribution matrix can directly control the shape of the material surface, indirectly adjust the blast furnace parameters. The same if the burden distribution matrix needs changed can be determined by blast furnace parameters. Blast furnace foremen determine the adjustment of the burden distribution matrix by observing the blast furnace conditions in the production. The gas utilization rate, the blast volume, the blast pressure of wind, the top pressure, the blast velocity, the permeability index, and the utilization coefficient are seven important blast furnace parameters. Based on the ELM algorithm, this paper uses the measured data to establish the model of blast furnace parameters and burden distribution matrixes. Compared with the probabilistic neural network (PNN) algorithm and the least squares support vector machine (LS-SVM) algorithm, the model established by ELM algorithm is more suitable to solve the problem of this paper.

2 Brief Introduction of Blast Furnace Parameters and Data Processing

At present, blast furnace environment is complex, and there are numerous parameters. Communicating with blast furnace foremen, then find several parameters which are not only concerned by blast furnace foremen, but also can be detected. They are used as input characteristics. When instruments measure blast furnace parameters, they will inevitably be disturbed by the environment, resulting in data with noise, so the data need de-noised, to reduce errors and improve the accuracy of the model.

2.1 The Relationship Between Burden Distribution Matrix and Blast Furnace Parameters

Burden distribution matrixes can change system of burden distribution, such as the development of center, the development of the edge or the development of center to control the development of edge [6], thus change the shape of the blast furnace burden surface, affect blast furnace parameters, for example: the gas utilization rate, the blast volume, the blast pressure, the top pressure, the blast velocity, the permeability index, the utilization coefficient and so on. Blast furnace parameters can also indirectly reflect the information of surface shape and determine whether the burden distribution matrix needs to be adjusted.

2.2 Data Processed

The blast furnace conditions can determine whether the burden distribution matrix needs to be adjusted, but the change of furnace conditions is difficult to predict before burden distribution, it is irregular, so the parameters belong to non-stationary signals. The contained noise in the blast furnace parameters do not belong to smooth white noise [7]. So the data need processed before classification. First, remove the null value, infinite value, special symbols and other obviously wrong samples, then process the data by wavelet transform method.

Wavelet transform is time-frequency analysis about signal with multi-resolution features, it can easily extract the original signal from mixed with a strong noise signal, thereby achieve non-smooth signal de-noising [8]. In this paper, take gas utilization rate data as an example, using wden () function in MATLAB to process the data. The wavelet de-noising quality evaluation index T is used as the composite evaluation index to determine the optimal parameters of the decomposition level and the threshold selection criterion in the wavelet transform [9]. The index T is linear combined by root mean square error and smoothness, by utilization coefficient of variation to determine the weight. The smaller the T value is, the better the effect of de-noising. Figure 1 is the trend of the evaluation index T with the decomposition level, under different threshold selection criteria.

From Fig. 1, it can be seen that when the decomposition level is the same, the T value of the heuristic threshold (heurure) is smaller than the T value of the fixed threshold (sqtwolog), the maximum minimum threshold (minimaxi), and the unbiased estimator (rigreure), so the heuristic threshold (heurure) is chosen as threshold selection criteria. From Fig. 1, it also can be seen that the evaluation index T is basically the same, when the decomposition level is greater than 6 layers, in other words, the de-noising effect is not obvious, so the decomposition level selects 6 layers. In summary, according to the data characteristics MATLAB wavelet toolbox function parameters are set as shown in Table 1.

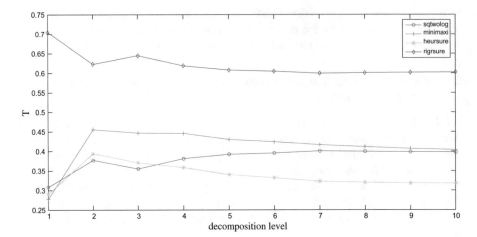

Fig. 1. The trend of the evaluation index T with the decomposition level under different threshold selection criteria

Table 1. The set of parameters

Threshold selection criteria	Threshold usage	Threshold processing changes with noise level	Decomposition level	Wavelet basis function
heursure	Soft threshold(s)	Does not change with noise level (one)	6 layers	Sym8

Figure 2 shows the gas utilization rate data before data processed, Fig. 3 shows the gas utilization rate data after data processed. Because the blast furnace parameters are non-stationary signal before date processed, so there will also be spikes in the signal after de-noising. Other parameters are processed in above method, such as: the blast volume, the blast pressure, the top pressure, the blast velocity, the permeability index, and the utilization coefficient.

3   Burden Distribution Matrix Classification Model Based on ELM

3.1   Extreme Learning Machine (ELM)

As a single hidden layer feed-forward neural network, the network structure of extreme learning machine can be shown in Fig. 4. It can be seen that the ELM includes input layer, hidden layer and output layer. The neurons in the layers of the network are connected in a fully connected manner. The specific idea of the algorithm is as follows.

Suppose given Q independent training samples $\{(X, T)|X = [x_1, x_2, \ldots, x_Q],$ $T = [t_1, t_2, \ldots, t_Q].\}$, $t_i = [t_{i1}, t_{i2}, \ldots, t_{im}]^T \in R$ stands for the target outputs

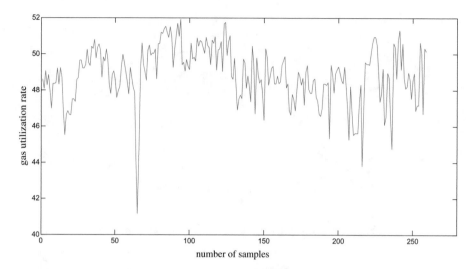

Fig. 2. Raw data of gas utilization rate

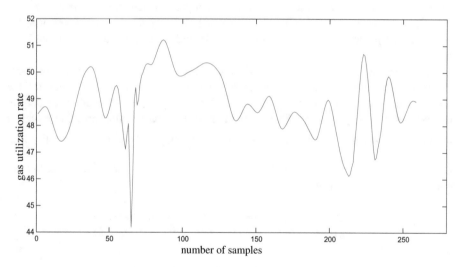

Fig. 3. Processed data of gas utilization rate

corresponding to the i th sample, and the output has m characteristics, $x_i = [x_{i1}, x_{i2}, \ldots, x_{in}]^T$ represents the i th sample with n characteristics. When the hidden layer has L nodes, the ELM network structure can be expressed as

$$y_j = \sum_{i=1}^{L} \beta_i g(\omega_i x_j + b_i), j = 1, 2, \ldots, Q \tag{2}$$

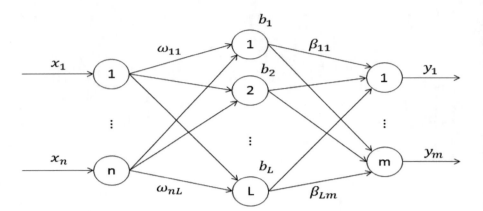

Fig. 4. The network structure of extreme learning machine

$g(\omega_i x_j + b_i)$ stands for the activation function of hidden layer, which is selected from sig, sin, hardlim and radbas generally; $\omega_i = [\omega_{i1}, \omega_{i2}, \ldots, \omega_{iL}]^T, i = 1, 2, \ldots, n$ stands for weight matrix of the hidden layer, w_{ij} represents the weight between the i th input layer neurons and the j th hidden layer neurons; b_i stands for the bias vector of the hidden layer; β stands for the weight between the hidden neurons and the output layer neurons.

Compared with the traditional neural network, the traditional neural network needs to adjust the input weight and the deviation value of hidden layer several times, but the input weight vector and the bias vector of the hidden layer are generated by random in the extreme learning machine. So the essential of extreme learning machine model is found a least squares solution β: $\left\|H\hat{\beta} - T\right\| = \min_{\beta}\|H\beta - T\| = 0$ of linear function $H\beta = T$. In other words $\hat{\beta} = H^{-1}T$, $H^{-1} = H^T(HH^T)^{-1}$ is the generalized inverse matrix of H. H is expressed as

$$H = \begin{bmatrix} g(w_1 x_1 + b_1) & g(w_2 x_1 + b_2) & \cdots & g(w_L x_1 + b_L) \\ g(w_1 x_2 + b_1) & g(w_2 x_2 + b_2) & \cdots & g(w_L x_2 + b_L) \\ \vdots & \vdots & \ddots & \vdots \\ g(w_1 x_Q + b_1) & g(w_2 x_Q + b_2) & \cdots & g(w_L x_Q + b_L) \end{bmatrix} \tag{3}$$

When solve the practical problems, the samples may have a multicollinearity problem, resulting in that HH^T is a singular matrix, Huang et al. [10] solved the problem by introducing a regular parameter C, eliminating the effect of matrix dissatisfaction on the classification results. The output expression of the model can be expressed as

$$y(x) = g(x)\beta = g(x)H^T\left(\frac{I}{C} + HH^T\right)^{-1} T \tag{4}$$

3.2   ELM Model

Whether the burden distribution matrix needs to be adjusted, that is a two classification problem essentially. Combined with the actual experience, seven important blast furnace parameters are chosen as input variables of input samples, they are the gas utilization rate, the blast volume, the blast pressure, the top pressure, the blast velocity, the permeability index, and the utilization coefficient. Whether the burden distribution matrix needs to be adjusted is the output variable. The specific steps are as follows:

(1) Make sure the input layer has 7 nodes and the output layer has a output node;
(2) The activation function is selected from several common activation functions such as sigmoid, sin, hardlim, radbas, and so on;
(3) Randomly generate the weight matrix w and the bias vector b;
(4) Enter the training samples, establish the network model, and calculate the hidden layer output matrix of the network;
(5) Get the output weight $\hat{\beta} = \boldsymbol{H}^T \left(\frac{I}{C} + \boldsymbol{H}\boldsymbol{H}^T\right)^{-1} T$;
(6) According to the output expression $Y = g(x)\beta = g(x)\boldsymbol{H}^T \left(\frac{I}{C} + \boldsymbol{H}\boldsymbol{H}^T\right)^{-1} T$, find out the output vector of the model.

4   Experimental Design and Result Analysis

This paper divides data into two groups according to whether the burden distribution matrixes need to be adjusted and labels, "0" means that the burden distribution matrixes do not need to be adjusted, "1" means that the burden distribution matrixes need to be adjusted. Select 256 groups actual measurement data, of which 51% belong to "1" class, 49% belong to "0" class. 80% of the data are used to train the model, and 20% of the data are used to be tested. The partially processed sample data and their classification conditions are shown in Table 2.

The training sample set and the testing sample set are randomly selected, In order to reduce the effect and increase the reliability of the forecast, 10 times 5 fold cross validation method is used to calculate the speed of the algorithm, the accuracy and the value of F1_score. The following describes the operation of 10 times 5 fold cross

Table 2. The partially processed sample data and their classification conditions

	Sample 1	Sample 2	Sample 3	Sample 4	Sample 5	Sample 6
Gas utilization rate	50.67	48.42	50.34	49.56	48.47	48.48
Blast volume	4122.69	3702.84	3742.64	4171.96	4421.40	4344.23
Blast pressure	329.18	291.74	298.12	315.76	329.07	328.85
Top pressure	173.14	155.12	155.33	165.03	179.76	180.21
Blast velocity	256.79	234.14	245.63	267.40	256.49	258.03
Permeability index	0.80	0.77	0.76	0.74	0.72	0.72
Utilization coefficient	2.12	2.10	2.24	2.27	2.14	2.05
Category	0	0	0	1	1	1

validation method. Firstly, the "0" and "1" sample data are divided into five parts. Secondly, take one part of "0" samples and one part of "1" samples as test set and the remaining four parts were taken as train set in turns, the mean of the five results is treated as the estimation of the algorithm accuracy [11]. Finally, perform 10 times 5 fold cross validation, then find the mean of results. This method can further improve the running speed, the measurement accuracy and the accuracy of F1_score.

4.1 Classification Model Based on ELM

Summarize the contents of the last section, we can see that the key to establish the ELM classification model is the determination of parameters. Whether the parameter setting is reasonable, that directly affects the prediction effect of the model. The important parameters in the ELM algorithm include the weight matrix of the hidden layer, the bias vector of the hidden layer and the output weight. However, the determination of the number of hidden layer nodes and the selection of the excitation function also need to determine by the algorithm speed, classification accuracy and robustness.

There are many evaluation indexes of ELM classification model, such as accuracy, precision, recall and F1-score [11]. Taking into account the F1-score contain the precision and recall, therefore, this paper determines the number of hidden layer nodes and the activation function by comparing the F1 scores of the model. F1-score is calculated as follows

$$F_1\text{- score} = \frac{2TP}{2TP + FP + FN} \tag{5}$$

In this model, TP represents the number of samples that correctly predictions for "0"; FP represents the number of samples that belong to "1" but are expected to be "0"; FN represents that the number of samples that belong to "0" but are expected to be "1". The larger the F1-score value is, the better the classifier is. Figure 5 shows the trend of the F1-score with the number of hidden layer nodes under different activation functions.

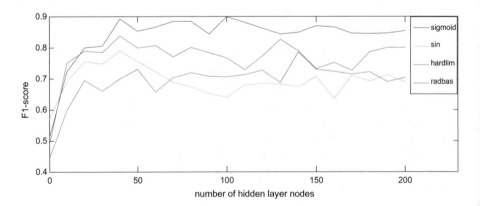

Fig. 5. F1-score trends with the nodes

From Fig. 5, it can be seen that when the number of hidden layer nodes is constant, the F1-score value of sigmoid function is higher than the F1-score value of other activation function, and many experiments have shown that the training time of sigmoid function, sin function, hardlim function and radbas function is about the same, so the sigmoid function is used as the activation function in this paper. Since the value of F1-score is basically stable after 50 nodes, comprehensive consideration, the number of hidden layer nodes is set as 60. The test results of 10 times 5 fold cross validation are shown in Fig. 6. The horizontal axis is the number of training, and the vertical axis is the F1-score value. The average of 10 times can be obtained F1-score was 0.86, and testing accuracy was 86%.

Fig. 6. The test results of 10 times 5 fold cross validation

4.2   Algorithm Comparison

There are many classification algorithms, in order to further prove the ELM model more suitable for the data, network models are established based on least squares support vector machine algorithm (LS-SVM) [12] and probabilistic neural network (PNN) [13]. The comparison results of accuracy, F1-score, training time and testing time are shown in Table 3.

Table 3. The comparison results of accuracy, F1-score, training time and testing time

	ELM	LS-SVM	PNN
Accuracy	86%	74%	73%
F1-score	0.86	0.75	0.70
Training time	0.0938	0.3281	0.73
Testing time	0.0045	0.0313	0.0781

Table 3 shows that the accuracy and F1-score of the ELM are higher than those of the LS-SVM model and the PNN model. And ELM model is faster than LS-SVM model and PNN model about training time and testing time. The ELM model is superior to the PNN model and LS-SVM model in the accuracy rate or the F1-score value. Therefore, considering the measured samples in this paper, the classification model based on the ELM algorithm is faster and more accuracy compared to PNN classification model and LS-SVM classification model. ELM algorithm model has a good learning ability and predictive ability, more suitable to solve the problem of burden distribution matrix classification.

5   Summary and Prospect

The determination of the burden distribution matrix is the key to achieve blast furnace automation and can reduce the unnecessary losses caused by some human factors. This paper is divided into three parts. First of all, the wavelet de-noising is carried out on the gas utilization rate, the blast volume, the blast pressure, the top pressure, the blast velocity, the permeability index and the utilization coefficient. Then, building classi-fication model based on the extreme learning machine, which can divide the burden distribution matrix into two categories, one is need to be adjusted, the other is not need to be adjusted, according to the gas utilization rate, the blast volume, the blast pressure, the top pressure, the blast velocity, the permeability index and the utilization coeffi-cient. Finally, the same function classification models based on the least squares support vector machine (LS-SVM) algorithm and probabilistic neural network (PNN) are created. Comparing with these models, the accuracy and F1-score of the ELM algorithm are relatively higher; the test time and train time is shorter; less model parameters are set in the model establishment process. It illustrates that the advantages of ELM model and proves the credibility of the classification results.

Acknowledgments. This work was supported by the National Natural Science Foundation of China (NSFC Grant No. 61333002 and No. 61673056).

References

1. Liu, Y.C.: Rules of Burden Distribution in Blast Furnace, 4th edn. Metallurgical Industry Press, Beijing (2012)
2. Yu, Z.J., Chen, L.K., Yang, T.J., Zuo, H.B., Guo, H.W.: New development of smelting expert system in No. 1 of Wuhan Steel. Res. Iron Steel
3. Zhu, Q.T., Cheng, S.S., Wei, Z.J., Guo, X.B.: Determination of the drop point of the charge. China Metall. **09**, 24–26 (2006)
4. Huang, G.B., Zhu, Q.Y., Siew, C.K.: Extreme learning machine: a new learning scheme of feedforward neural networks. In: IEEE International Joint Conference on Neural Networks, Proceedings, vol. 2, pp. 985–990 (2004)
5. Deng, C.W., Huang, G.B., Xu, J., Tang, J.X.: Extreme learning machines: new trends and applications. Sci. China (Inf. Sci.) **02**, 5–20 (2015)

6. Fu, S.G.: Development and application of blast furnace fabric matrix in Xinyu iron and steel. Iron-Mak. **06**, 31–34 (2013)
7. Gao, C.H., Jian, L., Chen, J.M., Sun, Y.X.: Data-driven modeling and predictive algorithm for complex blast furnace iron-making process. J. Autom. **35**(6), 725–730 (2009)
8. Chen, F., Cheng, X.M.: Signal de-noising technology based on wavelet transform and its implementation. Mod. Electron. Technol. **03**, 11–13 (2005)
9. Zhu, J.J., Zhang, Z.T., Kang, C.L., et al.: A reliable wavelet de-noising quality evaluation index. J. Wuhan Univ. Inf. Sci. Ed. **40**(5), 688–694 (2015)
10. Huang, G.B., Zhou, H., Ding, X., et al.: Extreme learning machine for regression and multiclass classification. IEEE Trans. Syst., Man, Cybern., Part B (Cybern.) **42**(2), 513–529 (2012)
11. Lo, M.J.: Evaluation method of classification algorithm based on ROC. Wuhan University of Science and Technology (2005)
12. Zhou, X.: Implementation and comparison of two classifiers based on LS-SVM algorithm. Comput. Knowl. Technol. **7**(29), 7281–7283 (2011)
13. Su, L., Song, X.D.: Realization and application of probabilistic neural network based on MATLAB. Comput. Mod. **11**, 47–50 (2011)

Hierarchical Pruning Discriminative Extreme Learning Machine

Tan Guo$^{(\boxtimes)}$, Xiaoheng Tan, and Lei Zhang

College of Communication Engineering,
Chongqing University, Chongqing, China
{tanguo,txh,leizhang}@cqu.edu.cn

Abstract. The extreme learning machine (ELM) provides efficient unified solutions for generalized single hidden layer feed-forward neural networks. Hierarchical learning based on ELM has now attracted lots of interests. This paper presents a hierarchical pruning discriminative ELM (H-PDELM) for feature learning and classification. The ELM pruning auto-encoder (ELM-PAE) is developed for unsupervised feature learning by promoting the output weights matrix to be row-sparse based on $l_{2,1}$-norm regularization. ELM-PAE can naturally distinguish and prune useless neurons in hidden layer to determine the structure of AE. Besides, we learn a flexible output weights matrix for supervised feature classification by relaxing the strict regression label matrix of ELM into a slack one for better generalization performance. H-PDELM performs layer-wise unsupervised feature learning using ELM-PAE, and conducts decision making by the flexible output weights matrix. The network of H-PDELM is compact with good generalization ability. Preliminary experiments on visual dataset show its effectiveness.

Keywords: Extreme learning machine (ELM) · Hierarchical learning
Auto-encoder · Neuron pruning · Label relaxation

1 Introduction

As an efficient learning algorithm for single-hidden-layer feed-forward neural networks (SFNNs), Extreme Learning Machine (ELM) has gained increasing interests in both theoretical research and practical applications [1–3]. Different from the other traditional learning algorithms, e.g., back propagation (BP)-based neural networks (NNs), a remarkable characteristic of ELM is that the parameters of hidden layers are randomly generated independent of training samples without fine-tuning. The parameters that need to be optimized are the connections (or weights) between the hidden layer and the output layer. The optimization of output weights matrix can be analytically solved, and closed-form solution is achieved. Theoretical studies have proven the learning capability, universal approximation capability, and generalization bound of ELM [2–4]. Also, ELMs have demonstrated to have excellent learning accuracy and speed in a variety of applications, such as semi-supervised and unsupervised learning [5], dimension reduction [6], domain adaptation [7, 8] and cost-sensitive learning [9].

© Springer Nature Switzerland AG 2019
J. Cao et al. (Eds.): ELM 2017, PALO 10, pp. 230–239, 2019.
https://doi.org/10.1007/978-3-030-01520-6_21

In recent years, lots of efforts have been made to exploit deep neural networks for representation learning [10]. ELMs are originally developed as an effective solution for single hidden layer shallow networks, which mainly focus on classification or regression tasks not for feature learning [2]. However, the performance of machine learning methods heavily depends on the features they have used. A favorable feature can uncover the underlying information of the observed data and intensely facilitate the progress of machine learning methods. Recent studies have demonstrated that NNs composed of multiple layers can learn representations of data with multiple levels of abstraction, which have remarkably improved the state-of-the-art in some computer vision problems [10]. Specifically, Kasun et al. [6] proposed a multilayer ELM (ML-ELM) architecture using ELM-based auto-encoder (ELM-AE) as its constructing unit. ML-ELM performs layer-wise unsupervised learning, and stacks on top of ELM auto-encoder to get a multilayer neural network. However, the encoded outputs are directly fed into the last layer without random feature mapping, which might not well exploit the advantages of universal approximation capability of ELM. Differently, Tang et al. [11] developed a new hierarchical ELM (H-ELM) by bridging self-taught feature extraction and supervised feature classification with random generated hidden weights. The method can be seen as a fusion of ELM-based sparse auto-encoder and the classic ELM, and outperforms the original ELM in many applications.

There are some problems should be noted in H-ELM. The ELM-based sparse auto-encoder in H-ELM can generate sparse and meaningful hidden features with l_1-norm penalty. There are three alternative network structures to represent the input features. As shown in Fig. 1, they are compressed, dimension invariant, and sparse representations. So the question arises: "How to choose the appropriate structure of ELM-based AE?" The problem can be transformed into an equivalent one that how to determine the number of neurons in hidden layer of AE. Meanwhile, there is little freedom in calculating the output weights matrix with strict regression labels matrix under classic ELM framework.

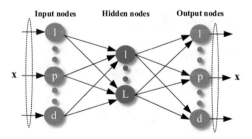

Fig. 1. Illustration for three structures of ELM-based AE. d > L, compressed representation. d = L, dimension invariant representation. d < L, sparse representation.

Focusing on these issues, we develop a multilayer neural network training scheme named hierarchical pruning discriminative extreme learning machine (H-PDELM). H-PDELM perform feature learning and classification in the same network. Each hidden layer weights matrix is calculated in a successive manner. The hierarchically computed outputs are then randomly projected for final supervised classification. To support this, we develop an ELM-based pruning auto-encoder (ELM-PAE) by introducing $l_{2,1}$-norm regularization into classic ELM-based auto-encoder (ELM-AE) for unsupervised representation learning. PELM-AE can automatically determine the appropriate number of hidden nodes by pruning useless ones. The structure of AE can be determined with little human intervention. Besides, we aim to learn a flexible output weights matrix to enlarge the margins between different classes as much as possible. To this end, we relax the strict regression label matrix into a slack one by introducing a non-negative label relaxation matrix, which can provide more freedom for the obtained output weights matrix. In brief, H-PDELM can automatically determine the number of hidden nodes required in each hidden layer, and the flexible output weights matrix is expected to have better generalization ability than classical ELM. In summary, we highlight the characteristics of the proposed method as follows:

(1) With ELM theory, we propose a new hierarchical learning scheme named hierarchical pruning discriminative extreme learning machine (H-PDELM). H-PDELM contains unsupervised feature learning followed by supervised feature classification.
(2) We develop an ELM pruning auto-encoder (ELM-PAE) for unsupervised feature learning by introducing $l_{2,1}$-norm regularization. ELM-PAE can naturally determine structure of AE for a compact but light network.
(3) For supervised feature classification, we aim to learn a flexible output weights matrix by relaxing the strict regression label matrix into a slack one for better generalization performance.

The remainder of this paper is organized as follows. In Sect. 2, we introduce the basic ELM and the hierarchical ELM (H-ELM). In Sect. 3, we present our hierarchical pruning discriminative extreme learning machine (H-PDELM). In Sect. 4, we report the experiments on visual data classification. Conclusions are drawn in Sect. 5.

2 ELM and Hierarchical ELM

For N arbitrary distinct samples $\{(\mathbf{x}_i, t_i)$, where $\mathbf{x}_i \in \Re^{\mathbf{D}}, t_i \in \Re^{\mathbf{m}}$, $i = 1,2\ldots N\}$. ELM is proposed for SLFNs and output function of ELM for SLFNs is

$$f_L(\mathbf{x}) = \sum_{i=1}^{L} \beta_i h_i(\mathbf{x}) = \mathbf{h}(\mathbf{x})\beta \tag{1}$$

Traditional learning algorithms proposed for feed-forward neural networks do not consider the generalization performance when they are proposed first time. ELM aims to reach better generalization performance by reaching both the smallest training error and the smallest norm of output weights [1].

$$argmin_{\beta \in \Re^{L \times d}} \frac{1}{2} \parallel \mathbf{H}\beta - \mathbf{X} \parallel^2 + \lambda \parallel \beta \parallel^2 \tag{2}$$

where $\mathbf{H}$ is the hidden layer output matrix (randomized matrix) for N training samples

$$\mathbf{H} = \begin{bmatrix} \mathbf{h}(\mathbf{x}_1) \\ \vdots \\ \mathbf{h}(\mathbf{x}_N) \end{bmatrix} = \begin{bmatrix} h_1(\mathbf{x}_1) & \cdots & h_L(\mathbf{x}_1) \\ \vdots & \vdots & \vdots \\ h_1(\mathbf{x}_N) & \cdots & h_L(\mathbf{x}_N) \end{bmatrix}_{N \times L}$$

and $\mathbf{T}$ is the training data target matrix:

$$\mathbf{T} = \begin{bmatrix} \mathbf{t}_1^{\mathrm{T}} \\ \vdots \\ \mathbf{t}_1^{\mathrm{T}} \end{bmatrix} = \begin{bmatrix} t_{11} & \cdots & t_{1m} \\ \vdots & \vdots & \vdots \\ t_{N1} & \cdots & t_{Nm} \end{bmatrix}_{N \times m}$$

The optimal solution is given by

$$\beta = \mathbf{H}^\dagger \mathbf{T} \tag{3}$$

where $\mathbf{H}^\dagger$ denotes the Moore–Penrose generalized inverse of matrix $\mathbf{H}$.

Extreme learning machine (ELM) is initially proposed for single hidden layer feed-forward neural networks. However, due to its shallow architecture, feature learning using ELM may not be effective to process natural data even with a large number of hidden nodes [11]. It is indicated that deep neural network composing of multiple layers are actually representation-learning methods with multiple levels of representation, and can approximate very complex functions efficiently [10]. For ELM-based hierarchical learning, Tang et al. [11] proposed a framework named Hierarchical ELM (H-ELM). H-ELM can be constructed by two parts: (1) unsupervised hierarchical feature representation by sparse ELM-based auto-encoder and (2) supervised feature classification by classic ELM. For the former part, sparse ELM-based auto-encoder is successively constructed to extract multilayer features of the input data. The formulation of sparse ELM-AE is

$$argmin_{\beta \in \Re^{L \times d}} \frac{1}{2} \parallel \mathbf{H}\beta - \mathbf{X} \parallel_F^2 + \lambda \parallel \beta \parallel_1 \tag{4}$$

The ELM-based sparse auto-encoder in H-ELM can learn sparse and meaningful hidden features. The encoded hierarchical features will be randomly mapped and then fed into the last layer for decision making like classic ELM. The introduction of random mapping before decision making can well utilize the universal approximation ability of ELM [13].

3 Hierarchical Pruning Discriminative ELM

This section will present our hierarchical pruning discriminative ELM (H-PDELM). H-PDELM contains two components, i.e., hierarchical unsupervised feature learning followed by supervised feature classification. For the former part, we develop a variation of ELM-based AE termed ELM-based pruning auto-encoder (ELM-PAE) by using $l_{2,1}$-norm regularization. Given training data $\mathbf{x}$, the output function of ELM based AE is formulated as

$$\mathbf{x} = \sum_{i=1}^{L} \boldsymbol{\beta}_i h_i(\mathbf{x}) = \mathbf{h}(\mathbf{x})\boldsymbol{\beta} \tag{5}$$

where $\boldsymbol{\beta} = [\boldsymbol{\beta}_1, \boldsymbol{\beta}_2 \ldots \boldsymbol{\beta}_L]^{\mathrm{T}} \in \Re^{L \times d}$ is the output weight matrix connecting the hidden layer and the output layer. If $\boldsymbol{\beta}_i$ ($i = 1, 2 \ldots L$) shrinks to be $\mathbf{0}$, the corresponding output $h_i(\mathbf{x})$ of i-th neuron will have no contribution for the final output. Thus, the i-th neuron can be pruned with little influence as shown in Fig. 2. This way, the output weights matrix $\boldsymbol{\beta}$ will be endowed with the function of hidden neurons selection by imposing row-sparse constraint. The sparsity of $\boldsymbol{\beta}$'s rows can be approximately measured by the sum of l_2-norm of $\boldsymbol{\beta}$'s rows. With the motivation, we have the following model

$$argmin_{\boldsymbol{\beta} \in \Re^{L \times d}} \frac{1}{2} \parallel \mathbf{H}\boldsymbol{\beta} - \mathbf{X} \parallel_F^2 + \lambda \sum_{i=1}^{L} \parallel \boldsymbol{\beta}_i \parallel_2 \tag{6}$$

where $\boldsymbol{\beta}_i$ is the i-th row of $\boldsymbol{\beta}$. Due to the fact that $\sum_{i=1}^{L} \sqrt{\sum_{j=1}^{m} \boldsymbol{\beta}_{ij}^2} = \sum_{i=1}^{L} \parallel \boldsymbol{\beta}_i \parallel_2 = \parallel \boldsymbol{\beta} \parallel_{2,1}$, the objection function (6) can be rewritten as

$$argmin_{\boldsymbol{\beta} \in \Re^{L \times d}} \frac{1}{2} \parallel \mathbf{H}\boldsymbol{\beta} - \mathbf{X} \parallel_F^2 + \lambda \parallel \boldsymbol{\beta} \parallel_{2,1} \tag{7}$$

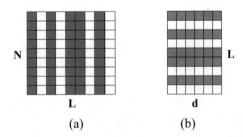

(a)   (b)

Fig. 2. An example for row-sparse output weight matrix $\boldsymbol{\beta}$. (a) is the hidden layer output matrix $\mathbf{H} \in \Re^{N \times L}$. The columns of $\mathbf{H}$ in blue are the selected hidden neurons for each sample. (b) is the obtained row-sparse output weight matrix $\boldsymbol{\beta} \in \Re^{d \times L}$, some rows of which shrink to be 0 (in blank), and therefore $\boldsymbol{\beta}$ is endowed with the ability of hidden neurons selection.

We take the derivation of the objective function and set it to be **0**, there is

$$\mathbf{H}^{\mathrm{T}}(\mathbf{H}\boldsymbol{\beta} - \mathbf{X}) + \lambda\boldsymbol{\Sigma}\boldsymbol{\beta} = 0 \tag{8}$$

where $\boldsymbol{\Sigma}$ is a diagonal matrix in $\Re^{L\times L}$ with the i-th diagonal component $\boldsymbol{\Sigma}_{ii}$ as $1/\parallel \boldsymbol{\beta}_i \parallel_2$. We can get the optimal solution as

$$\boldsymbol{\beta} = \left(\mathbf{H}^{\mathrm{T}}\mathbf{H} + \lambda\boldsymbol{\Sigma}\right)^{-1}\mathbf{H}^{\mathrm{T}}\mathbf{X} \tag{9}$$

Since $\boldsymbol{\Sigma}$ is related with $\boldsymbol{\beta}$, an iterative optimization method can be developed by acquiring $\boldsymbol{\Sigma}$ from $\boldsymbol{\beta}$ in former t-th iteration to update $\boldsymbol{\beta}$ of current $(t + 1)$-th iteration until the convergence condition is reached.

$$\boldsymbol{\beta}^{t+1} = \left(\mathbf{H}^{\mathrm{T}}\mathbf{H} + \lambda\boldsymbol{\Sigma}^t\right)^{-1}\mathbf{H}^{\mathrm{T}}\mathbf{X} \tag{10}$$

The obtained $\boldsymbol{\beta}$ is row-sparse, and the l_2-norm of rows of $\boldsymbol{\beta}$ indicates the importance of corresponding hidden neuron in information transmission. Therefore, one can prune the hidden neurons below certain threshold, and preserve the outputs of remaining hidden neurons. The output weights matrix will also be updated accordingly (Fig. 3).

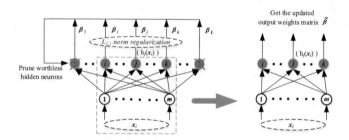

Fig. 3. The structure of ELM-based pruning auto-encoder (ELM-PAE). **Left**: An ELM-based auto-encoder is trained using $l_{2,1}$-norm regularization to get a row-sparse output weight matrix. **Right**: a more compact network is obtained by pruning hidden neurons of little value based on the row-sparse output weights matrix.

For the latter supervised feature classification part, we propose to learn a flexible output weights matrix to fully exploit the discriminative information in data. One problem for classic ELM is that the calculated output weight matrix $\boldsymbol{\beta}$ has little freedom when transforming the hidden layer output $\mathbf{H}$ into a strict label matrix. The design of strict regression targets is unfavorable to enhance the generalization capability of obtained network. A flexible $\boldsymbol{\beta}$ is desired to enlarge the margins between different classes as much as possible. In view of this goal, a technique called ε-dragging is introduced to force the regression targets of different classes moving along opposite directions such that the distances between classes can be enlarged [12]. Figure 4 shows the idea. For a two-class classification problem, the original regression label vector for

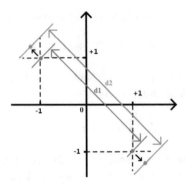

Fig. 4. An illustration for flexible output weights matrix learning. The regression label vectors of classic ELM are denoted as red points. After dragged to the position of blue ones, the distance between the regression vectors are enlarged. A flexible outputs weights matrix can be got to fully exploit the discriminative information in data for better generation performance.

samples of two classes are coded as [+1, −1] and [−1, +1] respectively. As shown in Fig. 4, the maximal distance between the two red regression labels is fixed as $2\sqrt{2}$. If we drag the two red regression vectors to the position of blue ones with proper dragging direction, the maximal distance between the two regression labels can be enlarged, that is, d2 > d1. This way, the learned β is expected to have better generalization ability since it can fully discover the discriminative information of training data. With this in mind, we have the following optimization model

$$argmin_{\beta,M \geq 0} \parallel H\beta - (T + B \odot M) \parallel_F^2 + \lambda \parallel \beta \parallel_F^2 \tag{11}$$

where the original label matrix is

$$T = \begin{bmatrix} t_1^T \\ \vdots \\ t_1^T \end{bmatrix} = \begin{bmatrix} t_{11} & \cdots & t_{1m} \\ \vdots & \vdots & \vdots \\ t_{N1} & \cdots & t_{Nm} \end{bmatrix}_{N \times m}$$

and $t_i = [-1 \ldots + 1 \ldots - 1]^T \in \Re^m$. Only the j-th element of t_i is +1, otherwise −1, which means the sample x_i comes from the j-th class. In (10), we initialize matrix B as T to record the dragging directions. Nonnegative matrix M is used to record the dragging values for each element in original label matrix T. $T + B \odot M$ is the relaxed regression targets. Model (10) can be solved in an iterative way by updating one variable with another one fixed until convergence [12].

In sum, the complete learning procedure for proposed H-PDELM model is shown in Fig. 5. The connecting weights matrix in the first two hidden layers of H-PDELM are successively calculated and stacked for high-level feature representation using ELM-PAE. The encoded hierarchical representations are then utilized for flexible output weights matrix calculation. The numbers of neurons in the hidden layers (expect the last hidden layer) are automatically determined by the developed ELM-PAE, and

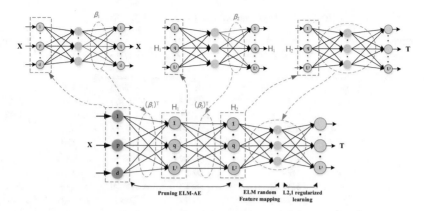

Fig. 5. The learning procedure of H-PDELM with three hidden layers. The weights matrix for the first two hidden layers of H-PDELM are successively calculated and stacked for high-level representation using ELM-PAE. The encoded hierarchical representation will then be utilized for final flexible output weights matrix calculation.

the flexible output weights matrix is expected to have better generalization ability than original ELM. As a result, H-PDELM network is compact with good generalization ability.

4   Experiments

In this section, we conduct experiments to show the performance of proposed H-PDELM. Several related methods are compared including ELM, DELM, PELM, H-ELM and proposed H-PDELM. Among them, DELM is formulated in (10) to learn a flexible output weights matrix, and PELM is developed by introducing $l_{2,1}$-norm regularization into ELM in a similar way with ELM-PAE. PELM can also prune useless hidden neurons for a more compact network. Preliminary experiments are conducted on COIL20 object dataset. The COIL20 dataset contains 20 objects. The images of each object were taken 5 degrees apart as the object is rotated on a turntable and each object has 72 images. The size of each image is 32×32 pixels, with 256 gray levels per pixel. Thus, each image is represented by a 1024-dimensional vector. We randomly select $i(5,10,20,$ and $30)$ samples from each class for training the network and the rest samples for testing. For each given image, we perform 10 times to randomly choose the training set.

Experimental results on the dataset are given in Table 1. For single hidden layer ELM models, classic ELM is fast in both training and testing. From the results, we observe that DELM outperforms ELM and PELM can achieve comparing performance with less hidden neurons and testing time. Besides, multilayer ELM based models such as H-ELM and H-PDELM, which construct deep neural network for feature learning, always outperform the single layer models. H-ELM and H-PDELM. Comparing with H-ELM, H-PDELM tends to perform better in testing accuracy and testing time. The reasons are twofold. Firstly, our ELM-PAE can automatically determine the structure

Table 1. Experimental results on COIL20 dataset

# samples per class	Methods	#hidden nodes in hidden layer	Testing Accuracy (%)	STD	Training time	Testing time
5	ELM	1600	81.58	1.3956	0.1536	0.2349
	DELM	1600	82.72	1.8666	4.2935	0.2282
	PELM	898.2	81.51	1.5787	9.9117	0.1836
	H-ELM	600-600-1600	85.89	2.3778	2.3214	0.3213
	H-PDELM	566.3-476.5-1600	87.07	1.6392	7.6250	0.2918
10	ELM	1600	88.19	1.3876	0.2309	0.2476
	DELM	1600	89.11	1.4128	5.0253	0.2261
	PELM	1206.2	88.12	1.4316	9.3096	0.2063
	H-ELM	600-600-1600	93.84	1.6607	2.5488	0.3194
	H-PDELM	600-599.8-1600	94.94	1.5863	8.1686	0.3139
20	ELM	1600	94.69	0.8441	0.2425	0.2027
	DELM	1600	94.87	0.6102	5.2622	0.2057
	PELM	1333	94.98	0.9517	9.8576	0.1884
	H-ELM	600-600-1600	98.47	0.8783	2.7969	0.2618
	H-PDELM	595.5-503.8-1600	98.76	0.8152	9.3244	0.2465
30	ELM	1600	97.46	0.6046	0.2955	0.1604
	DELM	1600	97.30	0.5021	5.7550	0.1673
	PELM	1455.6	97.70	0.4697	10.9133	0.1721
	H-ELM	600-600-1600	99.83	0.3234	2.9915	0.2218
	H-PDELM	599-465.6-1600	99.95	0.1506	10.2045	0.2014

of AE network by pruning useless hidden neurons, which will help reduce the network size and testing time. Secondly, the learnt flexible output weights matrix seems to have better generalization performance since it can fully exploit the discriminative information in data. H-PDELM network is compact with good generalization by taking the merits of ELM-PAE and flexible output weights matrix (Fig. 6).

Fig. 6. Example images from COIL20

5   Conclusions

This paper develops a hierarchical learning scheme named hierarchical pruning discriminative ELM (H-PDELM) for representation learning and classification. For representation learning, an ELM based pruning auto-encoder (ELM-PAE) is developed based on $l_{2,1}$-norm regularization. ELM-PAE can automatically determine the structure of the AE network by pruning valueless hidden neurons, which will also lead to a more compact network. Besides, we aim to learn a flexible output weights matrix to enlarge the margins between different classes as much as possible for better generalization ability. H-PDELM adopts ELM-PAE for hidden layer weights matrix learning followed by flexible output weights matrix for supervised feature classification. H-PDELM network seems to be compact with good generalization ability as preliminary experimental verifications show.

References

1. Huang, G., Huang, G.-H., Song, G., Youa, K.: Trends in extreme learning machines: a review. Neural Netw. **61**(C), 32–48 (2015)
2. Huang, G.-B., Zhou, H., Ding, X., Zhang, R.: Extreme learning machine for regression and multiclass classification. IEEE Trans. Syst. Man Cybern. B Cybern. **42**(2), 513–529 (2012)
3. Huang, G.-B., Zhu, Q.Y., Siew, C.K.: Extreme learning machine: theory and applications. Neurocomputing **70**(1–3), 489–501 (2006)
4. Huang, G.-B., Chen, L.: Convex incremental extreme learning machine. Neurocomputing **70**(16–18), 3056–3062 (2007)
5. Huang, G., Song, S., Gupta, J., Wu, C.: Semi-supervised and unsupervised extreme learning machines. IEEE Trans. Cybern. **44**(12), 2405 (2014)
6. Kasun, L.L., Yang, Y., Huang, G.B., et al.: Dimension reduction with extreme learning machine. IEEE Trans. Image Process. **25**(8), 3906 (2016)
7. Zhang, L., Zhang, D.: Robust visual knowledge transfer via extreme learning machine based domain adaptation. IEEE Trans. Image Process. **25**(10), 4959–4973 (2016)
8. Zhang, L., Zhang, D.: Domain adaptation extreme learning machines for drift compensation in E-nose systems. IEEE Trans. Instrum. Meas. **64**(7), 1790–1801 (2015)
9. Zhang, L., Zhang, D.: Evolutionary cost-sensitive extreme learning machine. IEEE Trans. Neural Netw. & Learn. Syst. (99), 1–16 (2015)
10. Lecun, Y., Bengio, Y., Hinton, G.: Deep learning. Nat. **521**(7553), 436–444 (2015)
11. Tang, J., Deng, C., Huang, G.B.: Extreme learning machine for multilayer perceptron. IEEE Trans. Neural Netw. & Learn. Syst. **27**(4), 809–821 (2016)
12. Guo, T., Zhang, L., Tan, X.: Neuron pruning-based discriminative extreme learning machine for pattern classification. Cogn. Comput. 1–15 (2017)
13. Huang, G.-B., Chen, L., Siew, C.-K.: Universal approximation using incremental constructive feedforward networks with random hidden nodes. IEEE Trans. Neural Netw. **17**(4), 879–892 (2006)

Mislabel Detection of Finnish Publication Ranks

Anton Akusok[1(✉)], Mirka Saarela[2], Tommi Kärkkäinen[2],
Kaj-Mikael Björk[3,6], and Amaury Lendasse[4,5]

[1] Arcada University of Applied Sciences, Helsinki, Finland
anton.akusok@arcada.fi
[2] University of Jyväskylä, Jyväskylä, Finland
[3] Arcada University of Applied Sciences, Helsinki, Finland
[4] Department of Mechanical and Industrial Engineering,
The University of Iowa, Iowa City, USA
[5] The Iowa Informatics Initiative, The University of Iowa, Iowa City, USA
[6] Hanken School of Economics, Helsinki, Finland

Abstract. The paper proposes to analyze a data set of Finnish ranks of academic publication channels with Extreme Learning Machine (ELM). The purpose is to introduce and test recently proposed ELM-based mislabel detection approach with a rich set of features characterizing a publication channel. We will compare the architecture, accuracy, and, especially, the set of detected mislabels of the ELM-based approach to the corresponding reference results in [1].

Keywords: ELM · Mislabel detection · Publication channel

1 Introduction

Finland, in the spirit of Norway and Denmark, introduced ranking system for academic publication channels (referring to scientific journals, conference series, book publishers etc.) called as *Jufo* (i.e. "Julkaisufoorumi" in Finnish, "Publication Forum" in English) in 2010, together with the renewed university legislation. The ranking of a publication channel, ranging from 0 (non-peer-reviewed) to 3 (most distinguished academic publication forums), is decided by a specially nominated panel of a particular scientific discipline. These panels decide the rankings based on their academic expertise in regular meetings. Because the rankings are directly linked to the allocated funding of the universities, there has been and is a lot of discussion about the fairness and objectivity of the ranks.

A versatile analysis of the 2015 Jufo-rankings was done in [1]. There, by using association rule mining, decision trees, and confusion matrices with respect to Norwegian and Danish ranks, it was shown that most of the expert-based rankings could be predicted and explained with machine learning methods. Moreover, it was found out that those publication channels, for which the Finnish expert-based rank is higher than the estimated one, are characterized by higher

© Springer Nature Switzerland AG 2019
J. Cao et al. (Eds.): ELM 2017, PALO 10, pp. 240–248, 2019.
https://doi.org/10.1007/978-3-030-01520-6_22

publication activity or recent upgrade of the rank. Hence, the outcomes of the system, the publication ranks, need to be assessed and evaluated regularly and rigorously.

Extreme Learning Machine (ELM), as proposed by Huang et al. [2,3], provides one of the key randomized neural network frameworks [4]. Probabilistic convergence analysis of the technique was provided in [5,6], where the necessity of repeated sampling of the feedforward kernel and the advantage of weight decay (ridge regression) were concluded. Here, to identify possibly mislabeled publication channel ranks, we apply the MD-ELM algorithm described and successfully tested in [7].

The rest of the paper is organized as following. Section 2 introduces the original dataset of *Jufo* rankings. The methodology, Sect. 3, describes the feature extraction process and summarizes the MD-ELM method. Section 4 explains the experimental setup, general prediction performance, and provides the comparison with the previous results in [1]. Section 5 summarizes the findings and describes the future research directions.

2   Data

The data for this study comes from two publicly available databases containing the Finnish publication source information and the actual national publication activity information.

1. *JuFoDB*: database of the Finnish publication forum, *JuFo*[1], which contains all nationally evaluated publication channels. Data was retrieved from this database in February 2015, so it describes the ranking situation after complete reevaluation round by the end of 2014.
2. *JuuliDB*: The publicly accessible database of *Juuli*[2] that contains all publications of Finnish researchers. Each publication channel in *JuFoDB* has a unique *Juuli* ID, through which all Finnish publications in that particular channel can be found. Data was retrieved from this database in September 2015, because only then all published work by the end of 2014 had been checked and included in the repository.

29,443 different publication channels with 33 attributes were retrieved from *JuFoDB* and 107,289 publications from *JuuliDB*. The Finnish expert-based *rank* of each publication channel as well as the Norwegian and Danish expert-based rankings can be obtained directly through the *JuFoDB* and also the three bibliometric indicators from Scopus, that is the *SJR*, the *SNIP* and the *IPP*, are featured. Moreover, through the link to *JuuliDB*, one can directly access the information of all researchers in Finland who have published in the particular channel.

---

[1] Available at http://www.tsv.fi/julkaisufoorumi/haku.php.
[2] Available at http://www.juuli.fi/?&lng=en.

The *panel* variable determines the list of experts[3] who have evaluated the publication channel and decided the Finnish expert-based rank. It basically indicates the research discipline of the publication channel. *Field, MinEdu field, Web of Science fields, Scopus fields* are further variables indicating the discipline of the publication channel. However, multiple linkings are possible for these variables and for some publication channels these linkings are not available at all. But each publication channel is attached to only one panel and the panel information is available for all publication channels except for 6,562 book publishers that have mostly been evaluated as rank 0 [1].

In addition to some more general data, such as the *title, subtitle, website, country of publication, language,* unique identifier (*ID*), *ISSN, Sherpa/Romeo code, starting year,* and *publisher,* the *JuFoDB* also provides information such as *abbreviation, title details, ISBN, DOAJ, end year, continued under the name* and *continued JuFo-rank.* The *evaluation history* provides information about the previous ranks in the system.

Similarly as in [1], the continuous variables are directly utilized as features and the categorical variables are transformed to own binary features for each category. All of the 29,443 publication channels have missing values for at least some of the 33 total variables. Hence, for utilizing all of available data in the analysis, one faces a significant sparsity problem [8]. Since the missing information was discovered as an important predictor of the Finnish expert-based rank in [1], we utilize here all the described variables as features plus for each variable the binary information whether it has an available value. Thus, for our final model we had 942 features (452 original + 400 added non-linear feature combinations).

3   Methodology

3.1   Feature Extraction

The original variables as described in the previous section were transformed into numerical features, either real-valued or binary ones. Each original feature has its own specific transformation into numerical format. The absence of a value, similarly to [1], is encoded with a separate binary variable for most features, as it provides valuable information (i.e., absence of a website of a poor quality conference).

The original features that are used for the analysis task, and their corresponding transformations are described below in Table 1. The results are notably missing Jufo-rankings for the previous years; those are omitted on purpose to make the rank prediction task unbiased by the previous decisions.

3.2   Mislabel Detection Using MD-ELM

The mislabel detection is based on the MD-ELM algorithm from [7]. The key idea is to include in a data set artificial mislabels, which then can be used as baseline

---

[3] See http://www.julkaisufoorumi.fi/en/publication-forum/panels.

Table 1. The list of original features and their numerical representations.

#	Feature	Meaning	Numerical representation
1	Level	Current Jufo ranking	An output variable with integer values in range $\{0, 1, 2, 3\}$
2	Title, Subtitle	Title and subtitle (if available) of the publication	Encoded in a Bag-of-Words representation, dimensionality reduced from 3700 to 30 by a Sparse Random Projection
3	Website	Website of the publication	Country code of the host represented in one-hot encoding with 117 binary variables (including *unknown*)
4	Type	Publication type (journal, conference, book series)	Represented in one-hot encoding with 3 binary variables; this feature has no missing values
5	ISSN	ISSN numbers of printed and online versions	A binary variable representing whether the publication has an ISSN, two variable total
6	StartYear	Start year of the publication	A logarithm of age of the publication, plus a binary variable representing missing value
7	Publication country	Country of publication	One-hot encoding of the publication origin with 114 binary variables, including the *unknown* origin
8	Publisher	Publisher of the series	One-hot encoding of 100 most popular publishers, plus *other* publisher
9	Language	Language of the publication	One-hot encoding of the publication language with 49 binary variables, including *undetermined*
10	ERIH-class	ERIH ranking of publications	One-hot encoding of the four available ranks, plus a *missing* rank
11	SJR SNIP IPP	Impact factors in three different systems	Three real-valued variables for the impact factors, plus three binary variables indicating the absence of an impact factor
12	DOAJ Sherpa/Romeo	Open access types	Eight binary variables: two for DOAJ levels, and six for the Sherpa/Romeo levels
13	Field	The field of study in Finnish classification	Ten binary variables for the ten fields, a publication may belong to multiple fields
14	MinEdu Field	The field of study according to the Ministry of Education classification	70 binary variables for the Ministry of Education fields, a publication may belong to several of them
15	Panel	The scientific panel that assigned a corresponding score	One-hot encoding of the panel number with 25 binary variables, including a *not available* panel
16	ISBN	ISBN numbers used by the publication	One variable representing the number of different ISBNs; can be zero

in a statistical detection of unknown mislabels using Welch's t-test and directly computable Leave-One-Out (LOO) cross-validation error (PRESS statistics). In this way, the MD-ELM algorithm detects samples whose original labels are likely incorrect.

More precisely, the MD-ELM analyses the changes in the LOO error of the model in response to randomly changing labels of a few training samples. If the new labels reduce the global LOO error, the mislabel score of those samples is increased. A small part of the samples, whose labels are randomly changed on purpose, create the control group called *artificial mislabels*. Scores of the artificially mislabeled samples help to determine whether the MD-ELM method succeeds, and define the stopping criterion.

The mislabel detection method uses Extreme Learning Machine as the powerful nonlinear prediction model with a fast LOO error. A practical implementation employs several ELM models with different sets of artificial mislabels, eliminating their possible impact on the results. The predicted originally mislabeled samples are samples with the mislabel score higher that the given quantile of a normal distribution fitted to all the scores.

4   Experimental Results

4.1   Prediction Performance

A successful MD-ELM method application requires a precise prediction model to work with. The prediction task uses features 2–16 from Table 1 as inputs and the feature 1 as the target output.

The dataset exhibit a strong class imbalance (see Table 2). The imbalance causes rank 3 to be completely neglected in the predictions, unless class balancing measures are taken.

Table 2. Number of data samples of each Jufo-rank.

Rank	0	1	2	3
#	5,743	20,503	2,329	668

The benchmark performance level is obtained with the Random Forest classifier. It achieves 89.3% test accuracy, but the predictions are biased due to the strong class imbalance as shown on Fig. 1. The smallest class 3 has only 18,6% correct predictions, while the largest class 1 is predicted correctly 98,4% of times.

Unfortunately, Random Forest model cannot be used in the Mislabeled Detection framework. So an Extreme Learning Machine was train instead. The input features consisted of the 542 numerical features derived from the data, 200 standard non-linear ELM neurons and another 200 Radial Basis Function neurons.

The output layer training proved difficult due to both class imbalance, and a high number of irrelevant linear features. The only successful model was an

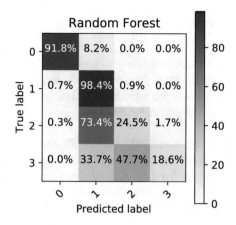

Fig. 1. Confusion matrix of Random Forest classifier on out-of-batch data.

ElasticNet linear classifier that combined L1 and L2 regularization, trained with the Stochastic Gradient Descend. The regularization strength parameter is found by a 5-fold stratified cross-validation, that keeps the proportion of samples from different classes equal between the folds. Additionally, the method performed class balancing by computing the corresponding sample weights.

The resulted ELM achieved 85% total accuracy, distributed much more equally among the classes as shown in Fig. 2. The resulting model selected only 289 input features out of the total of 942, reducing the data size for the MD-ELM method.

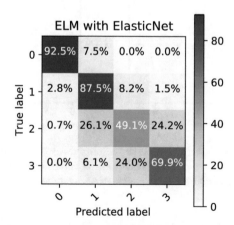

Fig. 2. Confusion matrix of an ELM model with an ElasticNet classifier output layer.

4.2   MD-ELM Performance

The MD-ELM method uses 289 best features selected in the prediction experiment. The method does not implement class balancing, so the scope of the experiment is limited to detecting incorrectly labeled samples of rank 3 using a dataset of 900 random samples from ranks 0, 1, 2 plus all the 668 samples of rank 3. Such reduced dataset has a smaller class imbalance, that does not negatively affect the results.

The final predictions are averaged over 10 different MD-ELM models. Each model uses its own dataset with different random samples of ranks 0, 1, 2, a random subset of 100 input features out of the available 289, and a different random subset of 3% artificially mislabeled samples. At each iteration of the method, two samples have their labels changes, one of which is always an original rank 3 sample.

The method continues until artificially mislabeled samples get an average score of 100. This takes 400,000 iterations. By that time, non-artificially mislabeled samples achieve an average mislabel score of only 19 with standard deviation of 28. The difference between the scores shows that MD-ELM methods succeeds at separating artificially mislabeled samples from the rest; it means that it should also succeed in detecting the originally mislabeled samples.

The mislabel scores of all the samples with the original rank 3 are shown on Fig. 3. A few outliers are clearly visible, together with other candidates to be the originally mislabeled samples. The analysis of these samples is presented below.

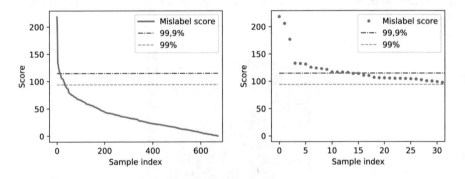

Fig. 3. Mislabel scores of samples with the original rank 3 averaged over 10 MD-ELM models; zoomed version on the right. The quantile values of 99% and 99.9% are shown by horizontal lines. Artificial mislabels achieve an average score of 100.

4.3   Characterization of Misclassified Publication Channels and Comparison to Earlier Results

As explained above, we concentrate only on misclassifications for the highest JuFo ranking, that is publication channels that were evaluated by the Finnish discipline experts as 3, but for which the automatic model suggested a lower rank.

We restrict our misclassification analysis here to this set because it also resembles the largest difference to the Danish and Norwegian systems that include only ranks 0, 1 and 2.

With a mislabelled score over 99% quantile of average scores, 34 publication channels were identified for which the Finnish expert-based ranking was 3 but the model suggested a different rank. However, 30 of these misclassifications could immediately be explained by the ranks in the Danish and Norwegian model, which evaluated these publication channels as 2, that is the highest rank in their systems.

The four remaining publication channels for which both, the automated model and the Danish and Norwegian systems, suggested a lower rank were *LIGHT: SCIENCE & APPLICATIONS*, *Etudes classiques*, *New German critique* (for all three of these journals, the rank has recently been updated to a higher one), and the *British medical journal*. The last one has a considerable higher publication activity: The average number of Finnish publications in JuFo rank 3 channels is 10.78 but the *British medical journal* has a total of 26 publications. All of these journals were also detected to be mislabelled in [1], but the misclassification could actually be explained. The three Scopus indicators had incorrectly not been included in *JuFoDB* for *LIGHT: SCIENCE & APPLICATIONS* and the *British medical journal*. These indicators could be manually found from Scopus and in both cases the indicators were so high that rank 3 actually seemed justified.

Although the methods utilized in here were very different from the ones utilized in [1], the main results obtained and the misclassification detected in here are to a large extend the same as the ones in [1]. Thus, we conclude that methodological triangulation [9,10] has strengthen our analysis results.

5   Conclusions

An extended version of the analysis of Finnish publication channel ranks was provided in this paper. Compared to the reference models in [1], we used here much more versatile set of features, with fully nonlinear ELM-based rank prediction model. The mislabel detection was based on the MD-ELM algorithm proposed in [7] and briefly recapitulated in Sect. 3.2.

In summary, the experimental results obtained and reported in Sect. 4.3 are very similar to the analysis results in [1]. In our future work, we intend to repeat the mislabel detection also for the other ranks, especially rank 2 for which the most suspicious publication channel quality misclassifications were identified in [1] and that, as explained above, actually contain the most misclassifications. The MD-ELM method will also be extended with a class balancing mechanism, allowing it to handle the whole original dataset.

References

1. Saarela, M., Kärkkäinen, T., Lahtonen, T., Rossi, T.: Expert-based versus citation-based ranking of scholarly and scientific publication channels. J. Informetrics **10**(3), 693–718 (2016)
2. Huang, G.B., Zhu, Q.Y., Siew, C.K.: Extreme learning machine: theory and applications. Neurocomputing **70**(1), 489–501 (2006)
3. Huang, G.B., Zhou, H., Ding, X., Zhang, R.: Extreme learning machine for regression and multiclass classification. IEEE Trans. Syst. Man Cybern. Part B (Cybern.) **42**(2), 513–529 (2012)
4. Gallicchio, C., Martin-Guerrero, J.D., Micheli, A., Soria-Olivas, E.: Randomized machine learning approaches: recent developments and challenges. In: ESANN 2017 Proceedings, European Symposium on Artificial Neural Networks, Computational Intelligence and Machine Learning, 26-28 April 2017, pp. 77–86. d-side Publi., Bruges (2017)
5. Liu, X., Lin, S., Fang, J., Xu, Z.: Is extreme learning machine feasible? a theoretical assessment (Part I). IEEE Trans. Neural Netw. Learn. Syst. **26**(1), 7–20 (2015)
6. Lin, S., Liu, X., Fang, J., Xu, Z.: Is extreme learning machine feasible? a theoretical assessment (Part II). IEEE Trans. Neural Netw. Learn. Syst. **26**(1), 21–34 (2015)
7. Akusok, A., et al.: MD-ELM: originally mislabeled samples detection using OP-ELM model. Neurocomputing **159**, 242–250 (2015)
8. Saarela, M., Kärkkäinen, T.: Analysing student performance using sparse data of core bachelor courses. JEDM-J. Educ. Data Mining **7**(1), 3–32 (2015)
9. Bryman, A.: Triangulation. In: The SAGE Encyclopedia of Social Science Research Methods. Sage Publications, Inc., pp. 1143–1144 (2004)
10. Denzin, N.: Strategies of Multiple Triangulation. The Research Act: A Theoretical Introduction to Sociological Methods, pp. 297–313 (1970)

Aviation Guide Gesture Recognition Using ELM with Multiscale CNN Features

Xiangyang Deng[1,2], Zhenyu Li[1], Dongshun Cui[3], Gaoming Huang[2],
Jiawen Feng[1], and Liming Zhang[1(✉)]

[1] Naval Aeronautical University, Yantai 264000, China
Xiangy.deng@gmail.com, zy_lee@outlook.com,
fengjiawen777@163.com, iamzlm@163.com
[2] Naval Engineering University, Wuhan 430000, China
[3] NanyangTechnological University, Singapore 639798, Singapore
dcui002@e.ntu.edu.sg

Abstract. Visual aided guide system (VaGS) is essential and important for the future UAVs, also for the mixed ground command and guide of both UAVs and manned fighters. The key module of VaGS is to recognize the commander's gestures and translate them into instructions. This paper introduces a new high-performance gesture recognition architecture for VaGS. It includes two main components: (1) a multiscale structure is adopted to conduct feature learning, which combined global human gesture features with local hands gesture features rather than an extremely deep neural networks; (2) the fused features are input into an extreme learning machine (ELM) for a classification procedure and then output the gesture instructions. Experimental results show that the proposed method can achieve a rather better expression in some standard datasets, and obtain an accuracy of 99.6% in an aviation guide gesture dataset with 40 classes built by ourselves, still a less training and forward time consumption.

Keywords: Aviation gesture recognition · ELM · Multiscale CNN
VaGS

1 Introduction

With the increasing utilization of UAVs, the ground mixed command and guide of UAVs and manned fighters together will become a common scene. To execute an efficient and safe guidance will greatly test the ground commanders and the mission control system of UAVs. Thus, a high-performance visual aided command and guide system (VaGS) can effectively assist the commanders to reduce the cognitive load, while helping the UAVs follow a correct guide directive.

The aviation guide gesture recognition is the critical part of VaGS. When compared with the general gesture recognition, it has the following difficulties:

(1) It often contains a number of gesture classes, and quite a few of them have similar features.
(2) The actions of commanders are often not holonomic, and usually have not a clear boundary between an action and the immediately following one.

© Springer Nature Switzerland AG 2019
J. Cao et al. (Eds.): ELM 2017, PALO 10, pp. 249–261, 2019.
https://doi.org/10.1007/978-3-030-01520-6_23

(3) The duration of an action is often short-time, which demands the gesture recognition processing has an extremely high computational efficiency.

Thus, an effective solution to aviation gesture recognition (AGR) problem applied in VaGS should have not only high recognition accuracy but also an extremely less time consumption in both training and recognition processes.

Generally, AGR, as a typical visual recognition task has two critical issues, should be seriously discussed, which are feature extraction and pattern classification. A commendable feature extraction method depends in large part on its representative capability. In the past few decades, a lot of image feature representation methods have been proposed, such as principal component analysis (PCA) [1], Bayesian subspace recognition [2], and linear discriminant analysis (LDA) [3], which are called subspace techniques. As well, enormous neighbor domain based techniques have also been presented, such as SIFT [4], HOG [5], LBP [6], etc. which obtain a rather better performance, and mainly extract the local feature representation generated from pixel brightness or color values. However, the aforementioned methods can seldom achieve a human level accuracy till the deep networks [7] occurred. Lately, the well-known image feature learning methods based on convolutional neural networks (CNN) beat human in some simple datasets [8]. They have recently been achieving state-of-the-art performance on a variety of pattern recognition tasks, most notably visual classification problems. However, CNN's feature representation depends seriously on the depth of networks. AlexNet is the champion of ILSVRC2012, and utilizes five convolutional layers and three pooling layers [9]. In 2014, Google proposed a network with 59 convolutional layers called GoogLeNet, which reduces the Top5 error to 6.73% [10] 。 However, the extremely deep networks require rather long time to train and forward computation. It is often unacceptable in engineering practice like AGR, which demands a compromise between high accuracy and time consumption, so that can obtain a good effect.

At the same time, visual classification problem also play a major role in solving AGR problem more precisely. The common network-based methods for classifying images mainly include softmax [11], support vector machine (SVM) [12], multi-layers perceptron (MLP) [13], radial basis function (RBF) [14], etc. And softmax and SVM have been widely used in classification applications. The softmax method is usually applied in deep learning networks conducted as output layers, but then, considering a large scale visual classification problem, multi-layers and large-scale parameters are also essential for a method with high classification accuracy. Due to the structure of mapping the data into a higher dimensional feature space through a nonlinear feature mapping function, SVM and its successors obtain a surprising classification capability in the past decades and are widely used in various classification tasks. Recently, the extreme learning machine (ELM) [15] has been proved to be a better solution to the classification problem, which can achieve a good balance between recognition accuracy and computational efficiency. ELM has some unique characteristics, i.e., simple and fast training, good generalization, and universal approximation and classification capability [16].

In this paper, a novel aviation gesture recognition architecture is proposed, which is able to combine the terrific representative capability of convolutional features with the outstanding generalization performance of ELM classifier. Firstly multi-channels CNN is trained using left-hand gesture pictures, right-hand gesture pictures and human gesture pictures, which have the same identified number and are derived from the same frame. Then, the multiscale CNN features are fed into a single hidden layer ELM for classification. Experimental results clarify that the proposed method could reach a comparable performance (99.6%) to the state-of-the-art approaches with less computation burden and shorter training time. The novelties of the proposed structure are categorized as follows:

(1) Employing a multiscale structure instead of extremely deep convolutional networks, which can be highly paralleled implemented with multi-processors or GPU, so as to extract the guide gesture features more effectively.
(2) Employing an extreme learning machine to fuse the global human gesture features and two types of local hand gesture features, which extremely improves the computation efficiency without reducing the accuracy.

The rest of this paper is organized as follows. Section 2 introduces the mechanism of CNN and ELM respectively and then describes proposed architecture method in details. Then experimental results and comparisons are shown in Sect. 3. Finally, the conclusion and future work are given in Sect. 4.

2 Theories of Foundation and the Proposed Structure

2.1 Convolutional Neural Networks

CNN is a type of feed-forward neural network that has been proven valuable in image processing. It is inspired by the human vision system, whose main part is the convolution operation. It is known as three specific characteristics, namely locally connected neurons, shared weight and spatial or temporal sub-sampling [17]. Generally, CNN can be considered to be made up of three main parts. Firstly, it has multiple hidden convolutional layers that process the output of the prior layer as receptive fields. Secondly, it contains local and global max pooling layers, which are introduced to reduce the number of free parameters and improve generalization. As a result, they form a hierarchical feature extractor that maps the original input images into feature vectors. Then the extracted features vectors are classified by the third part, that is, the fully-connected layers.

In convolutional layer, each neuron is connected locally to its inputs of the previous layer. The input and output of each layer are sets of arrays called feature maps, and its activation could be computed as the result of a nonlinear transformation.

$$a_{i,j} = \sigma(f \otimes x) = \sigma\left(\sum_{i'=1}^{I}\sum_{j'=1}^{J} F_{i',j'} \cdot x_{i+i',j+j'} + b\right)$$

Where $\mathbf{f}$ is the convolution kernel whose size is $I \times J$ and connects input feature map $\mathbf{x}$ to output feature map $\mathbf{a}$. $\sigma(\cdot)$ is a nonlinear activation function (usually sigmoid or hyperbolic tangent), b is the bias. $\otimes$ is a convolution operator.

The max-pooling layers aim to achieve spatial invariance by reducing the resolution of the feature maps obtained in convolution layer. It applies local pooling of feature maps using a max operation in the neighbourhood of the results of the convolution layer.

$$o_{i,j} = \max(y) = \max_{1 \leq i' \leq I', 1 \leq j' \leq J'} , y_{i+i'j+j'}$$

Where y is the $I' \times J'$ neuron patch connected with the neuron (i, j) in max pooling layer, $\max(\cdot)$ is the maximum response operator, $o_{i,j}$ is the output. Note that the maxpooling is non-overlapping and usually has a stride that equal to the size of the neuron patch, which ensures dimension reductions with the factor of I' and J' along each direction.

Different from other deep learning models, CNN is trained using back propagation because the amount of parameters is diminished significantly due to its combination of the three architectural characteristics. Hence, all the parameters are jointly optimized through back-propagation.

2.2 Extreme Learning Machine

ELM algorithms are a type of learning methods for single-hidden-layer feedforward neural networks (SFNNs), which was presented by Huang [15] and were widely applied in feature learning, clustering, regression, and classification, etc. [18]. The basic structure of ELM algorithms is a three layers neural network, containing an input layer, a hidden layer and an output layer (see Fig. 1).

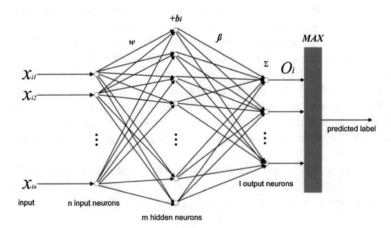

Fig. 1. ELM classifier architecture

The input weights between input and hidden layers are randomly assigned, and the output of hidden layers been seen as a type of feature mapping method of random projection, which is then full-connected to the output layer for linear classification.

In ELM algorithms, the connections between input and hidden layers implement a random feature mapping, and the weights of connections are not required to be changed. Therefore, only the output weights between hidden and output layers are trained. That is to say, layer-by-layer back-propagated tuning is not required in ELM. ELM algorithms can obtain an optimal and generalized solution for applied problems. Additionally, to take advantage of potential of ELM, multilayer implementation of ELM has been presented [19] and further research was performed on stacked deep network by using autoencoder technique [20]. ELM has also been extended in brain machine interfaces [21] and used for representational learning for big data [22].

Suppose that SLFNs with n hidden nodes can be represented by the following equation:

$$f_n(x) = \sum_{i=1}^{n} G_i(x, a_i, b_i) * \beta_i, a_i \in \mathcal{R}^d, b_i, \beta_i \in \mathcal{R}$$

Where $G_i(\cdot)$ denotes the ith hidden node activation function, a_i is the input weight vector connecting the input layer to the ith hidden layer, b_i is the bias weight of the ith hidden layer, and β_i is the output weight. For additive nodes with activation function g, G_i is defined as follows:

$$G_i(x, a_i, b_i) * \beta_i = g(a_i \times x + b_i)$$

and for RBF nodes with activation function g, G_i is defined as

$$G_i(x, a_i, b_i) * \beta_i = g(b_i \| x - a_i \|)$$

Let $L_2(X)$ be a space of functions f on a compact subset X in the d-dimensional Euclidean space R^d such that $|f|^2$ are integrable, that is, $\int_X |f(x)|^2 dx \mid < \infty$. For $u, v \in L^2(X)$, the inner product hu, vi is defined by

$$u, v = \int_X u(x)v(x)dx$$

The norm in $L_2(X)$ space is denoted as $\|\cdot\|$, and the closeness between network function f_n and the target function f is measured by the $L_2(X)$ distance:

$$\|f_L - f\| = \left(\int_X |f_n(x) - f(x)|^2 dx \right)^2$$

The ELM training algorithm can be summarized as follows [16]:

(1) Randomly assign the hidden node parameters: the input weights w_i and biases b_i, $i = 1,\. .., L$.
(2) Calculate the hidden layer output matrix H.
(3) Obtain the output weight

$$\beta = H^\dagger T$$

Where $T = [t_1, \ldots, t_N]^T$ and $H^\dagger$ is the Moore-Penrose (MP) generalized inverse of matrix H.

The orthogonal projection method [23] can be efficiently used for the calculation of MP inverse: $H^\dagger = (H^T H)^{-1} H^T$, if $H^T H$ is nonsingular; or $H^\dagger = H^T (H^T H)^{-1}$, if HH^T is nonsingular. According to the ridge regression theory [24], it was suggested that a positive value $1/\lambda$ be added to the diagonal of $H^T H$ or HH^T. The strategy can make the solution become stabler and has better generalization performance. That is, the output weight β can be calculated as follows:

$$\beta = H^T \left(\frac{1}{\lambda} + HH^T \right)^{-1} T$$

and correspondingly, the output function is:

$$f(x) = h(x)\beta = h(x) \left(\frac{1}{\lambda} + HH^T \right)^{-1} H^T T$$

or β can be calculated as follows:

$$\beta = \left(\frac{1}{\lambda} + HH^T \right)^{-1} H^T T$$

and the corresponding output function is:

$$f(x) = h(x)\beta = h(x)H^T \left(\frac{1}{\lambda} + HH^T \right)^{-1} T$$

2.3   Proposed Structure of ELM with Multiscale CNN Features

In most cases, the commander guides UAVs or manned fighters at a relatively long distance, while the automatic gesture recognition demands deeply the arm gestures or hand gestures. In other words, it can't obtain a good effect only based on the human gesture features, which is to be discussed in details in the subsequent experiment 4. Therefore, we should combine the global human features with local arm or hand features. In particular, the hand feature contains most of the semantic information implied in arms.

Thus, the proposed AGR method demonstrated in Fig. 2 includes two modules: (1) feature extraction module based on CNN and (2) classification module based on ELM. The features are extracted from each frame picture, which is divided into three components including human gesture picture, a local left-hand gesture sub-picture, and a local right-hand gesture sub-picture. Then we use ELM as a classifier, which is composed of an SFNN.

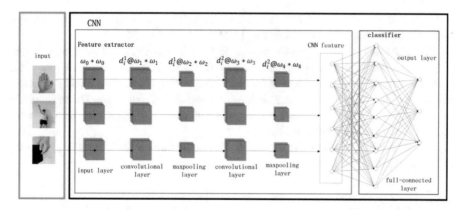

Fig. 2. The proposed architecture for aviation guide gesture recognition

The implementation includes two stages: (1) training and (2) recognition, which are described in Fig. 3.

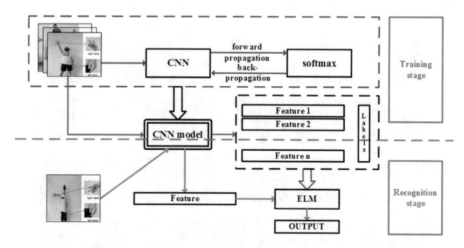

Fig. 3. How the proposed method to train and recognize

At the training stage, the back-propagation learning algorithm is used to train the CNN, which has multiple fully-connected layers to be used for classification. When all the training images organized as a serious of pitches have been fed into the CNN, finishing the training process, the fully-connected layers are removed from CNN. After that, the left CNN model is forward to generate the feature representation of a test image. It is the beginning of ELM's training stage, which uses ELM algorithm to estimate output weights of the SFNN given all feature date generated by CNN model in a batch learning mode. ELM algorithm admits two types of operations in the hidden nodes: (1) dot product and (2) kernel operation. Thus, the corresponding training algorithms are, respectively, denoted as ELM and kernel ELM.

The ELM randomly assigns the input weights while kernel ELM randomly assigns a set of kernels. The training inputs include a feature matrix X where each row is the feature vector of a training image and a class label vector Y where each entry indicates which sign class the training image belongs to, and X and Y are then fed to the ELM module to train the SFNN to make it encapsulate 40 categories of gesture. Details of the training process can be seen in Fig. 3.

At the recognition stage, the trained ELM module outputs the class label given the CNN feature of a test image.

3 Experiments

3.1 Aviation Guide Gesture Dataset

Aviation guide gesture generally includes several categories: skidding, steering, engine parking, etc. We choose 40 common guide gestures and take 300 pictures for each gesture. The 40 classes of aviation gesture are shown in Fig. 4.

Fig. 4. Aviation guide gesture dataset

There are a few difficulties in the AGR problem described in the aviation guide gesture dataset:

(1) Some of the commander's actions have the similar features, and it will result in discriminating them hardly. For example, the picture 29 and 30.

(2) When there is a busy circumstance, or when the commander is absent-minded sometimes, the guiding movements often become an incomplete gesture, which will lead to classification errors.

(3) If the commander is not always facing the video camera, or the observation angle is not good, the classification accuracy will be affected.

(4) Usually, the commander's actions conduct a short duration, so that the AGR processes demand high computational efficiency.

3.2 Experiments with Thomas Moeslund Hand Gesture Datasets

To evaluate the multiscale CNN model and ELM classifier, experiment 1 was performed using Thomas Moeslund Hand Gesture Database first, which consists of 2060 images of a hand performing the 24 static signs employed in the international sign language alphabet. We use the Caffe toolbox [25] to train the CNN feature extractor, running on the Ubuntu 14.04 system with Xeon(R) E5-2640v4 CPU (2.40 GHz), and 16 GB DDR4. Initial weights of the CNN are drawn from a uniform random distribution in the range [−0.01, 0.01]. The activation function of each neuron is sigmoid.

According to the principle of the proposed method, the feature extraction ability of multi-channels convolution neural network mainly comes from the different scales. Moreover, the selection of convolution kernel size is also an important influence factor, which is not fixed for different processing objects and different application scenarios. Therefore, this experiment includes six single-channel convolution neural networks (SCNNs) and six double-channels convolution neural networks(DCNN) with different convolution kernel sizes was performed to test the multiscale networks' capability. All the CNN models are tested with two classifiers, Softmax and ELM.

Table 1 shows the parameters for each experimental setup and the classification accuracies, MMCNN(3,5&5,5) indicates that the convolutional kernel sizes of the first channel are 3 × 3 and 5 × 5, and the kernel sizes of the second channel are 5 × 5 and 5 × 5. For all the experiments, the databases were divided into training and test, the selection was random, and it used 80% of the database for training, and 20% for testing. With Softmax Classifier, the SCNN(5 × 5) and DCNN(3 × 5 & 3 × 5) methods obtain better results, and DCNN(3 × 5 & 3 × 5) architecture gets the best results with ELM classifier. The best test error as shown is 0.5%.

For ELM, it is fed with CNN features. We choose the model with the best performance of SCNN and DCNN. Meanwhile, the different numbers of hidden nodes are chosen and the corresponding recognition accuracy is reported in Fig. 5. The recognition accuracy generally increases as there are more hidden nodes used. SCNN(7 × 7) reaches 99.5% with 500 hidden nodes, DCNN(3 × 5 & 3 × 5) reaches 99.5% with 500 hidden nodes. So far the highest recognition rate is 99.5%.

In the third experiment, a representative traditional gesture recognition algorithm and the CNN-based recognition methods were selected. Experiments were carried out on the Thomas Moeslund static gesture database. Based on the previous experimental results, the DCNN(7,5&5,5) with Softmax and ELM is selected, and the input image size is 32 × 32.

The results of the comparative experiments are shown in Table 2. The classification accuracy of the proposed DCNN + ELM in the Thomas Moeslund database is 99.52%, higher than that of the traditional CNN model.

Table 1. Results for the multi-scale experiments

Model	Kernel	Recognition rate(avg.)(%)	Model	Kernel	Recognition rate(avg.)(%)
Single channel CNN + Softmax	SCNN (5 × 5)	**98.1**	Double channel CNN + Softmax	DCNN (3 × 5&5 × 5)	97.8
	SCNN (3 × 5)	97.8		DCNN (7 × 5&5 × 5)	97.8
	SCNN (5 × 3)	97.1		DCNN (5 × 5&5 × 5)	97.9
	SCNN (7 × 5)	97.4		DCNN (3 × 5&7 × 5)	97.7
	SCNN (5 × 7)	97.7		DCNN (7 × 5&7 × 5)	97.8
	SCNN (7 × 7)	97.9		DCNN (3 × 5&3 × 5)	**98.1**
Single channel CNN + ELM	SCNN (5 × 5)	99.1	Double channel CNN + ELM	DCNN (3 × 5&5 × 5)	99.3
	SCNN (3 × 5)	99.1		DCNN (7 × 5&5 × 5)	99.2
	SCNN (5 × 3)	99.0		DCNN (5 × 5&5 × 5)	99.2
	SCNN (7 × 5)	**99.2**		DCNN (3 × 5&7 × 5)	99.3
	SCNN (5 × 7)	99.0		DCNN (7 × 5&7 × 5)	99.2
	SCNN (7 × 7)	98.8		DCNN (3 × 5&3 × 5)	**99.5**

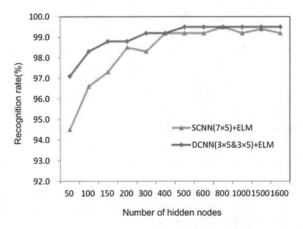

Fig. 5. ELM performance with different numbers of hidden nodes.

Table 2. Results for the different methods

Method	Accuracy(%)
Spatial Pyramid (BoW) [26]	85.32
bottom-up structured DCNN [27]	88.78
Tiled CNN [28]	90.48
Big and Deep MPCNN [29]	96.77
DCNN(7,5&5,5) + softmax	98.10
DCNN(7,5&5,5) + ELM	99.52

3.3   Experiments with Aviation Guide Gesture Dataset

Experiment 4 was performed to evaluate the multi-scale CNN with ELM classifier in Aviation Guide Gesture Dataset. First, we use body images only to train the CNN with one channel, the hit rate of each class is listed in Table 3. The average recognition accuracy is 87.3%. Then we take the CNN with three channels instead, and use local feature of hands as input with the body. The accuracy can reach 99.6% in this experiment.

Table 3. Recognition accuracy of each class with and without hands as input

Input	1	2	3	4	5	6	7	8	9	10	11
human only	100	100	100	65	93.5	93.44	100	100	57.4	95.1	80.3
human + hand	100	100	100	99.6	99.2	99.1	100	100	98.6	100	100
Input	12	13	14	15	16	17	18	19	20	21	22
human only	100	90.2	47.5	100	100	100	100	18	63.9	93.5	91.8
human + hand	100	99.3	98.6	100	100	100	100	99.7	98.9	99.1	98.9
Input	23	24	25	26	27	28	29	30	31	32	33
human only	98.4	78.7	100	86.9	93.4	85.2	72.1	72.1	100	54.1	60.7
human + hand	99.91	99.1	100	99.2	99.3	98.9	99.3	98.3	100	99.6	99.1
Input	34	35	36	37	38	39	40	mean			
human only	100	100	100	100	100	100	100	87.3			
human + hand	100	100	100	100	100	100	100	99.6			

Table 3 shows the comparison of the recognition rate between the proposed architecture and other recently reported results

4   Conclusion

In this paper, we designed a novel neural network structure for the aviation gesture recognition, which the input features are divided into global human features and two hands local features, based on a priori knowledge. Compared with the method of one-input deep convolution features extractor, we use three channels to train the feature extractor independently, which can directly do a multi-scale fusion of global human gesture features and local hands gesture features. Independent channel training can take advantages of the different scalar features more effectively without the task tuning. The multi-channel features' fusion into the ELM classifier can be regarded as an indirect semantic level blend. The proposed framework shows a terrific performance, and indicates that by means of applying appropriately the domain prior knowledge in algorithm designing, it can significantly improve the performance while keeping a more sophisticated structure. This is not only suitable for the classification of aviation guide gesture recognition, but also applies to some complex scene classification.

However, the aviation guide gesture recognition in ambiguous environments is a difficult project. In the future, we will try to design more effective features of signs for some contaminated conditions, such as bad weather, partial occlusions, viewpoint variations, rotations, and physical damages.

References

1. Turk, M., Pentland, A.: Eigenfaces for recognition. J. Cogn. Neurosci. **3**(1), 71–86 (1991)
2. Moghaddam, B., Jebara, T., Pentland, A.: Bayesian face recognition. Pattern Recognit. **33** (11), 1771–1782 (2000)
3. Ling, Y., Yin, X., Bhandarkar, S.M.: Sirface vs. Fisherface: recognition using class specific linear projection. In: Proceedings of International Conference on Image Processing 2003, ICIP 2003, vol. 2, pp. III-885–8. IEEE (2003)
4. Xie, S., et al.: Fusing local patterns of gabor magnitude and phase for face recognition. IEEE Trans. Image Process. **19**(5), 1349 (2010)
5. Huang, Z., et al.: An efficient method for traffic sign recognition based on extreme learning machine. IEEE Trans. Cybern. **47**(4), 920–933 (2016)
6. Wang, X., Han, T.X. Yan, S.: An HOG-LBP human detector with partial occlusion handling. In: IEEE, International Conference on Computer Vision IEEE, 32–39 (2010)
7. Hinton, G.E., Salakhutdinov, R.R.: Reducing the dimensionality of data with neural networks. Science **313**(5786), 504 (2006)
8. http://blog.kaggle.com/2015/01/02/cifar-10-competition-winners-interviews-with-dr-ben-graham-phil-culliton-zygmunt-zajac/
9. Krizhevsky, A., Sutskever, I., Hinton, G.E.: ImageNet classification with deep convolutional neural networks. In: International Conference on Neural Information Processing Systems Curran Associates Inc., pp. 1097–1105 (2012)
10. Szegedy, C., et al.: Going deeper with convolutions. Computer Vision and Pattern Recognition IEEE, pp. 1–9 (2015)
11. Bouchard, G.: Clustering and classification employing softmax function including efficient bounds. US, US 8065246 B2 (2011)
12. Guo, H., Wang, W.: An active learning-based SVM multi-class classification model. Pattern Recogn. **48**(5), 1577–1597 (2015)
13. Mirjalili, S., Mirjalili, S.M., Lewis, A.: Let a biogeography-based optimizer train your multi-layer perceptron. Information Sciences **269**(8), 188–209 (2014)
14. Tbarki, K., et al.: RBF kernel based SVM classification for landmine detection and discrimination. In: Image Processing, Applications and Systems IEEE (2017)

15. Huang, G.-B., Chen, L., Siew, C.-K.: Universal approximation using incremental constructive feedforward networks with random hidden nodes. IEEE Trans. Neural Netw. **17**(4), 879–892 (2006)
16. Huang, G.-B., Zhou, H., Ding, X., Zhang, R.: Extreme learning machine for regression and multiclass classification. IEEE Trans. Syst., Man, Cybern.-Part B: Cybern. **42**(2), 513–529 (2012)
17. Zeng, Y., et al.: Traffic sign recognition using extreme learning classifier with deep convolutional features. In: The 2015 International Conference on Intelligence Science and Big Data Engineering (IScIDE 2015), Suzhou, China (2015)
18. Deng, C., et al.: Extreme learning machines: new trends and applications. Sci. China Inf. Sci. **58**(2), 20301–020301 (2015)
19. Tang, J., Deng, C., Huang, G.B.: Extreme learning machine for multilayer perceptron. IEEE Trans. Neural Netw. Learn. Syst. **27**(4), 809–821 (2016)
20. Zhou, H., et al.: Stacked extreme learning machines. IEEE Trans. Cybern. **45**(9), 2013 (2015)
21. Chen, Y., Yao, E., Basu, A.: A 128-channel extreme learning machine-based neural decoder for brain machine interfaces. IEEE Trans. Biomed. Circuits Syst. **10**(3), 679–692 (2016)
22. Kasun, L.L.C., et al.: Representational learning with ELMs for big data. Intell. Syst. IEEE **28** (6), 31–34 (2013)
23. Rao, C.R., Mitra, S.K.: Generalized inverse of matrices and its applications, pp. 601–620 (1971)
24. Hoerl, A.E., Kennard, R.W.: Ridge regression: biased estimation for nonorthogonal problems. Technometrics **12**(1), 55–67 (1970)
25. Jia, Y., et al.: Caffe: convolutional architecture for fast feature embedding. In: ACM International Conference on Multimedia ACM, pp. 675–678 (2014)
26. Lazebnik, S., Schmid, C., Ponce, J.: Beyond bags of features: spatial pyramid matching for recognizing natural scene categories. In: IEEE Computer Society Conference on Computer Vision and Pattern Recognition IEEE Computer Society, pp. 2169–2178 (2006)
27. Yamashita, T., Watasue, T.: Hand posture recognition based on bottom-up structured deep convolutional neural network with curriculum learning. In: IEEE International Conference on Image Processing IEEE, pp. 853–857 (2015)
28. Le, Q.V., et al.: Tiled convolutional neural networks. In: International Conference on Neural Information Processing Systems Curran Associates Inc., pp. 1279–1287 (2010)
29. Abdel-Hamid, O., Deng, L., Yu, D.: In: Exploring Convolutional Neural Network Structures and Optimization Techniques for Speech Recognition. INTERSPEECH 2013, p. 1173-5 (2013)
30. Sermanet, P., Lecun, Y.: Traffic sign recognition with multi-scale convolutional networks. In: International Joint Conference on Neural Networks IEEE, pp. 2809–2813 (2011)
31. Ciresan, D., et al.: A committee of neural networks for traffic sign classification. **42**(4), 1918–1921 (2011)

Facial Age Estimation with a Hybrid Model

Zhan-Li Sun$^{(\boxtimes)}$, Nan Wang, Ru-Xia Ban, and Xia Chen

School of Electrical Engineering and Automation, Anhui University, Hefei, China
zhlsun2006@126.com

Abstract. How to accurately estimate facial age is still an intractable task because of insufficiency of training data. In this paper, a hybrid model is proposed to estimate facial age by means of extreme learning machine (ELM) and label distribution support vector regressor (LDSVR). In the proposed method, the bio-inspired features are adopted to estimate the facial age due to its prominent performance. In order to improve the accuracy and decrease the computation burden, the ratio of feature's between-category to within-category sums of squares (BW) is designed as a criterion to select features. To define the category of each sample, the training data is divided into several sets according to age group. Different virtual class labels are assigned to the samples of each set, respectively. Given the reduced data, a multiple-input-single-output ELM regression model is established to estimate the facial ages. Moreover, a label distribution support vector regressor is adopted to estimate facial age based on a multiple-input-multiple-output regression model. After obtaining the outputs of ELM and LDSVR, a linear weighting strategy is devised to compute the final estimation of facial age. Experimental results on a well known facial image database demonstrates the feasibility and efficiency of the proposed hybrid model.

Keywords: Facial age estimation · Extreme learning machine
Label distribution support vector regressor · Feature selection

1 Introduction

The task of facial age estimation is to automatically estimating human ages by exploring the various features of facial images [1–3]. Due to its wide applications, such as safety and security, facial age estimation has become an important topic of face image processing. In computer vision and pattern recognition, facial age estimation remains a challenge problem because of several negative factors.

The work was supported by a grant from National Natural Science Foundation of China (No. 61370109), a key project of support program for outstanding young talents of Anhui province university (No. gxyqZD2016013), a grant of science and technology program to strengthen police force (No. 1604d0802019), and a grant for academic and technical leaders and candidates of Anhui province (No. 2016H090).

J. Cao et al. (Eds.): ELM 2017, PALO 10, pp. 262–270, 2019.
https://doi.org/10.1007/978-3-030-01520-6_24

For the same person, the photos taken at different years may have a relative large appearance variation. Moreover, the available facial images sometimes are few because the time span is long and the image acquisition is difficult in the past.

As an effective and efficient single-hidden layer feedforward neural networks (SLFNs), extreme learning machine (ELM) has been developed [4] and widely applied in the various fields [5–8]. Instead of precise prediction, a multiple feature based ELM approach was proposed in [9] to classify face images into one of several pre-defined age-groups. By considering as an ordinal ranking-oriented issue, a multifeature extreme ordinal ranking machine was presented in [10] for facial age estimation.

Inspired by the appearance similarity of faces at close ages, an adaptive label distribution learning algorithm was proposed in [11,12] by assigning a label distribution to each face image. In [13], facial age estimation was divided into three stages, i.e. age grouping, age estimation within age groups, and decision fusion for final age estimation. In [14], the crowd opinion is here expressed by the distribution of ratings given by a sufficient amount of people. A label distribution support vector regressor (LDSVR) is proposed to predict the pre-release crowd opinion.

Considering the superiority of ELM and LDSVR, a hybrid model is proposed in this paper to accurately estimate facial age. In the proposed method, the widely used biologically inspired features (BIF) are adopted as the training data. The BIF has shown good performance in facial age estimation by simulating the primate visual system [15]. In order to improve the recognition accuracy and to decrease the computation burden, the ratio of feature's between-category to within-category sums of squares (BW) is designed as a feature selection criterion. As the category labels of facial ages are not known in the regression model, a strategy of virtual labels assignment are devised to compute BW. Finally, a linear weighting model is presented to fuse the outputs of ELM and LDSVR.

The remainder of the paper is organized as follows. In Sect. 2, we present the proposed hybrid model-based facial age estimation algorithm. Experimental results and the related discussions are given in Sect. 3, and our concluding remarks are presented in Sect. 4.

2   Methodology

Figure 1 shows the flowchart of the hybrid model for facial age estimation. There are three main steps in the proposed method: feature selection with BW, facial age estimation with ELM and LDSVR, and multi-model fusion with a weighting strategy. A detailed description of these three steps is presented in the following subsections.

2.1   Feature Selection with BW

Assume that there are N training samples $\mathbf{x}_i$ $(i = 1, \cdots)$ with n-dimensional features, i.e. $\mathbf{x}_i = (x_{i1}, x_{i2}, \cdots, x_{in})^T \in R^n$. For the training data, the goal of

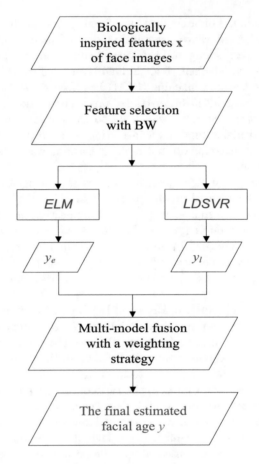

Fig. 1. Flowchart of the hybrid model for facial age estimation.

BW is to reduce the feature dimension by selecting those with large between-category distances and small within-category distances. In the regression model of ELM or LDSVR, the outputs are the estimated ages, not the category labels of facial ages. In order to derive the category labels, a strategy of virtual class division is devised according to the age interval. In a relative small age interval, the appearances of face images are close for the same individual. Thus, we divide the face image into m intervals, and assign the images in each intervals with different labels.

For the jth feature, the within-category distance s_{kj}^{w} of the kth class can be computed between the samples $\mathbf{x}_i$ and the mean μ_{kj}, i.e.,

$$s_{kj}^{w} = \sum_{i=1}^{n_k}(\mathbf{x}_{ij} - \mu_{kj})^2, \quad \mathbf{x}_i \in \mathbf{X}_k, \tag{1}$$

where n_k is the sample number of the kth class. Similarly, the between-category distance s_{kj}^b can be calculated as,

$$s_{kj}^b = (\mu_{kj} - \mu_j)^2, \tag{2}$$

where μ_j is the mean of the jth feature. Given s_{kj}^w and s_{kj}^b, the ratio of feature's between-category to within-category sums of squares can be given by

$$\lambda_j = \frac{\sum_{k=1}^m s_{kj}^b}{\sum_{k=1}^m s_{kj}^w}, j = 1, \cdots, n. \tag{3}$$

Finally, the features with large λ_j values are selected for facial age estimation.

2.2 ELM Regression Model

Assume that there are N samples $(\mathbf{x}_i, y_i)$, where $(\mathbf{x}_i = (x_{i1}, x_{i2}, \cdots, x_{in})^T \in R^n$ is an n-dimensional feature vector, and y_i is the actual age of the ith sample. For the ith sample, the ELM regression model of facial age estimation can be mathematically modeled as [4],

$$\sum_{i=1}^{\tilde{N}} \beta_i g_i(\mathbf{x}_j) = \sum_{i=1}^{\tilde{N}} \beta_i g(\mathbf{w}_i \cdot \mathbf{x}_j + b_i) = \tilde{y}_j, j = 1, \cdots, N, \tag{4}$$

where $\tilde{N}$ is the number of hidden nodes, $g(x)$ is the activation function. The vector $\mathbf{w}_i = (w_{i1}, \cdots, w_{in})^T$ is the weight vector connecting the ith hidden node and the input nodes, β_i is the weight vector connecting the ith hidden node and the output node, and b_i is the threshold of the ith hidden node. The operator $\mathbf{w}_i \cdot \mathbf{x}_j$ denotes the inner product of $\mathbf{w}_i$ and $\mathbf{x}_j$.

The N equations of (4) can be written compactly as,

$$\mathbf{H}\beta = \mathbf{y}, \tag{5}$$

where the hidden layer output matrix

$$\mathbf{H}(\mathbf{w}_1, \cdots, \mathbf{w}_{\tilde{N}}, b_1, \cdots, b_{\tilde{N}}, \mathbf{x}_1, \cdots, \mathbf{x}_N)$$
$$= \begin{bmatrix} g(\mathbf{w}_1 \cdot \mathbf{x}_1 + b_1) & \cdots & g(\mathbf{w}_{\tilde{N}} \cdot \mathbf{x}_1 + b_{\tilde{N}}) \\ \vdots & \cdots & \vdots \\ g(\mathbf{w}_1 \cdot \mathbf{x}_N + b_1) & \cdots & g(\mathbf{w}_{\tilde{N}} \cdot \mathbf{x}_N + b_{\tilde{N}}) \end{bmatrix}_{N \times \tilde{N}} \tag{6}$$

is called the hidden layer output matrix of the neural network. The activation function used in this paper is the sigmoidal function,

$$g(x) = \frac{1}{1 + e^{-x}}, \tag{7}$$

which has been demonstrated to be an effective activation function of ELM. Finally, the smallest norm least squares solution of the linear system (5) can be given by,

$$\beta = \mathbf{H}^\dagger \mathbf{y} \tag{8}$$

where $\mathbf{H}^\dagger$ denotes the Moore-Penrose generalized inverse operation of $\mathbf{H}$.

After training, we can get the input weight $\mathbf{w}_i$, output weight β, and the threshold b_i. Given the test data, we can compute the corresponding hidden layer output matrix $\mathbf{H}$, and the estimated ages $\tilde{\mathbf{y}}$,

$$\tilde{\mathbf{y}} = \mathbf{H}\beta. \tag{9}$$

2.3   Label Distribution Support Vector Regressor

The basic idea of LDSVR is to fit a sigmoid function to each component of the label distribution simultaneously by a support vector machine [14]. Suppose the label distribution $\mathbf{d}$ of the instance $\mathbf{x}$ is modeled by an element-wise sigmoid vector,

$$d = \frac{1}{1 + \exp(-\mathbf{W}\varphi(\mathbf{x}) - \mathbf{b})}. \tag{10}$$

Then, we can generalize the single-output SVR by minimizing the sum of the target functions on all dimensions,

$$\Gamma(\mathbf{W}, \mathbf{b}) = \frac{1}{2} \sum_{j=1}^{c} \| \mathbf{w}^j \|^2 + C \sum_{i=1}^{n} L(u_i), \tag{11}$$

where w^j is the transpose of the j-th row of $\mathbf{W}$ and $L(u_i)$ is the loss function for the i-th example.

2.4   Multi-model Fusion with a Weighting Strategy

Denote y_e and y_l as the estimated facial ages by ELM and LDSVR, respectively. The final estimation value can be given by a linear weighting of y_e and y_l, i.e.,

$$y = \lambda_e y_e + \lambda_l y_l, \tag{12}$$

where λ_e and λ_l denote the weighting coefficients.

3   Experimental Results

3.1   Experimental Data

As one of the largest publicly available longitudinal face databases, the MORPH database is used here to verify the proposed method. The MORPH database contains 55,000 images of more than 13,000 people within the age ranges of 16 to 77 [16]. As an example, Fig. 2 shows the face images of one subject in the MORPH database.

In order to measure the estimation performance, the mean-square error ϵ between the estimated values $\tilde{y}_i$ and the true values y_i is used as the performance index, i.e.,

$$\epsilon = \frac{1}{N} \sqrt{\sum_{i=1}^{N} (y_i - \tilde{y}_i)^2}. \tag{13}$$

Smaller estimation errors means that the estimated facial ages are more close to the true values, i.e. the estimation results are more accurate.

Fig. 2. The face images of one subject in the MORPH database.

3.2   Experimental Results of the MORPH Database

After sorting, how to determine the number of features to be selected is still a difficult problem for BW. To simplify the computation, in experiments, we first set the percentages of features to be selected as 10%, 20%, 30%, $\cdots$, 100%, respectively. Figure 3 shows the estimation errors of one set of training data, validation data, and testing data, and the corresponding training time with different percentages of features. After comparing the validation error, we found that a good result can generally be achieved when the percentage of features is set as 50%.

For ELM, one problem is to design an appropriate network size, i.e. to determine the number of hidden neurons. When the number of the hidden neurons is equal to N, the SLFNs can approximate these N samples with zero error. This means that N can be assumed to be the largest value of $\tilde{N}$. Since N is relatively large, it is impossible to verify every possible value in the interval $[1, N]$. To simplify the computation, in experiments, $\tilde{N}$ is selected as $\{100, 200, 300, \cdots, N\}$ in sequence. As one example, Fig. 4 shows the estimation errors of one set of training data, validation data, and the testing data, and the corresponding training time with different number of hidden neurons. Then, the number with the relative small validation error is selected as the optimal parameter. Considering both the validation error and the training time, the parameter $\tilde{N}$ is set as

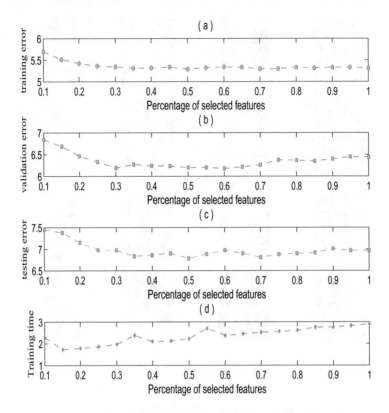

Fig. 3. The estimation errors of one set of training data, validation data, and the testing data, and the corresponding training time with different percentages of features.

800 for ELM. In terms of the validation errors, the weighting coefficients λ_e and λ_l are set as 0.7 and 0.3, respectively.

Table 1. The estimation errors of five trials and the corresponding mean and standard deviation $(\mu \pm \sigma)$ for the MORPH database of different methods.

	1	2	3	4	5	$\mu \pm \sigma$
IIS-LDL	11.3013	10.6564	10.6685	10.7293	10.9090	10.8529 ± 0.2703
ELM	6.7667	5.9171	6.7145	6.4369	6.3744	6.4419 ± 0.3391
LD-SVR	7.2189	6.1217	7.1823	6.4551	6.4947	6.6945 ± 0.4843
HM	6.7256	5.7587	6.6577	6.1317	6.0753	6.2698 ± 0.4112

In order to verify the effectiveness of the hybrid model (HM), we compare it with two recently reported methods, i.e. IIS-LDL [11,12] and LD-SVR [14]. Moreover, as a comparison, the estimation results of a single ELM regressor [4] are also presented here. In experiments, a 5-fold cross validation is performed

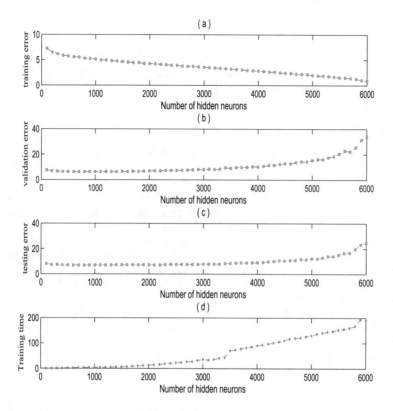

Fig. 4. The estimation errors of one set of training data, validation data, and the testing data, and the corresponding training time with different number of hidden neurons.

on the MORPH database to investigate the robustness of the various methods. Considering the balance, the sample number of each set is the same for every age. Initially, the first set is used as the testing data, and the remained four sets are adopted as the training data. Successively, the set from the second to the last is adopted as the testing data in sequence. Table 1 shows the estimation errors of five trials and the corresponding mean and standard deviation ($\mu \pm \sigma$) for the MORPH database of different methods.

From Table 1, we can see that the estimation errors of ELM and LD-SVR are obviously higher than that of IIS-LDL. Therefore, ELM and LD-SVR have a better estimation performance than IIS-LDL. Further, it can be seen that the estimation errors of ELM are slightly lower than LD-SVR. In general, the performances of these two methods are close each other. Nevertheless, we can see that the estimation errors of HM are all lower than that of ELM and LD-SVR in five trials. Therefore, the proposed method is effective and feasible.

4   Conclusions

In this paper, an effective approach is proposed to estimate facial age by combining ELM and LDSVR. The experimental results demonstrated the effectiveness and feasibility of the proposed method.

References

1. Wu, Z.D., Yuan, J., Zhang, J.W., Huang, H.X.: A hierarchical face recognition algorithm based on humanoid nonlinear least-squares computation. J. Ambient. Intell. Humanized Comput. **7**(2), 229–238 (2016)
2. Iqbal, K., Odetayo, M.O., James, A.: Face detection of ubiquitous surveillance images for biometric security from an image enhancement perspective. J. Ambient. Intell. Humanized Comput. **5**(1), 133–146 (2014)
3. Lozano-Monasor, E., López, M.T., Vigo-Bustos, F., Fernández-Caballero, A.: Facial expression recognition in ageing adults: from lab to ambient assisted living Lozano-Monasor. J. Ambient. Intell. Humanzed Comput. **8**(4), 567–578 (2017)
4. Huang, G.B., Zhu, Q.Y., Siew, C.K.: Extreme learning machine: theory and applications. Neurocomputing **70**, 489–501 (2006)
5. Arigbabu, O.A., Mahmood, S., Ahmad, S.M.S., Arigbabu, A.A.: Smile detection using hybrid face representation. J. Ambient. Intell. Humanized Comput. **7**(3), 415–426 (2016)
6. Sun, Z.L., Au, K.F., Choi, T.M.: A hybrid neuron-fuzzy inference system through integration of fuzzy logic and extreme learning machines. IEEE Trans. Syst., Man, Cybern. Part B **37**(5), 1321–1331 (2007)
7. Sun, Z.L., Choi, T.M., Au, K.F., Yu, Y.: Sales forecasting using extreme learning machine with applications in fashion retailin. Decis. Support Syst. **46**(1), 411–419 (2008)
8. Sun, Z.L., Wang, H., Lau, W.S., Seet, G., Wang, D.W.: Application of BW-ELM model on traffic sign recognition. Neurocomputing **128**, 153–159 (2014)
9. Sai, P.K., Wang, J.G., Teoh, E.K.: Facial age range estimation with extreme learning machines. Neurocomputing **149**, 364–372 (2015)
10. Zhao, W., Wang, H., Huang, G.B.: Multifeature extreme ordinal ranking machine for facial age estimation. Math. Probl. Eng. **2015**(1), Article ID 840840 (2015)
11. Geng, X., Yin, C., Zhou, Z.H.: Facial age estimation by learning from label distributions. IEEE Trans. Pattern Anal. Mach. Intell. **35**(10), 2401–2412 (2013)
12. Geng, X., Wang, Q., Xia, Y.: Facial age estimation by adaptive label distribution learning. In: International Conference on Pattern Recognition, pp. 4465–4470 (2014)
13. Liu, K.H., Yan, S.C., Kuo, C.C.J.: Age estimation via grouping and decision fusion. IEEE Trans. Inf. Forensics Secur. **10**(11), 2408–2423 (2015)
14. Geng, X., Yin, C., Zhou, Z.H.: Pre-release prediction of crowd opinion on movies by label distribution learning. In: International Joint Conference on Artificial Intelligence, pp. 3511–3517 (2015)
15. Guo, G.G., Mu, G.W., Fu, Y., Huang, T.S.: Human age estimation using bio-inspired features. In: IEEE Conference of Computer Vision and Pattern Recognition, pp. 112–119 (2009)
16. Ricanek, K., Tesafaye, T.: MORPH: a longitudinal image database of normal adult age-progression. In: IEEE International Conference on Automatic Face and Gesture Recognition, pp. 341–345 (2006)

Seizure Prediction for iEEG Signal with Bag-of-Wave Model and Extreme Learning Machine

Song Cui[1,2], Lijuan Duan[1,3(✉)], Yuanhua Qiao[4], and Xing Su[1]

[1] Faculty of Information Technology,
Beijing University of Technology, Beijing 100124, China
ljduan@bjut.edu.cn
[2] Beijing Key Laboratory on Integration and Analysis of Large-scale Stream
Data, Beijing 100124, China
[3] Beijing Key Laboratory of Trusted Computing, National Engineering
Laboratory for Critical Technologies of Information Security Classified
Protection, Beijing 100124, China
[4] College of Applied Sciences, Beijing University of Technology,
Beijing 100124, China

Abstract. Long-term epileptic seizure prediction has potential to transform epilepsy care and treatment. However, the accuracy of seizure prediction is still difficult to satisfy the requirement of application. In this paper, a seizure prediction system is proposed based on Bag-of-Wave Model and Extreme Learning Machine. To get the representation of segments in iEEG signals, interictal codebook and preictal codebook are constructed by clustering algorithm. Histogram features are then extracted by projecting waves within the sliding window on two codebooks. In the end, classifying the feature with ELM into interictal phase and preictal phase. Experiments are operated on Kaggle Seizure Prediction Challenge dataset, which show the proposed approach is effective in seizure prediction.

Keywords: Signal analysis · Seizures prediction · Bag-of-Wave
iEEG · Extreme learning machine

1 Introduction

In neurological disease, epilepsy influence almost 1% people in the world [1]. Epileptic patients are tortured by unexpected seizure attacks, which seriously reduce their quality of life. Antiepileptic drugs failed to control seizures in 20%–30% patients. In addition, most of antiepileptic drugs have side effects in different levels. If the seizures can be predicted, on one side, treatments can be implemented to prevent the further damage during seizure attack. And patients can take drugs or other treatments to interfere the occurrence of seizures. Moreover, the study on seizure prediction can help us to further understand the mechanism of epilepsy. With the further research on epilepsy, evidences on clinical researches show that symptoms can be detected to predict the occurrence of seizures [2–5]. In recent years, a new option for epileptic patients to control seizure

© Springer Nature Switzerland AG 2019
J. Cao et al. (Eds.): ELM 2017, PALO 10, pp. 271–281, 2019.
https://doi.org/10.1007/978-3-030-01520-6_25

attack is closed-loop stimulation. The seizures control device can change the continues stimulation into focus stimulation before seizure by using a long-term seizure prediction. Therefore, the side effect of continuous stimulation and the energy consumption of the device would be reduced remarkably. Electroencephalogram (EEG) signals are the most direct reflection of the signal transduction in the brain. However, the symptoms of epilepsy in EEG signals are still unclear. It is crucial to find relationship between EEG signals and epilepsy seizures.

Seizure prediction is to determine whether the seizure is going to occur. In clinical study, the process of a seizure can be divide into four phases: interictal, preictal, ictal and postictal [6]. Interictal takes the longest duration in these phase. In the opposite, ictal is drowning in long period of peace time. But, ictal could be formed in long period of time before ictal is reflect in epilepsy performance. In seizure prediction, symptoms of seizure are seen to be contained in preictal EEG. The challenge of seizure prediction is to detect the symptoms in EEG signals. Generally, there is no need to detect ictal and postictal phases. Therefore, seizure prediction problem is converted into the classification of interictal and preictal phase. In recent years, studies on intracranial EEG (iEEG) recordings have found their usage in treatment of seizures. An over loop seizure control device can be implanted into the brain. When the advisory device predicts the seizure coming, the patient can receive treatment to prevent the seizure [7]. Following the traditional signal processing, approaches are proposed on the basis of feature extraction and classification [8–10]. Different temporal domain and frequent domain features are applied into seizure prediction, such as spike rate [11], spectral power [12], wavelet energy [13] and so on [14–17]. Williamson et al. [18] combine patient-specific machine learning and spatiotemporal correlation features to predict seizures. To detect 25 min prior to each seizure, it achieves sensitivity varies between 86% and 95%. Mirowski et al. [19] integrate 6 linear and non-linear bivariate features into a pseudo image, and classify the images with convolution neural networks (CNN). It is the first time that deep learning algorithm is used in seizure prediction, but the network is only used as a classifier. In 2016, Shiao et al. [20] extract features by combining signal in time, power spectrum and correlation matrix, and use SVM as classifier. The system achieves robust prediction of seizures from iEEG data. Overall, there are two aspects in feature extraction becomes more valuable, (1) correlation between electrodes; (2) ensemble of different features. Bag-of-Words (BoWs) are firstly applied to text classification [21], and widely used in computer vision [22, 23]. Inspired by BoWs feature extraction, the phase of an intracranial electroencephalography (iEEG) signal in seizure can be expressed by the changes of waves. The proposed Bag-of-Wave (BoWav) model measures the correlation in temporal and spatial by unsupervised learning. It provides different features to represent EEG signals.

In this paper, we apply BoWav model and ELM to achieve long-term seizure prediction through iEEG signal. First, we construct two different codebooks from interictal and preictal iEEG signals with clustering algorithm. Second, BoWav features are extracted for every segment of iEEG signal by projecting to two codebooks. Third, concatenating the feature extracted and classifying by ELM algorithm.

The contribution of this paper includes: (1) we proposed BoWav feature extraction for iEEG feature expression; (2) we proposed a framework based on BoWav feature extraction and ELM classifier for seizure prediction; (3) experiments show the proposed approach is effective and we visualized some features for further analysis.

The remainder of the paper is organized as follows. The method for seizure prediction is demonstrated in Sect. 2. The details of experiments are presented in Sect. 3. In the end, the conclusion and discussion are given in Sect. 4.

2   Method for Seizure Prediction

In this section, we first introduce the framework of the proposed seizure prediction algorithm. After that, the details of BoWav model and feature classification are described separately.

2.1   Framework Description

EEG signals measure the rapid voltage changes on electrodes. They are continuous on time and hard to understand. Therefore, feature extraction is important before classification. We try to find basic components from iEEG in order to express long-term signals. By clustering algorithm, waves in similar shape are categorized into same class. It makes waves in the same class can be represented by their cluster center. Therefore, we create our codebooks, which are consist of cluster centers, from iEEG signals. In seizure prediction, interictal and preictal codebooks are constructed from training samples. After that, signals in sliding window are represented as two histograms of codebooks. Then, feature of long-term EEG signal is extracted by concatenating the histogram of each window by time. In the end, the BoWav feature is classified by ELM. The framework of the seizure prediction is presented in Fig. 1.

2.2   BoWav Model for EEG Feature Extraction

In the proposed framework, BoWav model is proposed to express the state of the brain in a period of time. In a local segment of EEG, the temporal order and electrodes position are ignored. On the other hand, the correlation between signals on different electrodes are considered. It is because if there are similar waves on different electrodes, the related class on histogram would be increased.

The procedures are described as follows:

(1) Codebook construction. EEG signals from interictal and preictal contains different kinds of waves. Therefore, we construct interictal codebook D_{inter} and preictal codebook D_{pre} from different training data. Each sample is first divided into small segments according to the length of component wave L_w. To reduce the number of segments, we select a certain number of segments randomly from them. The selected segments are then clustered with K-means. Each cluster center c_k is seen as a word in codebook D.

(2) BoWav feature extraction. The feature of a sample can be extracted from the concatenate of the histogram in a sliding window S_t. The sliding window S_t contains L_S seconds of voltage sampling on every electrodes. Euclidian distance between wave w_i in S_t and cluster centers c_k is calculated to determine the

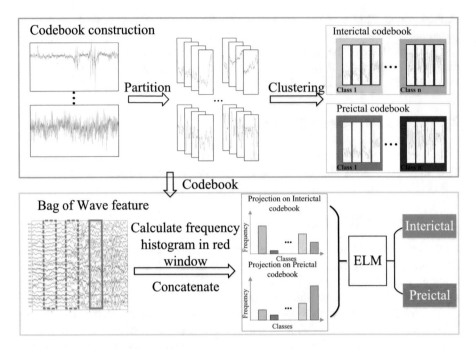

Fig. 1. Framework of proposed seizure prediction

category of w_i. If the nearest cluster center of w_i is c_k, k-th bin in histogram H is added. Suppose the number of electrodes is N_E, waves are slide with length L_{stride}. There will be $N_E \times \left(\frac{L_S - L_w}{L_{stride}} + 1 \right)$ waves projected onto the codebook D. The histogram H can be formulated in Eq. (1).

$$
H = \begin{bmatrix} h_1 \\ \vdots \\ h_K \end{bmatrix}, \ h_k = \sum_k w_{ik},
$$

$$
w_{ik} = \begin{cases} 1 & if \ k = \min_k ||w_i - c_k||^2 \\ 0 & else \end{cases},
\tag{1}
$$

$$
i = 1, \cdots, N_E \times \left(\frac{L_S - L_w}{L_{stride}} + 1 \right), \ k = 1, \cdots, K.
$$

In this way, two histogram features can be obtained with D_{inter} and D_{pre}. The feature of a sample is presented by concatenating two histogram features and features in every slide window S_t.

2.3   Classifier with ELM

To learn the pattern to distinguish interictal and preictal iEEG signals, extreme learning machine is used to classify BoWav feature. ELM is an effective learning algorithm based on single-hidden layer feed forward networks (SLFN). It first proposed by Huang et al. [24], and has been widely applied in many fields [25–27] because of its efficiency.

A classic ELM contains one hidden layer, the weights between input nodes and hidden nodes are generated with distributions with random parameters. And weights from hidden nodes to outputs are optimized by a close-form solution. For input $x \in \mathbb{R}^n$, the output of hidden layer is presented as $h(x) = [g(w_1, b_1, x), \cdots, g(w_L, b_L, x)]$, where $(w_i, b_i), i = 1, \cdots, L$ are generated from stochastic distribution. The structure of ELM is present in Fig. 2.

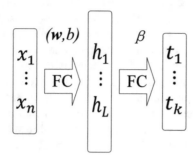

Fig. 2. ELM network structure

The output of a SLFNs is shown as follows:

$$f(x) = \sum_{i=1}^{L} \beta_i g_i(x) = h(x)\beta. \tag{2}$$

The β is learned from the optimization of Eq. (3)

$$\begin{aligned} min & \quad \tfrac{1}{2}||\beta||^2 + C \sum_{i=1}^{N} \xi_i^2, \\ s.t. & \quad h(x_i)\beta = t_i - \xi_i, i = 1, \cdots, N, \end{aligned} \tag{3}$$

where t_i is the label of the i-th data.

Thus, the learned β is formulated in Eqs. (4) and (5).

$$\beta = \left(H^T H + \frac{I}{C} \right)^{-1} H^T T, \tag{4}$$

when the dimension of input layer N is larger than hidden node number L and

$$\beta = H^T \left(H^T H + \frac{I}{C} \right)^{-1} T, \tag{5}$$

when $N < L$.

3 Experiments and Discussions

3.1 Dataset

We use the dataset published for Seizure Prediction Challenge on Kaggle. The dataset contains hours of iEEG signal recorded from dogs with naturally epilepsy occurrence and patients with iEEG monitoring before surgery. In the canine iEEG collection, the electrode arrays were implanted in five dogs, each of them contains 16 electrodes and sampled at 400 Hz. On epilepsy patients, the electrodes are placed by the considerations of clinical. The iEEG data are collected with 5000 Hz sampling from two patients, they are settled 15 electrodes and 24 electrodes, respectively. The preictal data are sampled before 10 s of seizure onset and 5 min to prevent the influence of early ictal activity. The interictal data are collected from time which is more than 4 h before or after seizures. They are collected 1 h of signal each time, and divided into 10 min clips.

The dataset we use is just part of Kaggle seizure prediction which has label on it. The details of the dataset are shown in Table 1. We have further clipped every 10 min into 10 segments on patient samples, which means 1 min in a sample. We use mean performance of 5-fold cross validation to get our final results. Models for dog seizure prediction and patient seizure prediction are separated.

Table 1. Details of Kaggle seizure prediction dataset

Subject	Channels	Interictal (10 min)	Preictal (10 min)
Dog 1	16	480	24
Dog 2	16	500	42
Dog 3	16	1440	72
Dog 4	16	804	97
Dog 5	15	450	40
Patient 1	15	50	18
Patient 2	24	42	18

3.2 Parameter Selection

As for the iEEG collected from patients, it has a small number of samples and a high sampling frequency in each samples. The raw iEEG data take up more than 40 GB. We first down sampling the data to a sampling frequency of 500 Hz. It is still a frequency with a medium resolution. For canine data, we preserve the sampling frequency of 400 Hz. We select 4000 segments of 200 ms from interictal and preictal samples separately. These segments are then clustered into 100 classes, which construct interictal codebook and preictal codebook. On patient signal, signals have been clipped into 1 min samples. We use a sliding window with 4 s. The time step for sliding window is 2 s. And stride in sliding window is set to 200 ms. Therefore, the dimension of the features for patient sample is $\left(\frac{60-4}{2} + 1\right) \times (2 \times 100) = 5980$. On canine signal, we use a sliding window with 3 s, and also a 2 s time step. There is also no overlap of wave matching in sliding window. The dimension of features for canine samples is 39600.

3.3   Evaluation

We evaluate the performance with AUC and Sensitivity (SS). AUC score is calculated from Eq. (6).

$$AUC = \frac{\sum_{ins_i \in positiveclass} rank_{ins_i} - \frac{M \times (M+1)}{2}}{M \times N}, \tag{6}$$

where M is the number of positive samples, and N is the number of negative samples. AUC is a common score to measure the performance of classifier. Sensitivity is also called true positive rate or recall. It measures the percentage of positive that are correctly identified, which is more important for medical treatment. In seizure prediction, a good sensitivity is very important. It is calculating with Eq. (7).

$$TPR(SS) = \frac{TP}{P} = \frac{TP}{TP + FN}, \tag{7}$$

where TP is the number of true positive, FN is the number of false positive.

3.4   Visualization

In this section we visualize the waves in codebook. The codebooks for patients are shown in Fig. 3. As shown in the codebooks, waves extracted from preictal samples contain more high frequency components. It is consistent with the clinical experience. Different from spike which is often used for seizure detection, each wave in codebook is a broader period of signal. The size of wave is larger than the common concept of spikes. Each wave in the codebook presents a local variance in some electrodes. In BoWav feature, these local variances constitute the status of brain in a small period of time. Machine learning approaches are promising in finding the changes in the status of brain.

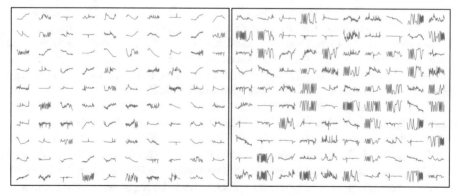

    (a) interictal codebook   (b) preictal codebook

Fig. 3. Visualization of codebooks

We also visualize the distribution of each clips in BoWav features with t-SNE. The visualization results are shown in Fig. 4. A 3D visualization is used in our experiment. In Fig. 4, two viewpoints are given in (a) and (b) separately. Blue points are data from interictal segments, and yellow points present preictal segments. All data points are divided into two parts, matching two patients in the dataset. It reflects that different patients have their specific processes in the development of seizure. The data from interictal and preictal are close to points in the same category. It means that the BoWav features have projected the original data into a more separable space.

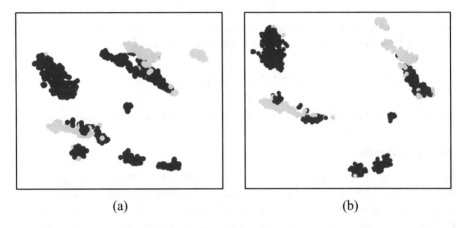

(a)   (b)

Fig. 4. t-SNE visualization of patient dataset

3.5   Prediction Results

The Sensitivity and AUC result presented are average value of 5-fold cross validation.

As comparison we also experiment three different features, Spectral relative power (SRP), Spectral entropy (SE), and Coherence mean (CM) and the combination of these three features on Kaggle dataset. The code [28] for the extraction of these features is from publish program on Github[1]. For samples in Kaggle dataset, we extracted these feature from every 1 min. Therefore, the dimension of SRP, SE and CM feature for 10 min' segment is 10, and the dimension of the combination feature is 30. After these features, we classify the samples with Kernel ELM. Results are shown in Table 2. The proposed BoWav feature achieves high performance on Kaggle dataset.

We also try to use different classifiers for seizure prediction. In our experiment, Logistic regression, SVM, ELM, Kernel ELM are implemented for prediction. Results are shown in Table 3. As shown in Table 3, results on Kernel ELM are higher than other classifiers. SVM also presents an accurate prediction result. For ELM, the number of hidden nodes are 500 and sigmoid activation function is selected. For Kernel ELM, regularization coefficient is 2^1, kernel function is set to be linear kernel, the number of kernel parameter is 500.

---

[1] https://github.com/otoolej/qEEG_feature_set.

Table 2. Result comparison between different features

Feature	Dog		Patient	
	Sensitivity (%)	AUC	Sensitivity (%)	AUC
Spectral_relative_power (SRP)	6.08	0.5099	55.95	0.5814
Spectral_entropy (SE)	5.30	0.5158	55.56	0.5932
Coherence_mean (CM)	10.59	0.5111	53.05	0.5771
SRP + SE + CM	13.28	0.5132	64.74	0.6307
BoWav	100	1	98.89	0.9944

Table 3. Result comparison between different classifiers

Classifier	Dog		Patient	
	Sensitivity (%)	AUC	Sensitivity (%)	AUC
Logistic regression	77.74	0.7988	71.12	0.7061
SVM	100	1	98.06	0.9877
ELM	96.59	0.9851	92.51	0.9358
Kernel ELM	100	1	98.89	0.9944

4 Conclusions

In this paper, we proposed a framework for long-term seizure prediction. A new Bag-of-Wave feature is extracted for the representation of EEG signal. Then, ELM algorithm is used as classifier for further prediction. From the visualization of codebook extracted and BoWav feature, the proposed feature shows consistence distribution with clinical experience. Experimental results show the proposed feature is better than other traditional features. The seizure prediction task is still limited by the individual difference of EEG signal. It is promising to combine BoWav feature with other feature to eliminate the influence brought by individual difference in the future.

Acknowledgements. This research is supported in part by NSFC [grant nos. 61672070, 61370113, and 61572004], the Beijing Municipal Natural Science Foundation [grant no. 4152005], the Science and Technology Program of Tianjin [grant no. 15YFXQGX0050], and the Science and Technology Planning Project of Qinghai Province [grant no. 2016-ZJ-Y04].

References

1. Shorvon, S.D., Goodridge, D.M.: Longitudinal cohort studies of the prognosis of epilepsy: contribution of the National General Practice Study of Epilepsy and other studies. Brain **136** (11), 3497–3510 (2013)
2. Rajna, P., et al.: Hungarian multicentre epidemiologic study of the warning and initial symptoms (prodrome, aura) of epileptic seizures. Seizure **6**(5), 361–368 (1997)
3. Mormann, F., et al.: Seizure prediction: the long and winding road. Brain **130**(2), 314–333 (2006)

4. Schulze-Bonhage, A., et al.: Seizure anticipation by patients with focal and generalized epilepsy: a multicentre assessment of premonitory symptoms. Epilepsy Res. **70**(1), 83–88 (2006)
5. Brinkmann, B.H., et al.: Crowdsourcing reproducible seizure forecasting in human and canine epilepsy. Brain **139**(6), 1713–1722 (2016)
6. Moghim, N., Corne, D.W.: Predicting epileptic seizures in advance. PLoS One **9**(6), e99334 (2014)
7. Gadhoumi, K., et al.: Seizure prediction for therapeutic devices: a review. J. Neurosci. Methods **260**, 270–282 (2016)
8. Sackellares, J.C., et al.: Predictability analysis for an automated seizure prediction algorithm. J. Clin. Neurophysiol. **23**(6), 509–520 (2006)
9. Kuhlmann, L., et al.: Patient-specific bivariate-synchrony-based seizure prediction for short prediction horizons. Epilepsy Res. **91**(2), 214–231 (2010)
10. Aarabi, A., He, B.: Seizure prediction in hippocampal and neocortical epilepsy using a model-based approach. Clin. Neurophysiol. **125**(5), 930–940 (2014)
11. Li, S., et al.: Seizure prediction using spike rate of intracranial EEG. IEEE Trans. Neural Syst. Rehabil. Eng. **21**(6), 880–886 (2013)
12. Bandarabadi, M., et al.: Epileptic seizure prediction using relative spectral power features. Clin. Neurophysiol. **126**(2), 237–248 (2015)
13. Gadhoumi, K., Lina, J.-M., Gotman, J.: Seizure prediction in patients with mesial temporal lobe epilepsy using EEG measures of state similarity. Clin. Neurophysiol. **124**(9), 1745–1754 (2013)
14. Park, Y., et al.: Seizure prediction with spectral power of EEG using cost-sensitive support vector machines. Epilepsia **52**(10), 1761–1770 (2011)
15. Cook, M.J., et al.: Prediction of seizure likelihood with a long-term, implanted seizure advisory system in patients with drug-resistant epilepsy: a first-in-man study. Lancet Neurol. **12**(6), 563–571 (2013)
16. Zheng, Y., et al.: Epileptic seizure prediction using phase synchronization based on bivariate empirical mode decomposition. Clin. Neurophysiol. **125**(6), 1104–1111 (2014)
17. Korshunova, I., et al.: Towards improved design and evaluation of epileptic seizure predictors. IEEE Trans. Biomed. Eng. **65**(3), 502–510 (2017)
18. Williamson, J.R., et al.: Seizure prediction using EEG spatiotemporal correlation structure. Epilepsy Behav. **25**(2), 230–238 (2012)
19. Mirowski, P., et al.: Classification of patterns of EEG synchronization for seizure prediction. Clin. Neurophysiol. **120**(11), 1927–1940 (2009)
20. Shiao, H.-T., et al.: SVM-based System for prediction of epileptic seizures from iEEG signal. IEEE Trans. Biomed. Eng. **64**(5), 1011–1022 (2017)
21. Sriram, B., et al.: Short text classification in twitter to improve information filtering. In: Proceedings of the 33rd International ACM SIGIR Conference on Research and Development in Information Retrieval. ACM (2010)
22. Filliat, D.: A visual bag of words method for interactive qualitative localization and mapping. In: IEEE International Conference on Robotics and Automation 2007. IEEE (2007)
23. Zhang, Y., Jin, R., Zhou, Z.-H.: Understanding bag-of-words model: a statistical framework. Int. J. Mach. Learn. Cybern. **1**(1-4), 43–52 (2010)
24. Huang, G.-B., Zhu, Q.-Y., Siew, C.-K.: Extreme learning machine: a new learning scheme of feedforward neural networks. IEEE International Joint Conference on Neural Networks 2004 Proceedings, vol. 2. IEEE (2004)
25. Huang, G.-B., et al.: Extreme learning machine for regression and multiclass classification. IEEE Trans. Syst., Man, Cybern., Part B (Cybern.) **42**(2), 513–529 (2012)

26. Zong, W., Huang, G.-B.: Face recognition based on extreme learning machine. Neurocomputing **74**(16), 2541–2551 (2011)
27. Tang, J., Deng, C., Huang, G.-B.: Extreme learning machine for multilayer perceptron. IEEE Trans. Neural Netw. Learn. Syst. **27**(4), 809–821 (2016)
28. Toole, J.M.O., Boylan, G.B.: NEURAL: quantitative features for newborn EEG using Matlab. arXiv preprint (2017). arXiv:1704.05694

A Robust Object Tracking Approach with a Composite Similarity Measure

Shu-Heng Ma, Zhan-Li Sun$^{(\boxtimes)}$, and Cheng-Gang Gu

School of Electrical Engineering and Automation, Anhui University, Hefei, China
zhlsun2006@126.com

Abstract. How to achieve a robust performance remains an intractable problem in the various object tracking algorithms due to some unfavorable factors, e.g. occlusions, appearance change, etc. In this paper, a robust object tracking approach is proposed based on a composite similarity measure. Experimental results on several challenging sequences demonstrate the effectiveness and feasibility of the proposed method.

Keywords: Object tracking · Sparse collaborative model
Adaptive structural local sparse appearance model

1 Introduction

Given the initial regions of objects, the task of visual tracking is to estimate their locations in the consecutive video frames. Due to wide applications, e.g. video surveillance, human-computer interaction, etc., object tracking has long been an important topic in the computer vision field [1,2]. Nevertheless, object tracking remains a challenging topic nowadays because of many unfavorable factors, such as, cluttering background, similar objects and appearance changes caused by pose, illumination, occlusion and motion [3,4].

So far, many works have been developed to alleviate the negative effectiveness of these unfavorable factors. In [5], the hierarchical dense structures on an undirected hypergraph was exploited in handling long-term occlusions or distinguishing spatially close targets with similar appearance in crowded scenes. In order to avoid the gap between the label prediction and the object position, an adaptive visual object tracking was presented in [6] based on structured output prediction.

In order to alleviate the negative effectiveness caused by the drastic appearance change, a sparse collaborative model (SCM) was proposed in [7] by exploiting both the holistic templates and the local representations. The experiments

The work was supported by a grant from National Natural Science Foundation of China (No. 61370109), a key project of support program for outstanding young talents of Anhui province university (No. gxyqZD2016013), a grant of science and technology program to strengthen police force (No. 1604d0802019), and a grant for academic and technical leaders and candidates of Anhui province (No. 2016H090).

© Springer Nature Switzerland AG 2019
J. Cao et al. (Eds.): ELM 2017, PALO 10, pp. 282–291, 2019.
https://doi.org/10.1007/978-3-030-01520-6_26

on several challenging videos demonstrated that the tracker performed favorably against several state-of-the-art algorithms. Specifically, an adaptive structural local sparse appearance (ASLSA) model was presented in [8] by exploiting both partial information and spatial information of the target. Nevertheless, we found that the stability of ASLSA is still needed to be improved. Inspired by the complementary of the similarity measures in ASLSA and SCM, a composite similarity measure is proposed in this paper to improve the stability of object tracking. Experimental results on several challenging videos demonstrate the effectiveness and feasibility of the proposed algorithm.

The remainder of the paper is organized as follows. In Sect. 2, we present the proposed composite similarity measure based object tracking algorithm. Experimental results and related discussions are given in Sect. 3, and our concluding remarks are presented in Sect. 4.

2   Methodology

There are three main components in the proposed method: Compute the likelihood value of ASLSA, compute the confidence value of SCM, design the composite similarity measure. A detailed description of these three parts is presented in the following subsections.

2.1   The Likelihood Value of ASLSA

Given the first several frames, a set of target templates are first sampled around the target region. Then, the best template for each frame is searched by means of kd-tree function. Furthermore, an initial dictionary $\mathbf{D}$ is constructed by dividing each template into N local patches with a smaller size. For each local patch $\mathbf{a}_i$ of a target candidate, the sparse representation model can be formulated as [8],

$$\min_{\mathbf{b}_i} \| \mathbf{a}_i - \mathbf{D}\mathbf{b}_i \|_2^2 + \lambda \| \mathbf{b}_i \|_1$$
$$s.t. \ \mathbf{b}_i \geq 0 \tag{1}$$

where $\mathbf{b}_i$ denotes the sparse coding coefficients of local patchs. The constraint $\mathbf{b}_i \geq 0$ means that all the elements of $\mathbf{b}_i$ are nonnegative. Assume that there are n candidate target templates in one frame, a so-called histogram can be given by,

$$\rho = [\mathbf{B}_1, \mathbf{B}_2, \cdots, \mathbf{B}_n]. \tag{2}$$

Given a predefined threshold ε_0, the smoothing operator can be given by,

$$\mathbf{o}_i = \begin{cases} \varepsilon_i < \varepsilon_0, \\ \text{otherwise,} \end{cases} \tag{3}$$

Further, the smoothing operation is performed as,

$$\mathbf{s} = \rho \odot \mathbf{o}, \tag{4}$$

where $\odot$ denotes the element-wise multiplication. After smoothing, the normalized weighting coefficient of i-th local patch can be computed as,

$$\mathbf{v}_i = \frac{1}{\mathbf{C}} \sum_{k=1}^{n} \mathbf{s}_i^{(k)}, i = 1, 2, \cdots, N, \tag{5}$$

Further, the sum h_i of $\mathbf{v}_i$ is used as likelihood value of the ASLSA model.

2.2 The Confidence Value of SCM

The target on the first frame is manually marked. The positive templates $\mathbf{Y}^+$, mostly covered by the target, are sampled around the target position. On the contrary, the negative templates $\mathbf{Y}^-$, mostly covered by the background, are sampled slightly away from the target position.

For the current frame, M candidates are drawn around the estimated target position in the previous frame with a particle filter. The sparse decomposition of one candidate ($\mathbf{c}$) that has not been processed by the image smoothing can be given as follows [7],

$$\min_{\mathbf{x}} \| \mathbf{c} - \mathbf{Y}\mathbf{x} \|_2^2 + \lambda \| \mathbf{x} \|_1, \tag{6}$$

where $\mathbf{Y} = [\mathbf{Y}^+, \mathbf{Y}^-]$. $\mathbf{x}$ is the corresponding sparse code of one candidate. The reconstruction error e^+ with the foreground template set and the reconstruction error e^- with the background template set $\mathbf{Y}^+$ can be computed as

$$e^+ = \|\mathbf{c} - \mathbf{Y}^+\mathbf{x}^+\|_2^2 \tag{7}$$

and

$$e^- = \|\mathbf{c} - \mathbf{Y}^-\mathbf{x}^-\|_2^2, \tag{8}$$

respectively. After obtaining e^+ and e^-, the confidence value c_y of the candidate $\mathbf{c}$ can be defined as,

$$c_y = \exp(-(e^+ - e^-)/\sigma). \tag{9}$$

2.3 The Composite Similarity Measure

After obtaining the confidence value c_i and the similarity of histograms h_i, the composite similarity measure of the i-th candidate is defined as,

$$p_i = c_i h_i. \tag{10}$$

In terms of (10), we can obtain the composite similarity measure values of all candidates. After sorting, the candidate with the maximum composite similarity measure value is selected as the target region of the current frame.

3 Experiments

The performance of the proposed composite similarity measure (CSM) model is evaluated on some typical video sequences, which contains several challenging situations in object tracking, such as heavy occlusion, motion blur, in-plane and out-of-plane rotation, large illumination change, scale variation, complex background, etc [7]. Figure 1 shows some typical frames with unfavorable factors in some video sequences.

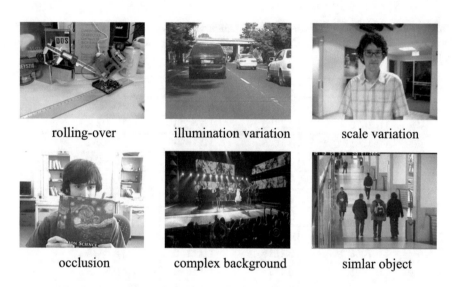

| rolling-over | illumination variation | scale variation |

| occlusion | complex background | simlar object |

Fig. 1. Some typical frames with unfavorable factors in 9 video sequences.

For comparison, we carry out the experiments with three state-of-the-art object tracking algorithms, i.e. the SCM algorithm [7], the compressive tracking algorithm [9], and the L1 Tracker using accelerated proximal gradient approach (L1APG) algorithm [10]. In experiments, the parameters of the various methods are set as the default values in the program packages provided by the corresponding authors. Two widely used performance indices, the overlapping rate (r_o) and the center location error (ϵ_c), are used here to evaluate the object tracking performance.

All simulations were conducted in the MATLAB environment running on an ordinary personal computer with double 3.0-GHZ CPU and 4-GB memory. As the problem concerned in this paper is the instability caused by the random sampling of the training sets, ten trials are performed for each sequence by setting different random seeds, i.e. from 1 to 10. For a identical random seed, the training sets are invariable when the trials are repeated in the MATLAB environment.

Take the sequence *woman_sequence, board, car11, faceocc2* for example, Tables 1, 2, 3, 4, 5, 6, 7 and 8 show the overlapping rates and the center-location errors of ten trials with different random seed number (SN), and the corresponding mean and standard deviation values ($\mu \pm \sigma$).

Tables 9 and 10 show the mean and standard deviation values ($\mu \pm \sigma$) of ten trials of the overlapping rates and the center-location errors, respectively. For each sequence, the best result and the second-best result are highlighted in red and blue, respectively. We can see that the performance of CT is obviously not as

Table 1. The overlapping rates of the sequence *woman_sequence* for ten trials with different random seed number (SN), and the corresponding mean and standard deviation values ($\mu \pm \sigma$).

SN	SCM	CT	L1APG	CSM
1	0.181	0.1599	0.0632	0.8471
2	0.1675	0.1709	0.0626	0.8008
3	0.1691	0.1648	0.1639	0.8264
4	0.1549	0.15	0.0663	0.8415
5	0.1774	0.1609	0.0643	0.8317
6	0.1509	0.1329	0.1633	0.8269
7	0.1792	0.1684	0.0578	0.1819
8	0.1771	0.1748	0.1816	0.8331
9	0.7053	0.1516	0.1766	0.8478
10	0.4925	0.1715	0.1684	0.8469
$\mu \pm \sigma$	0.2555 ± 0.1881	0.1606 ± 0.0128	0.1168 ± 0.0572	0.7684 ± 0.2066

Table 2. The center-location errors of the sequence *woman_sequence* for ten trials with different random seed number (SN), and the corresponding mean and standard deviation values ($\mu \pm \sigma$).

SN	SCM	CT	L1APG	CSM
1	160.298	109.1778	164.6071	2.7047
2	163.8442	108.3892	163.4712	3.0109
3	174.8962	109.9774	129.9975	2.9717
4	164.2194	108.5402	143.87	2.3016
5	132.9356	113.8328	157.4566	2.4298
6	121.5435	112.3248	114.8274	2.4898
7	114.5318	113.7015	148.2793	147.3537
8	126.2511	114.0808	129.0217	2.6586
9	8.013	110.4557	122.6223	2.7344
10	42.2027	105.1306	128.8145	2.1774
$\mu \pm \sigma$	120.8736 ± 55.1158	110.5611 ± 2.9178	140.2968 ± 17.7173	17.0833 ± 45.7732

Table 3. The overlapping rates of the sequence *board* for ten trials with different random seed number (SN), and the corresponding mean and standard deviation values $(\mu \pm \sigma)$.

SN	SCM	CT	L1APG	CSM
1	0.3856	0.4998	0.1564	0.7501
2	0.1505	0.4586	0.1553	0.7471
3	0.3917	0.6297	0.1357	0.7259
4	0.3198	0.6573	0.1442	0.8047
5	0.6621	0.4944	0.0682	0.8233
6	0.6357	0.4145	0.0835	0.8036
7	0.6973	0.6151	0.0563	0.744
8	0.1771	0.562	0.0634	0.6926
9	0.2587	0.5602	0.1417	0.8113
10	0.2757	0.5349	0.0605	0.7668
$\mu \pm \sigma$	0.3954 ± 0.2018	0.5427 ± 0.0777	0.1065 ± 0.0433	0.7669 ± 0.0426

Table 4. The center-location errors of the sequence *board* for ten trials with different random seed number (SN), and the corresponding mean and standard deviation values $(\mu \pm \sigma)$.

SN	SCM	CT	L1APG	CSM
1	89.9184	54.7441	136.6233	7.2938
2	150.7434	61.7923	133.2281	7.2287
3	97.4459	41.9543	141.5647	7.5512
4	89.2823	34.9	136.4032	8.2863
5	27.207	56.4613	262.2831	8.6862
6	53.4269	76.861	227.5954	7.8609
7	19.584	42.2834	268.0604	7.4576
8	140.1907	51.1819	262.053	7.7596
9	120.6125	43.0808	140.1248	8.2287
10	116.3416	56.8747	270.8578	7.7051
$\mu \pm \sigma$	90.4753 ± 44.8857	52.0134 ± 12.1706	197.8794 ± 64.6518	7.8058 ± 0.4692

good as other three methods. In general, the tracking results of SCM and CSM are better than that of CT and L1APG. Moreover, we can see that the tracking results of 9 video sequences are relative good either for CSM. Thus, CSM is more robust than other methods to the instability caused by the random sampling.

Table 5. The overlapping rates of the sequence *car11* for ten trials with different random seed number (SN), and the corresponding mean and standard deviation values ($\mu \pm \sigma$).

SN	SCM	CT	L1APG	CSM
1	0.8085	0.6196	0.5815	0.8521
2	0.8059	0.3325	0.5818	0.8182
3	0.8057	0.0127	0.5714	0.8291
4	0.8057	0.1451	0.5577	0.8589
5	0.7918	0.1171	0.8082	0.8425
6	0.7784	0.2877	0.5669	0.8205
7	0.8196	0.029	0.5835	0.8432
8	0.8011	0.0127	0.5622	0.8352
9	0.7835	0.5299	0.5661	0.8222
10	0.7972	0.6875	0.8136	0.8393
$\mu \pm \sigma$	0.7997 ± 0.0123	0.2774 ± 0.2576	0.6193 ± 0.1014	0.8361 ± 0.0137

Table 6. The center-location errors of the sequence *car11* for ten trials with different random seed number (SN), and the corresponding mean and standard deviation values ($\mu \pm \sigma$).

SN	SCM	CT	L1APG	CSM
1	1.6212	5.4742	18.432	1.387
2	1.6882	14.361	18.3046	1.904
3	1.6465	75.3598	18.8686	1.6666
4	1.6463	24.0521	18.8058	1.2889
5	1.7431	57.2918	1.4526	1.586
6	1.7028	38.2836	18.4684	1.9459
7	1.6365	98.5651	19.6962	1.6686
8	1.629	82.9618	18.6205	1.7872
9	1.6842	8.511	18.3701	1.583
10	1.6787	4.3394	1.4961	1.3846
$\mu \pm \sigma$	1.6677 ± 0.0384	40.92 ± 35.2552	15.2515 ± 7.2721	1.6202 ± 0.2214

Table 7. The overlapping rates of the sequence *faceocc2* for ten trials with different random seed number (SN), and the corresponding mean and standard deviation values $(\mu \pm \sigma)$.

SN	SCM	CT	L1APG	CSM
1	0.8271	0.5969	0.3287	0.8471
2	0.8331	0.6428	0.3362	0.5348
3	0.8403	0.4241	0.3595	0.8297
4	0.8299	0.5566	0.3478	0.8275
5	0.8409	0.5881	0.3601	0.5975
6	0.8342	0.5881	0.3426	0.8078
7	0.8371	0.5623	0.3221	0.8229
8	0.8412	0.5846	0.3321	0.8363
9	0.8359	0.6155	0.3196	0.8387
10	0.8393	0.61	0.3363	0.823
$\mu \pm \sigma$	0.8359 ± 0.0048	0.5769 ± 0.0593	0.3385 ± 0.0141	0.7765 ± 0.1124

Table 8. The center-location errors of the sequence *faceocc2* for ten trials with different random seed number (SN), and the corresponding mean and standard deviation values $(\mu \pm \sigma)$.

SN	SCM	CT	L1APG	CSM
1	4.4289	16.3885	26.3159	3.4739
2	4.4233	12.1551	16.0066	44.9475
3	4.2452	31.462	17.181	3.9354
4	4.4714	15.9244	23.7691	4.2859
5	4.2178	15.7652	17.9755	24.7366
6	4.4093	14.8789	24.3291	5.1295
7	4.2501	19.1515	19.5144	4.5343
8	4.2421	15.6997	18.5083	3.9274
9	4.393	14.4874	19.8671	3.7261
10	4.2348	15.6047	23.743	4.175
$\mu \pm \sigma$	4.3316 ± 0.1009	17.1517 ± 5.318	20.721 ± 3.5296	10.2872 ± 13.798

Table 9. The mean and standard deviation values ($\mu \pm \sigma$) of ten trials of the overlapping rates for 9 sequences.

Sequence	SCM	CT	L1APG	CSM
woman_sequence	0.2555 ± 0.1881	0.1606 ± 0.0128	0.1168 ± 0.0572	0.7684 ± 0.2066
faceocc2	0.8359 ± 0.0048	0.5769 ± 0.0593	0.3385 ± 0.0141	0.7765 ± 0.1124
ThreePastShop2cor	0.3445 ± 0.2879	0.1881 ± 0.0212	0.3908 ± 0.0761	0.7284 ± 0.2742
singer1	0.8755 ± 0.0063	0.3336 ± 0.0149	0.201 ± 0.0242	0.7954 ± 0.0172
car4	0.8998 ± 0.006	0.2221 ± 0.0189	0.2593 ± 0.1078	0.9037 ± 0.0032
car11	0.7997 ± 0.0123	0.2774 ± 0.2576	0.6193 ± 0.1014	0.8361 ± 0.0137
davidin300	0.7167 ± 0.1722	0.432 ± 0.0781	0.2743 ± 0.1224	0.734 ± 0.1477
board	0.3954 ± 0.2018	0.5427 ± 0.0777	0.1065 ± 0.0433	0.7669 ± 0.0426
stone	0.5496 ± 0.1665	0.3454 ± 0.0534	0.5942 ± 0.1035	0.417 ± 0.2092

Table 10. The mean and standard standard deviation values of the center location errors ($\mu \pm \sigma$) of five methods for 9 sequences.

Sequence	SCM	CT	L1APG	CSM
woman_sequence	120.8736 ± 55.1158	110.5611 ± 2.9178	140.2968 ± 17.7173	17.0833 ± 45.7732
faceocc2	4.3316 ± 0.1009	17.1517 ± 5.318	20.721 ± 3.5296	10.2872 ± 13.798
ThreePastShop2cor	52.6075 ± 26.5033	59.0003 ± 10.7181	22.2953 ± 14.5313	14.8578 ± 26.2953
singer1	3.2055 ± 0.2527	15.5628 ± 2.8598	149.7278 ± 11.0075	5.0352 ± 0.5956
car4	3.6346 ± 0.4437	202.9943 ± 27.3357	163.797 ± 81.5155	3.8665 ± 0.2762
car11	1.6677 ± 0.0384	40.92 ± 35.2552	15.2515 ± 7.2721	1.6202 ± 0.2214
davidin300	11.2227 ± 17.8841	19.4869 ± 11.4169	65.6158 ± 26.3384	11.2276 ± 24.2121
board	90.4753 ± 44.8857	52.0134 ± 12.1706	197.8794 ± 64.6518	7.8058 ± 0.4692
stone	20.2255 ± 51.5028	26.1649 ± 11.4479	9.4231 ± 10.9066	31.5113 ± 45.4465

4   Conclusions

In this paper, a robust object tracking approach is proposed based on a composite similarity measure. Moreover, the experimental results on 9 typical video sequences have demonstrated the effectiveness and feasibility of the proposed CSM model.

References

1. Oh, H., Shiraz, A.R., Jin, Y.: Morphogen diffusion algorithms for tracking and herding using a swarm of kilobots. Soft Comput. **22**(6), 1833–1844 (2018)
2. Ahmed, S.A., Dogra, D.P., Kar, S.: Unsupervised classification of erroneous video object trajectories. Soft Comput. **8**, 1–19 (2017)
3. Lin, L., Lin, W., Huang, S.: Group object detection and tracking by combining RPCA and fractal analysis. Soft Comput. **22**(1), 231–242 (2018)
4. Cabido, R., Montemayor, A.S., Pantrigo, J.J.: High performance memetic algorithm particle filter for multiple object tracking on modern GPUs. Soft Comput. **16**(2), 217–230 (2012)
5. Wen, L., Lei, Z., Lyu, S., Li, S.Z., Yang, M.H.: Exploiting hierarchical dense structures on hypergraphs for multi-object tracking. IEEE Trans. Pattern Anal. Mach. Intell. **38**(10), 1983–1996 (2016)
6. Hare, S., et al.: Struck: structured output tracking with kernels. IEEE Trans. Pattern Anal. Mach. Intell. **38**(10), 2096–2109 (2016)
7. Yang, M.H., Lu, H., Zhong, W.: Robust object tracking via sparsity-based collaborative model. In: IEEE Conference on Computer Vision and Pattern Recognition, pp. 1838–1845 (2012)
8. Jia, X.: Visual tracking via adaptive structural local sparse appearance model. In: IEEE Conference on Computer Vision and Pattern Recognition, pp. 1822–1829 (2012)
9. Zhang, K.H., Zhang, L., Yang, M.H.: Real-time compressive tracking. In: Proceedings of the 12th European conference on Computer Vision, pp. 864–877 (2012)
10. Bao, C., Wu, Y., Ling, H., Ji, H.: Real time robust L1 tracker using accelerated proximal gradient approach. In: IEEE Conference on Computer Vision and Pattern Recognition, pp. 1830–1837 (2012)

Target Coding for Extreme Learning Machine

Dongshun Cui[1,2(✉)], Kai Hu[3], Guanghao Zhang[2], Wei Han[2],
and Guang-Bin Huang[2]

[1] Energy Research Institute @ NTU (ERI@N), Interdisciplinary Graduate School,
Singapore, Singapore
[2] School of Electrical and Electronic Engineering, Nanyang Technological University,
Singapore, Singapore
{dcui002,gzhang009,hanwei,egbhuang}@ntu.edu.sg
[3] College of Information Engineering, Xiangtan University, Xiangtan, China
kaihu@xtu.edu.cn

Abstract. Target coding is an indispensable part of supervised learning. Currently, the assumption of the most popular target coding like one-hot (one-of-K) coding is that targets are independent. However, this assumption is limited due to the complex relationship between targets. In this paper, we will explore the effects of kinds of target coding methods on the performance of Extreme Learning Machine Classifiers (ELM-C). Linearly independent coding (e.g., one-of-k coding, Hadamard coding) and linearly dependent coding (e.g., ordinal coding, binary coding) are analyzed and compared. The experimental results on OCR letter dataset show that different target coding will indeed affect the performance of the same classifier.

Keywords: Target coding · Extreme Learning Machine
ELM classifier · Hadamard coding

1 Introduction

Feature extraction/learning and Target coding are two essential parts for supervised learning, especially for classification problems. With years of research, there are lots of work have been done on how to extract effective features using ELMs [1,2]. However, little work has been done on target coding for ELM. By far, most people adopt one-of-k target coding methods instantly when they use ELM classifiers [3], while others adopt the numeric coding methods to simplify the computation.

One-of-k (one-hot) coding comes from electronics, and there is only one "hot" value in the coding list. It is introduced into the modern data science by [4], and the one "hot" value is replaced of '1' for simplicity reasons. One-of-k coding

D. Cui—This work is supported by Delta Joint Lab.

J. Cao et al. (Eds.): ELM 2017, PALO 10, pp. 292–303, 2019.
https://doi.org/10.1007/978-3-030-01520-6_27

is the most frequently used target coding method for machine-learning based classification applications in recent years [5].

In this paper, we have explored the effects of one-of-k coding methods on ELM classifiers. Besides, motivated by the properties of orthogonality and equal weight (each coding has the same number of non-zero elements) of Hadamard coding [6–8], we have compared its effects with one-of-k coding. Two simple coding methods ordinal coding and binary coding have also been compared to show the effective of one-of-k coding and haramard coding. To the best of our knowledge, this is the first discussion on target coding for ELM.

The structure of this paper is as follows. We give a brief introduction on ELM-C and target coding in Sect. 2. Detailed analysis of target coding methods for ELM classifiers are explained in Sect. 3 and results are presented in Sect. 4. Finally, Sect. 5 summarizes our contribution and the future work.

2 Preliminaries

Extreme Learning Machine for Classification. ELM is a single (hidden) layer forward neural network which has been proposed in [9]. The weights $(\boldsymbol{W})$ and bias $(\boldsymbol{B})$ between the input layer and hidden layer are randomly assigned, while only the weights $(\boldsymbol{\beta})$ between the hidden layer and output layer need to be computed by solving a linear equation. This network has been proved to have good generalization performance with high training and test speed with remarkable classification and regression ability [10–14].

Assume there are N_1 training samples $\boldsymbol{X}_1$, and the feature dimension of each sample is N_I. We use $\boldsymbol{Y}_1$ to denote the label matrix which consists of coding vectors of the training target and the length of each coding vector is N_O. The number of hidden nodes is N_H, then the dimension of $\boldsymbol{W}$ and $\boldsymbol{B}$ is $N_I \times N_H$ and $N_1 \times N_H$ respectively. Noted that each row in $\boldsymbol{B}$ is a duplicate of the first row. Now, with the full-connected single-layer forward network, the output of hidden nodes is

$$\boldsymbol{H}_1 = g(\boldsymbol{X}_1 \boldsymbol{W} + \boldsymbol{B}), \tag{1}$$

where $g(*)$ is an activation function.

Training this network is equivalent to solving a linear system

$$\boldsymbol{Y}_1 = \boldsymbol{H}_1 \boldsymbol{\beta}, \tag{2}$$

and the minimum norm least squares solution is

$$\boldsymbol{\beta} = \boldsymbol{H}_1^\dagger \boldsymbol{Y}_1. \tag{3}$$

Here † is the Moore-Penrose inverse operator, and the dimension of $\boldsymbol{\beta}$ is $N_H \times N_O$.

For the N_2 testing samples $\boldsymbol{X}_2$, similar steps are performed, and the output of hidden nodes are

$$\boldsymbol{H}_2 = g(\boldsymbol{X}_2 \boldsymbol{W} + B), \tag{4}$$

and the predicted labels are

$$\widehat{\boldsymbol{Y}}_2 = \boldsymbol{H}_2 \boldsymbol{\beta}. \tag{5}$$

Here, $\widehat{Y}_2$ is the predicted output of the test samples. To evaluate the classification accuracy, $\widehat{Y}_2$ should be compared with Y_2, and more details will be discussed on performance evaluation in the following sections.

To avoid the singularity of the matrix H, a unified learning network of ELM is proposed in [12] which minimizes the training error, and the norm of the output weight matrix by introducing a regularization factor C. The target is to minimize $\frac{1}{2}\|\boldsymbol{\beta}\|^2 + \frac{1}{2}C\sum_{i=1}^{N}|\boldsymbol{\xi}_i|^2$. Here, $\boldsymbol{\xi}_i$ is the i-th training error. The solution of this function is

$$\boldsymbol{\beta} = \begin{cases} \boldsymbol{H}^T\left(\frac{\boldsymbol{I}}{C} + \boldsymbol{H}_1\boldsymbol{H}_1^T\right)^{-1}\boldsymbol{Y}_1, N_1 < N_H \\ \left(\frac{\boldsymbol{I}}{C} + \boldsymbol{H}_1\boldsymbol{H}_1^T\right)^{-1}\boldsymbol{H}^T\boldsymbol{Y}_1, N_1 \geq N_H \end{cases} \tag{6}$$

Target Coding. Coding techniques have been developing for many years in the field of information science [15,16], and various methods have been proposed for different propose like data compression, forward error correction, and encryption. Recently, coding for targets has gained a growing attraction from machine learning community, especially the supervised learning like classification [17,18]. The detailed definition of target coding is given in [8], and here we just restate it briefly as following.

A target coding (code) $\mathcal{T}$ is defined as a matrix whose dimension is $n \times l$, where n is the number of unique categories and l is the length of a vector whose elements come from an integer set $\mathcal{S}$ (*alphabet set*). Each item in $\mathcal{S}$ is called a *symbol*, so every element in $\mathcal{T}$ is a symbol.

For example, we have a dataset which contains three instances. These three instances are divided into 2 groups, and their categories are 'dog', 'cat', 'dog' in turn. Assume $\mathcal{S} = \{0,1\}$, and $l = 4$. Then one possible assignment of target coding is

$$\mathcal{T} = \begin{bmatrix} 0,1,0,1 \\ 1,0,1,0 \\ 0,1,0,1 \end{bmatrix}, \tag{7}$$

so the coding of 'dog' is $[0,1,0,1]$, and 'cat' is $[1,0,1,0]$.

3   Target Coding for ELM's Classification

From Eqs. 3 and 5, we have

$$\widehat{Y}_2 = H_2 H_1^\dagger Y_1. \tag{8}$$

Substitute Eqs. 1 and 4 into Eq. 8, we can obtain

$$\widehat{Y}_2 = [g(X_2W + B)]\,[g(X_1W + B)]^\dagger\,Y_1. \tag{9}$$

Assume $M = [g(X_2W + B)]\,[g(X_1W + B)]^\dagger$, then M is determined by the training samples X_1, test samples X_2, weight matrix W and bias matrix B. For ELM, W and B are randomly generated. So M will be a determined matrix once we know the training and test samples.

$$\widehat{Y}_2 = MY_1. \tag{10}$$

For the classification problem of a test set, one sample is a *true sample* when its predicted label is equal to its true label. Otherwise, it is a *false sample*. The accuracy of the classifier is

$$Accuracy = \frac{\#TS}{\#TS + \#FS}, \tag{11}$$

where $\#TS$ and $\#FS$ represent the number of true samples and false samples respectively.

For ELM classification, the true label matrix and predicted label matrix of a test set are denoted as $\boldsymbol{Y}_2$ and $\widehat{\boldsymbol{Y}}_2$. According to Eq. 11, we can derive that the test accuracy of an ELM classifier is

$$Accuracy = \frac{\sum_{i=1}^{N} \mathfrak{B}\left(y_{2_i}, \mathfrak{S}\left(\boldsymbol{Y}_2, \widehat{y}_{2_i}\right)\right)}{N_2}, \tag{12}$$

where $\mathfrak{B}$ is a logical function and $\mathfrak{S}$ is a nearest-neighbor search function. Their analytical formulas are

$$\mathfrak{B}(\boldsymbol{v_1}, \boldsymbol{v_2}) = \begin{cases} 1, \boldsymbol{v_1} = \boldsymbol{v_2} \\ 0, \boldsymbol{v_1} \neq \boldsymbol{v_2}, \end{cases} \tag{13}$$

and

$$\arg\min_{y_2'} \mathfrak{S}(\boldsymbol{Y}_2, \widehat{y}_{2_i}) = \{y_2' \in \boldsymbol{Y}_2 \mid D(y_2', \widehat{y}_{2_i}) = \min_{y_2''} D(y_2'', \widehat{y}_{2_i})\}, \tag{14}$$

where $\boldsymbol{v_1}$ and $\boldsymbol{v_2}$ are arbitrary equal-length vectors, and $D(\cdot)$ is a distance matrix.

There are many kinds of linearly dependent coding. Actually, a group of vectors must be linearly dependent if the dimension of the vector L_C is smaller than a number of vectors N_C. The most popular linearly dependent coding are ordinal coding and binary coding.

Ordinal Coding. Ordinal coding means each class of target is represented as an ordinal integer, shown in Table 1. This is the simplest way to encode targets, and ordinal coding can not only be used for classification but also for regression.

Table 1. Ordinal coding for different classes

Classes	Ordinal coding
#1	1
#2	2
#3	3
#⋯	⋯
#N	n

Binary Coding. Binary coding indicates that each class of target is represented as a binary number. An example is shown in Table 2, where the dimension of

Table 2. Two-bit binary coding for different classes

Classes	Two-bit binary coding
#1	0 0
#2	0 1
#3	1 0
#4	1 1

each binary coding is two, and the amount is four. Binary coding can be used for single-label classification and multi-label classification.

There are also many methods for linearly independent coding, and the two most widely adopted coding are one-of-k coding and Hadamard coding.

One-of-k Coding. Assume the dimension of one target code vector is K, then one-of-k coding means the k-th element of the vector of the k-th class is one and other elements are all zero. An simple example is shown in Table 3.

Table 3. One-of-k coding for different classes

Classes	One-of-k coding
#1	1 0 0 0 $\cdots$ 0
#2	0 1 0 0 $\cdots$ 0
#$\cdots$	$\cdots$
#N	0 0 0 0 $\cdots$ 1

Table 4. Hadamard coding for different classes

Classes	Hadamard coding
#1	1 1 1 1
#2	1 0 1 0
#3	1 1 0 0
#4	1 0 0 1

Hadamard Coding. The Hadamard code (shown in Table 4) is one of the popular error-correcting codes which are usually adopted in communication systems. Hadamard code is generated by using the Hadamard matrix $\mathcal{H}$, and the dimension (denoted as n) of $\mathcal{H}$ should be not less than the amount (denoted as m) of code we need. Each element of a Hadamard matrix is either -1 or $+1$, and $\mathcal{H}\mathcal{H}_T = nI$, where T is a transpose operator and I is a unit matrix. One way to produce a new Hadamard matrix is to embed an old Hadamard matrix into a specified template which is,

$$\mathcal{H}_{new} = \begin{bmatrix} \mathcal{H}_{old} & \mathcal{H}_{old} \\ \mathcal{H}_{old} & -\mathcal{H}_{old} \end{bmatrix}.$$

After we have the relationship of $\mathcal{H}_{new}$ and $\mathcal{H}_{old}$, we also need some initialization kernels. Here here we only list three cases: $n \in \{p * 2^q, n \geq 2 | p \in [1, 12, 20], q \in \mathbb{N}\}$, and the three kernels are:

$$\mathcal{H}_2 = \begin{bmatrix} + & + \\ + & - \end{bmatrix},$$

$$\mathcal{H}_{12} = \begin{bmatrix}
+ & + & + & + & + & + & + & + & + & + & + & + \\
+ & - & + & - & + & + & + & - & - & - & + & - \\
+ & - & - & + & - & + & + & + & - & - & - & + \\
+ & + & - & - & + & - & + & + & + & - & - & - \\
+ & - & + & - & - & + & - & + & + & + & - & - \\
+ & - & - & + & - & - & + & - & + & + & + & - \\
+ & - & - & - & + & - & - & + & - & + & + & + \\
+ & + & - & - & - & + & - & - & + & - & + & + \\
+ & + & + & - & - & - & + & - & - & + & - & + \\
+ & + & + & + & - & - & - & + & - & - & + & - \\
+ & - & + & + & + & - & - & - & + & - & - & + \\
+ & + & - & + & + & + & - & - & - & + & - & -
\end{bmatrix},$$

and

$$\mathcal{H}_{20} = \begin{bmatrix}
+ & + & + & + & + & + & + & + & + & + & + & + & + & + & + & + & + & + & + & + \\
+ & - & - & + & + & - & - & - & - & - & + & - & + & - & + & + & + & + & - & - & + \\
+ & - & + & + & - & - & - & - & + & - & + & - & + & + & + & + & - & - & + & - \\
+ & + & + & - & - & - & - & + & - & + & - & + & + & + & + & - & - & + & - & - \\
+ & + & - & - & - & - & + & - & + & - & + & + & + & + & - & - & + & - & - & + \\
+ & - & - & - & - & + & - & + & - & + & + & + & + & - & - & + & - & - & + & + \\
+ & - & - & - & + & - & + & - & + & + & + & + & - & - & + & - & - & + & + & - \\
+ & - & - & + & - & + & - & + & + & + & + & - & - & + & - & - & + & + & - & - \\
+ & - & + & - & + & - & + & + & + & + & - & - & + & - & - & + & + & - & - & - \\
+ & + & - & + & - & + & + & + & + & - & - & + & - & - & + & + & - & - & - & - \\
+ & - & + & - & + & + & + & + & - & - & + & - & - & + & + & - & - & - & - & + \\
+ & + & - & + & + & + & + & - & - & + & - & - & + & + & - & - & - & - & + & - \\
+ & - & + & + & + & + & - & - & + & - & - & + & + & - & - & - & - & + & - & + \\
+ & + & + & + & + & - & - & + & - & - & + & + & - & - & - & - & + & - & + & - \\
+ & + & + & + & - & - & + & - & - & + & + & - & - & - & - & + & - & + & - & + \\
+ & + & + & - & - & + & - & - & + & + & - & - & - & - & + & - & + & - & + & + \\
+ & + & - & - & + & - & - & + & + & - & - & - & - & + & - & + & - & + & + & + \\
+ & - & - & + & - & - & + & + & - & - & - & - & + & - & + & - & + & + & + & + \\
+ & - & + & - & - & + & + & - & - & - & - & + & - & + & - & + & + & + & + & - \\
+ & + & - & - & + & + & - & - & - & - & + & - & + & - & + & + & + & + & - & -
\end{bmatrix}.$$

So, for $p = 1$, $\mathcal{H}_4 = [\mathcal{H}_2, \mathcal{H}_2; \mathcal{H}_2, -\mathcal{H}_2]$, $\mathcal{H}_8 = [\mathcal{H}_4, \mathcal{H}_4; \mathcal{H}_4, -\mathcal{H}_4]$, ...
For $p = 12$, $\mathcal{H}_{24} = [\mathcal{H}_{12}, \mathcal{H}_{12}; \mathcal{H}_{12}, -\mathcal{H}_{12}]$, $\mathcal{H}_{48} = [\mathcal{H}_{24}, \mathcal{H}_{24}; \mathcal{H}_{24}, -\mathcal{H}_{24}]$, ...
For $p = 20$, $\mathcal{H}_{40} = [\mathcal{H}_{20}, \mathcal{H}_{20}; \mathcal{H}_{20}, -\mathcal{H}_{20}]$, $\mathcal{H}_{80} = [\mathcal{H}_{40}, \mathcal{H}_{40}; \mathcal{H}_{40}, -\mathcal{H}_{40}]$, ...

In order to be consistent with the above target coding method, we map '+1' and '−1' in the Hadamard matrix we produced to '0' and '1' respectively. Besides, we delete the elements in the first column and the first row because they are all the same. So the dimension of each Hadamard code is $n - 1$. A greedy search method introduced in [8] is implemented in our experiments.

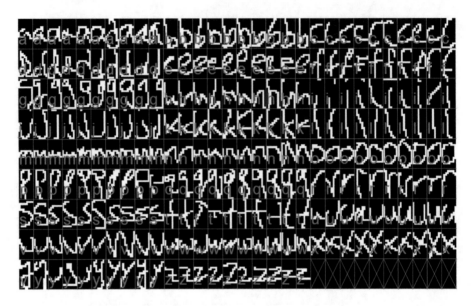

Fig. 1. Some Samples from the OCR letter dataset. Each row contains two letters' 10 samples respectively, and the ground truth of each letter is also marked in red color.

4   Experiments

We adopt OCR letter dataset to show the effects of the four above-mentioned target coding methods on ELM classification. OCR letter database is collected by Rob Kassel [19], and it contains 52152 samples of 26 letters (a–z). Each letter in this database is normalized and stored as a 16×8 binary image. Some examples from OCR database are shown in Fig. 1), from which we can see that letters are difficult to recognize due to the large and varied deformation (e.g., partial occlusion and rotation) (Fig. 2).

In the experiments, we feed the same feature data (raw pixel values) into the classification network, and all the parameters in the ELM classifier are kept consistently, including activation function, the number of hidden nodes and regularization factor. The activation function is uniformly set to the Sigmoid function, which is defined as

$$Sig(x) = \frac{1}{1 + e^{(-x)}}. \tag{15}$$

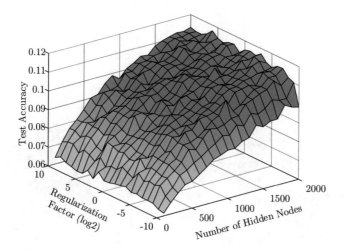

Fig. 2. Test accuracy with ordinal target coding on OCR letter dataset using ELM.

The number of hidden nodes ranges from 100 to 2000 with a step of 100, and the regularization factor is set to 2^i where i goes from -10 to 10 by a unit step.

We follow the default 10-fold cross-validation protocol to train and test the four target coding methods we have introduced in Sect. 3.

Here we have six corresponding results for binary target coding, and they are shown in Fig. 3. We can see that the performance becomes better as n increases.

Previously, we have introduced that each vector from one-of-k target coding only has one non-zero element. So for OCR letter database, the one-of-k target coding corresponds to a 26-dimension identity matrix, and each row vector stands for a different letter respectively. The results of one-of-k target coding are shown in Fig. 4, and the accuracy increases as the number of hidden nodes rises.

Similar to the operation on binary target coding, we also selected six values for Hadamard target coding. After we obtained the corresponding Hadamard code, we randomly selected 26 coding vectors and assigned them to the 26 letters. The classification results with Hadamard target coding by using ELM on OCR letter database are shown in Fig. 5.

We summarized the best Results of the above experiments in Table 5, from which we can see that

- Also as linearly dependent coding, binary method performs better than ordinal method.
- For linearly independent coding, one-of-k target coding and Hadamard target coding are evenly matched.
- Linearly independent coding methods are better than linearly dependent coding methods in these experiments.

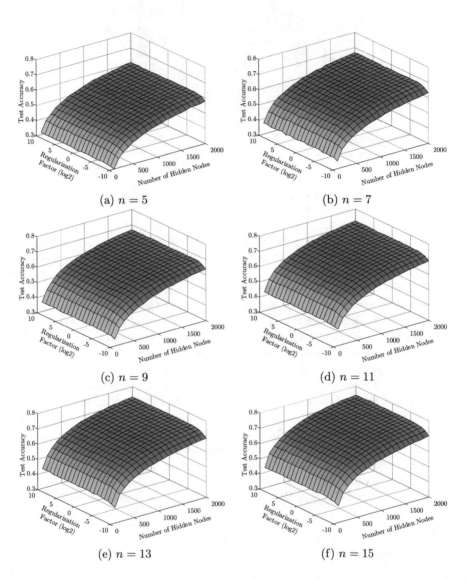

(a) $n = 5$

(b) $n = 7$

(c) $n = 9$

(d) $n = 11$

(e) $n = 13$

(f) $n = 15$

Fig. 3. Test accuracy with binary target coding on OCR letter dataset using ELM.

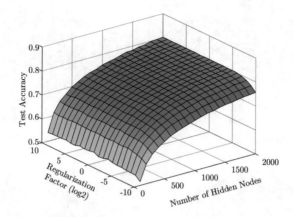

Fig. 4. Test accuracy with one-of-k target coding on OCR letter dataset using ELM.

Table 5. Comparison of different target coding methods on training accuracy and test accuracy

Target coding method		Training accuracy	Test accuracy
ordinal		0.1187	0.1172
one-of-k		0.8381	0.8033
Hadamard	n = 31	0.8378	0.8045
	n = 39	0.8380	0.8043
	n = 47	0.8382	0.8047
	n = 63	0.8381	0.8043
	n = 79	0.8375	0.8038
	n = 127	0.8376	0.8040
Binary	n = 5	0.6421	0.6088
	n = 7	0.6967	0.6697
	n = 9	0.7072	0.6705
	n = 11	0.7594	0.7250
	n = 13	0.7666	0.7331
	n = 15	0.7821	0.7484

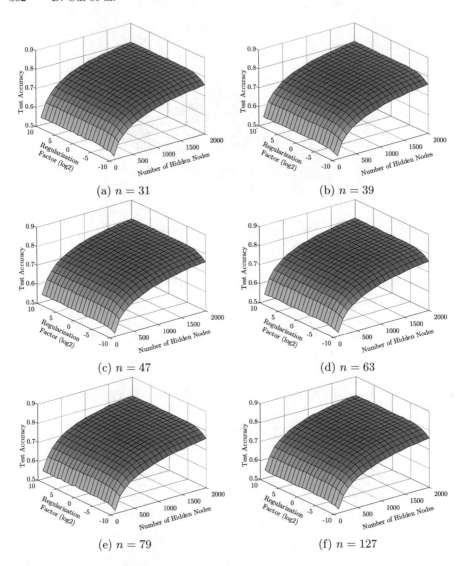

(a) $n = 31$

(b) $n = 39$

(c) $n = 47$

(d) $n = 63$

(e) $n = 79$

(f) $n = 127$

Fig. 5. Test accuracy with Hadamard target coding on OCR letter dataset using ELM.

5   Conclusion

In this paper, we have explored the effectiveness of target coding on ELM classifiers. In particular, we have analyzed two linearly dependent coding (ordinal coding and binary coding) and two linearly independent coding (one-of-k coding and Hadamard coding). Experiments have been implemented on OCR letter database to show the different effects of different target coding, and discussion is given based on the results. We have shown that different target coding methods indeed affect the performance of ELM classifiers.

References

1. Kasun, L.L.C., Yang, Y., Huang, G.-B., Zhang, Z.: Dimension reduction with extreme learning machine. IEEE Trans. Image Process. **25**(8), 3906–3918 (2016)
2. Cui, D., Kasun, L.L.C., Huang, G.-B., Zhang, G., Han, W.: Elmnet: feature learning using extreme learning machines. In: International Conference on Image Processing, pp. 1857–1861. IEEE (2017)
3. Wang, S., Deng, C., Lin, W., Huang, G.-B., Zhao, B.: NMF-based image quality assessment using extreme learning machine. IEEE Trans. Cybern. **47**(1), 232–243 (2017)
4. Omlin, C.W., Giles, C.L.: Stable encoding of large finite-state automata in recurrent neural networks with sigmoid discriminants. Neural Comput. **8**(4), 675–696 (1996)
5. Huang, G., Huang, G.-B., Song, S., You, K.: Trends in extreme learning machines: a review. Neural Netw. **61**, 32–48 (2015)
6. Golomb, S.W., Baumert, L.D.: The search for hadamard matrices. Am. Math. Monthly **70**(1), 12–17 (1963)
7. MacWilliams, F.J., Sloane, N.J.A.: The Theory of Error Correcting Codes. Elsevier (1977)
8. Yang, S., Luo, P., Loy, C.C., Shum, K.W., Tang, X.: Deep representation learning with target coding. In: AAAI, pp. 3848–3854 (2015)
9. Huang, G.-B., Zhu, Q.-Y., Siew, C.-K.: Extreme learning machine: theory and applications. Neurocomputing **70**(1), 489–501 (2006)
10. Huang, G.-B., Siew, C.-K.: Extreme learning machine: RBF network case. In: ICARCV 2004 8th Control, Automation, Robotics and Vision Conference, vol. 2, pp. 1029–1036. IEEE (2004)
11. Huang, G.-B., Wang, D.H., Lan, Y.: Extreme learning machines: a survey. Int. J. Mach. Learn. Cybern. **2**(2), 107–122 (2011)
12. Huang, G.-B., Zhou, H., Ding, X., Zhang, R.: Extreme learning machine for regression and multiclass classification. IEEE Trans. Syst. Man Cybern. Part B (Cybern.) **42**(2), 513–529 (2012)
13. Wang, B., Zhao, W., Du, Y., Zhang, G., Yang, Y.: Prediction of fatigue stress concentration factor using extreme learning machine. Comput. Mater. Sci. **125**, 136–145 (2016)
14. Cui, D., Huang, G.-B., Liu, T.: Smile detection using pair-wise distance vector and extreme learning machine. In: 2016 International Joint Conference on Neural Networks (IJCNN), pp. 2298–2305. IEEE (2016)
15. Hamming, R.W.: Coding and Theory. Prentice-Hall (1980)
16. Rappaport, T.S., et al.: Wireless Communications: Principles and Practice, vol. 2. Prentice Hall PTR, New Jersey (1996)
17. Hsu, D.J., Kakade, S.M., Langford, J., Zhang, T.: Multi-label prediction via compressed sensing. In: Advances in Neural Information Processing Systems, pp. 772–780 (2009)
18. Cisse, M.M., Usunier, N., Artieres, T., Gallinari, P.: Robust bloom filters for large multilabel classification tasks. In: Advances in Neural Information Processing Systems, pp. 1851–1859 (2013)
19. Kassel, R.: The MIT on-line character database. http://ai.stanford.edu/~btaskar/ocr

Deformable Surface Registration with Extreme Learning Machines

Andrey Gritsenko[1,2(✉)], Zhiyu Sun[1,3], Stephen Baek[1,3], Yoan Miche[4], Renjie Hu[1,2], and Amaury Lendasse[1,2,5]

[1] Department of Mechanical and Industrial Engineering,
The University of Iowa, Iowa City, USA
`andrey-gritsenko@uiowa.edu`
[2] Iowa Informatics Initiative, The University of Iowa, Iowa City, USA
[3] Center for Computer-Aided Design, The University of Iowa, Iowa City, USA
[4] Bell Labs, Nokia, Espoo, Finland
[5] Arcada University of Applied Sciences, Helsinki, Finland

Abstract. One of the most important open problems in the field of computer-aided design and computer graphics is the task of surface registration for non-isometric cases. One of the approaches of addressing surface registration problem is to find the point-wise correspondence between surfaces using state-of-the-art shape descriptors. This paper introduces an improvement to this approach by means of Extreme Learning Machines. The ELM model is trained to distinguish pairs of corresponding points from non-corresponding ones on the dataset with highly non-isometric distortions between models. The proposed method is compared with original shape descriptors. The results show the increase of accuracy in surface registration task, and also reveal the bottleneck of the state-of-the-art shape descriptors.

Keywords: Surface registration · Deformable registration
Non-isometric distortion · Spectral descriptors
Non-strict classification · Similarity measure · Distance metric
3D mesh · Extreme learning machines · Computer graphics

1 Introduction

In computer-aided design and computer graphics, a surface registration task is a task of finding for each point on one figure a matching point on another figure. If one figure cannot be obtained from another through rigid transformations (isometry) or scaling, then the task is said to be *deformable registration of surfaces*, and it is one of the fundamental problems in computer-aided design and computer graphics. Successful solution of this task has a wide range of applications such as shape interpolation [1,2], geometry transfer [3], statistical shape analysis and modeling [4,5], and so on. Finding the point-wise correspondence between surfaces is the most straightforward approach to solve successfully the registration task. So far, the research in this area is mainly focused on the development

© Springer Nature Switzerland AG 2019
J. Cao et al. (Eds.): ELM 2017, PALO 10, pp. 304–316, 2019.
https://doi.org/10.1007/978-3-030-01520-6_28

of shape descriptors that would improve the performance of the shape retrieval task, while the correspondence problem itself was solved using simple similarity metric [6–9]. Spectral descriptors, based on the spectrum of the Laplace-Beltrami operator [10], are the most prominent shape descriptors due to their desirable properties, like isometry invariance. Spectral descriptors have proved themselves efficient is such tasks as shape retrieval [11,12], mesh segmentation [13,14] and isometric matching [15], compared to other types of descriptors [11,16].

However, despite their properties, spectral descriptors are suitable to perform surface registration tasks only for near-isometry cases, i.e. they can provide correct similarity measure only for models with small metric distortions between them (e.g., consecutive surface scans of the same individual performing a motion), but not very suitable for deformable registration tasks (e.g., registering a skinny person to an obese person). The reason is, spectral descriptors can perfectly represent the local geometry and therefore they are highly sensitive to the metric distortion. In this paper a novel approach is introduced – instead of focusing on the development of a new shape descriptor, we propose to learn a similarity metric by means of the Extreme Learning Machines model, optimal for the point-wise correspondence task.

2   Methodology

2.1   Spectral Descriptors

One of the conventional state-of-the-art shape descriptors is a family of spectral descriptors, originally proposed in [17] and since then been a subject for improvements and modifications [12,18–20]. This type of shape descriptors is based on the spectrum of the Laplace-Beltrami operator and utilizes its eigendecomposition [21]. Spectral descriptors become a conventional method for shape description because they inherit properties of the Laplace-Beltrami operator, among which isometry invariance is one of the most desirable [19,22], since a lot of deformations in real-world can be characterized as an isometry or a near-isometry.

The first spectral descriptor was proposed by Reuter *et al.* in 2006 [17]. This spectral descriptor is an ascending sequence of eigenvalues of the Laplace-Beltrami operator per se. Though it is the first spectral shape descriptor, it is omitted from the experiments described in this paper due to its simplicity. Instead, the following three conventional spectral descriptors are used.

Global Points Signature is a spectral descriptor proposed in [18] and defined at each point x on the surface as a vector containing the eigenfunctions of different modes scaled by the corresponding eigenvalues:

$$\text{GPS}(x) = \left[\frac{\phi_1(x)}{\sqrt{\lambda_1}}, \frac{\phi_2(x)}{\sqrt{\lambda_2}}, \ldots, \frac{\phi_n(x)}{\sqrt{\lambda_n}} \right]^\top, \tag{1}$$

where λ_k and ϕ_k are the k-th eigenvalue and eigenfunction of the Laplace-Beltrami operator defined on the manifold, respectively. In the original paper [18], it was suggested to use $n = 25$ eigenvalues.

Heat Kernel Signature is a more sophisticated spectral descriptor based on the different heat diffusion characteristics according to the geometric shape of the surface. In [19] it has been shown that heat diffusion process on a manifold can be formulated using the spectrum of the Laplace-Beltrami operator, and serve as a shape descriptor. More specifically, the amount of heat diffused from a point x to a point y on a 2D manifold during a certain time t can be calculated as

$$H_t(x,y) = \sum_{k=0}^{\infty} e^{-\lambda_k t}\phi_k(x)\phi_k(y), \tag{2}$$

where λ_k and ϕ_k are the k-th eigenvalue and eigenfunction of the Laplace-Beltrami operator, respectively. When defining a Heat Kernel Signature spectral descriptor, for each point x on the surface heat kernel values $H_t(x,x)$ were calculated, measuring the amount of heat that stays at point x after different amount of time $t_1, t_2, \ldots, t_n$:

$$\text{HKS}(x) = [H_{t_1}(x,x), H_{t_2}(x,x), ..., H_{t_n}(x,x)]^{\top}. \tag{3}$$

In [19], it was suggested to use the first 300 eigenvalues and eigenvectors for the approximation of Eq. (2). Also, the recommended number of time samples is $n = 100$, sampled uniformly from the logarithmic scale over the time interval from $4\ln 10/\lambda_{300}$ to $4\ln 10/\lambda_2$.

Wave Kernel Signature is a spectral shape descriptor that follows a similar idea to the HKS, replacing the heat equation with the Schrödinger wave equation: the WKS represents the average probability of measuring a quantum mechanical particle at a specific location [20]. To achieve this, the solution of the Schrödinger wave equation is given by means of the Laplace-Beltrami operator's spectrum:

$$\psi_E(x,t) = \sum_{k=0}^{\infty} e^{i\lambda_k t}\phi_i(x)f_E(\lambda_k), \tag{4}$$

where i is the imaginary number and f_E^2 is an energy probability distribution with expectation value E, in practice approximated by the log-normal distribution

$$f_E^2 = e^{\frac{-(\rho-\ln\lambda_k)^2}{2\sigma^2}}, \tag{5}$$

where ρ is the energy scale. The physical meaning of the l_2 norm $\|\psi_E(x,t)\|^2$ is the probability of measuring the particle at a point x at a certain time t. To get rid of time parameter as it has no straightforward interpretation in the characteristics of the shape, the average probability is calculated:

$$P_\rho(x) = \lim_{T\to\infty} \frac{1}{T}\int_0^T \|\psi_E(x,t)\|^2 dt = \sum_{k=0}^{\infty} \phi_k^2(x)f_E^2(\lambda_k). \tag{6}$$

The Wave Kernel Signature is then defined as a vector of the probability values, calculated for different energy scales $\rho_1, \rho_2, \ldots, \rho_n$:

$$\text{WKS}(x) = [P_{\rho_1}(x), P_{\rho_2}(x), \ldots, P_{\rho_n}(x)]^\top. \tag{7}$$

Again, it is recommended to use the first 300 eigenvalues to approximate Eq. (6), and to uniformly sample $n = 100$ energy scale values over an interval from $\ln(\lambda_1)$ to $\ln(\lambda_{300})$ [20].

In addition to the general properties of spectral descriptors, like invariance to the isometry, proportionality to a deformation or a metric change, HKS and WKS are both known to be multiscale in a sense that they inherently capture both the local and global shape characteristics through different time scales.

2.2 Extreme Learning Machines

The ELM algorithm was originally proposed by Huang et al. in [23–26]. The method is proven to be an universal approximator given an enough number of hidden neurons [27–32]. It works as follows: Consider a set of N distinct samples $(\mathbf{x}_i, \mathbf{t}_i)$ with $\mathbf{x}_i \in \mathbb{R}^d$ and $\mathbf{t}_i \in \mathbb{R}^c$ – target variable, where c is the dimensionality of regression function in regression problems or number of classes in classification problems [33–36].

Assuming the ELM model would perfectly approximate the data, it can be modeled as:

$$\sum_{i=1}^{L} \beta_i \phi(\mathbf{w}_i \mathbf{x}_j + b_i) = \mathbf{t}_j, \ j \in [1, N]. \tag{8}$$

This can be concisely written as $\mathbf{H}\beta = \mathbf{T}$, where

$$\mathbf{H} = \begin{pmatrix} \phi(\mathbf{w}_1 \mathbf{x}_1 + b_1) & \cdots & \phi(\mathbf{w}_L \mathbf{x}_1 + b_L) \\ \vdots & \ddots & \vdots \\ \phi(\mathbf{w}_1 \mathbf{x}_N + b_1) & \cdots & \phi(\mathbf{w}_L \mathbf{x}_N + b_L) \end{pmatrix}, \tag{9}$$

$$\beta = (\beta_1^\top \ldots \beta_L^\top)^\top, \ \mathbf{T} = (\mathbf{t}_1^\top \ldots \mathbf{t}_N^\top)^\top. \tag{10}$$

Finding the output weights β from the hidden layer outputs $\mathbf{H}$ and targets $\mathbf{T}$ is a linear regression problem. In the general case of $N \neq d$, a minimum L^2-norm solution is given by the Moore-Penrose generalized inverse, or pseudo-inverse, of the matrix $\mathbf{H}$ denoted as $\mathbf{H}^\dagger$ [37].

In the following, a binary classification problem is considered. In this case, the target variable is encoded as a binary variable that takes a value of 0 for one class and 1 for another.

3 Experiments

3.1 Dataset

For the experiments, the *Dyna* dataset [38], which is composed of over 40,000 scans of ten human subjects with a wide range of body shapes (see Fig. 1), is used to generate data samples. Originally, the body scans were collected by using a custom-built multi-camera active stereo system by capturing 14 different motions (e.g. running, jumping, shaking, etc.) at 60 frames per second. The system outputs 3D meshes with the average number of vertices around 150,000. To reduce the number of vertices and bring it to the same number over all scans, a template mesh was conformed to each of the scanned meshes resulting in a new dataset with 3D meshes containing only 6,980 vertices. More important, the topology of each mesh becomes compatible to each other, and the vertices with the same index get to correspond to each other geometrically. This information is used later as the ground truth for the point-wise correspondence task. Finally, over all meshes of the *Dyna* database, for each point the spectral descriptors are computed in a way described in Sect. 2.1.

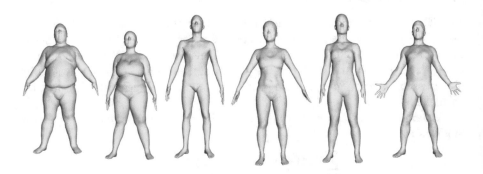

Fig. 1. Examples of different 3D models in *Dyna* dataset, spanning a range of body shapes

From the resulting dataset, two subsets (training and test sets) are formed in the following way. The training set is formed by 10,000 batches. Each batch contains 512 pairs of vertices from two randomly selected models, resulting in the training set having 5,120,000 samples. Vertices for pairs are also randomly selected and satisfy the following condition: 256 of pairs are pairs of matching vertices, i.e. vertices have the same indices on meshes, and 256 pairs are pairs of non-matching vertices. This scheme allows to have duplicates of vertices in the scope of one batch, while meshes that were used once for training set are removed from the pool of possible choices in the future. This means, that the dataset is divided in two almost equal parts: the first part used for the training (10,000 batches × 2 models) and the rest of the data is used for the test.

3.2   Learning Similarity Metric with ELM

The goal of the presenting approach is to perform registration task between two models, more particular find point-to-point correspondence, by calculating the measure of distance between points. It is proposed to train an Extreme Learning

Machines model to learn an appropriate similarity metric with desirable suitability for deformable registration task, directly from original spectral descriptors. In order to do so, the data from *Dyna* dataset at first is been altered in a following way: for each pair of vertices from two registering models a new data sample is created by merging values of spectral descriptors corresponding to these vertices: $x_i = [sd_i^1 sd_i^2]$, where $sd_i^{1,2}$ are values of corresponding spectral descriptors for the i-th pair of vertices from the first and second 3D models, respectively. This trick halves the size of the data, but doubles its dimensionality, and makes it suitable to be used by the ELM model to learn similarity metric.

Now, the goal of the proposed ELM model that is to learn an optimal similarity metric for spectral descriptors in the original space can be achieved. In this case, it becomes intuitive to assign target values in the following way:

$$y_k = \begin{cases} 0 & v_k^1 \equiv v_k^2 \\ \text{rand}[0.8 \ldots 1] & v_k^1 \neq v_k^2, \end{cases} \tag{11}$$

where v_k^1 and v_k^2 represent the k-th pair of vertices from the first and second 3D models, respectively. As a result, Extreme Learning Machines model is trained to directly compute the measure of similarity between a pair of shape descriptors by means of nonlinear distance metric. It should be noted, that for pairs of non-matching vertices a random value between 0.8 and 1 is been assigned to the target variable, instead of strictly assigning 1 as a target variable that is a standard procedure of performing a binary classification with ELM (as well as other methods that convert classification problem into regression). The choice of this more relaxed classification is done under practical reasoning that even non-matching pairs might have a highly similar geometry, e.g. a point on the left thumb would probably have a highly similar local geometry with a symmetric point on the right thumb. However, technically they represent a non-matching pair. Obviously, such conflicting examples can confuse the ELM model if appear in the training set. To avoid this, a certain relaxation has been added to the model by assigning a random value between 0.8 and 1 to y_k of non-matching vertices. The correct matching pairs are still kept to $y_k = 0$.

The threshold value is defined equal to 0.5, so that a pair of vertices with a similarity measure less than or equal to 0.5 should be considered as a matching pair, otherwise, as a non-matching pair.

3.3   Model Selection

The main parameter in the ELM model that influence its performance is the number of neurons in the hidden layer. In order to estimate the optimal number of hidden neurons, validation is performed. When constructing the validation subset 1,000 batches is used and the same policy as for the construction of training subset is applied. The analysis of the validation subset results has shown that the model does not overfit even when the model reaches the maximum number of hidden neurons that is set equal to 40,000. The maximum possible number of hidden neurons depends on the amount of available RAM in the system [39], and for the current ELM implementation a system with 256 GB of RAM has been used.

In order to evaluate the performance of the ELM model the combination of three parameters is used: mean square error (MSE), false-positive rate (FPR) – ratio of the *negative* samples that were predicted as *positive*, and false-negative rate (FNR) – ratio of the *positive* samples that were predicted as *negative*. The left subfigure of Fig. 2 presents the performance of the trained ELM model for the validation subset, in terms of three evaluation criteria. The right subfigure shows how the number of the hidden neurons influence the performance of the ELM model for the training and validation sets, in terms of combination of evaluation criteria

$$\varepsilon = \mathrm{MSE} + \mathrm{FPR} + \mathrm{FNR}. \tag{12}$$

It can be observed, that the overall error continues to decrease and the model never overfits, even when the number of hidden neurons reaches the maximum value. This behavior of ELM model is common for Big Data problems with huge number of data samples, and the size of the model, i.e. number of hidden neurons, is restricted only by the size of the available memory [39]. Based on the analysis of the validation results and this reasoning, it is suggested to use the maximum possible number of hidden neurons, which is equal to 40,000 for the used system.

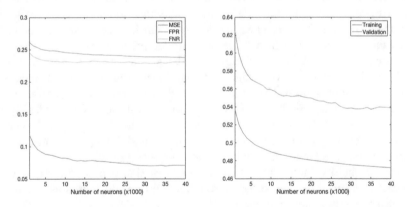

Fig. 2. Influence of the number of hidden neurons in the ELM model on its performance. The relationship between the number of hidden neurons and values of the evaluation parameters (MSE, TNR, TPR), computed for the validation set, is shown on the left. The decrease of the error ε (see Eq. 12) for both training and validation sets is depicted on the right

4   Results

Performance of the proposed approach implemented by the means of Extreme
Learning Machines is tested in a series of experiments, compared to conventional
spectral descriptors, that are GPS, HKS and WKS, using Euclidean distance as a
distance metric. In the presented experiments only Heat Kernel Signature spectral descriptor is used to build ELM model (hereinafter, referred as ELMHKS).

For each experiment, a pair of meshes is randomly selected from the test
subset of *Dyna* dataset. For a given pair of 3D meshes, all pairwise distances
between vertices are calculated, using Euclidean distance in case of conventional
spectral descriptors, and distance metric learned by ELM model for ELMHKS.
From the resulting distance matrix, for each vertex from the first 3D model a
matching pair is defined as a vertex from the second 3D model with the smallest distance between spectral descriptors, in other words, a matching vertex is
found as a nearest neighbor. This pair of matching vertices is said to be correctly defined if it satisfies a condition: the geodesic distortion between vertices
should not be more than 10% of the shape diameter, otherwise, it is rejected.
For ELMHKS, a pair should also satisfy an additional condition of the distance
not being over the threshold value equal to 0.5.

The results of 6 performed experiments with different 3D models are presented both in terms of absolute number of correctly determined matching pairs
(see Table 1) and visually depicting them (see Fig. 3).

Table 1. Statistics of the number of correct matching pairs for the results presented
in Fig. 3. For each experiment, Model 1 is presented by a yellow figure, and Model 2
by a blue one. The two leftmost entries of each row show IDs of corresponding models

Model 1 (yellow)	Model 2 (blue)	GPS	HKS	WKS	ELMHKS
00455	00228	0	211	669	**1,228**
00165	00170	147	2,209	2,124	**2,769**
00090	00266	84	1,145	1,328	**2,718**
00168	00374	25	627	1,072	**2,350**
00233	00302	206	455	1,162	**1,892**
00515	00214	17	1,474	**1,943**	1,633

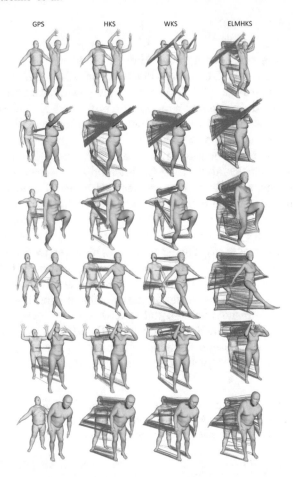

Fig. 3. Visualization of the matches within the geodesic distortion of 10% of the shape diameter. Each row shows a comparison between conventional spectral descriptors using Euclidean distance, namely, GPS, HKS and WKS, and HKS descriptor using distance metric learned by ELMHKS model, respectively

5   Conclusions and Future Work

From the results presented in Sect. 3.2, it can be observed that accuracy of performing the registration task for pairs of models with large metric distortions can be increased by introducing a reliable distance metric robust to non-isometric shape deformations. To obtain this, the Extreme Learning Machines model is suggested to learn the appropriate distance metric directly for the conventional spectral descriptors, by an example of Heat Kernel Signature. As it can be seen from the Table 1, for 5 out of 6 experiments using ELMHKS model indeed allows to improve the accuracy of the deformable registration task drastically, in terms of absolute number of correctly defined matching pairs. Though, it should also be noted that in the last experiment using distance metric learned by ELMHKS model allows to discover less matching pairs than using Euclidean distance for Wave Kernel Signature descriptor. Nevertheless, when analyzing results of this experiment visually we can notice that matching pairs discovered by ELMHKS model cover larger area than any of the conventional spectral descriptors (see Fig. 3), thus, still considering it as an improvement. Essentially, the similar results can be observed for other experiments as well. This is, in fact, because some of the body parts such as hands, feet, and faces, are near-isometry across the models. In this reason, the matching results of the conventional spectral descriptors are mostly concentrated around those body parts, whereas the matching results of the ELMHKS model are distributed widely over the entire body area, even including highly non-isometric areas.

Although, in general learning a distance metric results in the increase of accuracy of the registration task for 3D models with non-isometric deformations, there are ways for further improvements discussed below.

Dimensionality of Spectral Descriptors. In Sect. 2.1, the recommended dimensionality of conventional spectral descriptors is mentioned, as suggested by the authors in the original papers [18–20]. Though these recommendations were given in order to improve shape retrieval by spectral descriptors, they may be not optimal for surface registration task, especially for cases with non-isometric deformations. Finding intrinsic dimensionality of conventional spectral descriptors may result in learning similarity measure more suitable for deformable registration task, thus further improving its accuracy.

Relaxed Classification. When designing ELMHKS model (see Sect. 3.2), it was proposed to relax the strict classification problem for non-matching pairs by assigning a random value from the interval of $[0.8, \ldots, 1]$ to the target variable. Obviously, this strategy is not optimal and more sophisticated ways could be considered, e.g. make the assigned value depend on the geodesic distortion between vertices.

Geodesic Distortion as Matching Criteria. When performing experiments, particularly, when defining if a pair of vertices is a matching pair, a certain condition has been introduced. According to that condition, if geodesic distance between a pair of vertices exceeds 10% of the shape diameter then this pair of vertices is rejected. This condition becomes possible to be implemented due to the fact that all models in *Dyna* dataset have been preliminary registered to a template model. In the real world problems the ground truth template model is usually not known, or geodesic distance could not be calculated. In this case, instead of searching for a matching vertex across the whole model, we could at first search for a region on a model that most likely contains a desired matching vertex, i.e. perform a generalized task of body parts detection.

Finding Matching Pair. In a set of described experiments (see Sect. 3), no sophisticated algorithm for the calculation of the matching pairs has been used, such as the ones presented in [6,40]. Instead, distances between spectral descriptors (both Euclidean and the one learned by ELMHKS model) were directly compared, such that a more straight forward nearest neighbor search could be done on the quality of the metric. Using more sophisticated comparison algorithms could improve the overall performance of the deformable registration task.

Finally, an interesting phenomenon can be observed from the visualization of pairs of matching vertices (see Fig. 3). Regardless of the overall number of matching vertices between two models, there are certain regions on the models that consistently have lower number of matching vertices. These regions can be described as regions with relatively simple geometry (e.g. belly, loin, tights, shoulders). This phenomenon denotes a limitation of spectral descriptors that are designed to focus on capturing local geometry features. In conclusion, it should be stated that successful performance of surface registration task with non-isometric deformations should not rely solely on spectral descriptors but implement other techniques that will compensate the noted limitation of spectral descriptors and strengthen the reliability of the proposed methods.

References

1. Baek, S.Y., Lim, J., Lee, K.: Isometric shape interpolation. Comput. Graph. **46**(1), 257–263 (2015)
2. Kilian, M., Mitra, N.J., Pottmann, H.: Geometric modeling in shape space. In: ACM SIGGRAPH 2007 Papers, SIGGRAPH 2007. ACM, New York (2007)
3. Sumner, R.W., Popović, J.: Deformation transfer for triangle meshes. ACM Trans. Graph. (TOG) **23**(3), 399–405 (2004)
4. Allen, B., Curless, B., Popović, Z.: The space of human body shapes: reconstruction and parameterization from range scans. ACM Trans. Graph. (TOG) **22**(3), 587–594 (2003)
5. Baek, S.Y., Lee, K.: Parametric human body shape modeling framework for human-centered product design. Comput.-Aided Des. **44**(1), 56–67 (2012)
6. Litman, R., Bronstein, A.M.: Learning spectral descriptors for deformable shape correspondence. IEEE Trans. Pattern Anal. Mach. Intell. **36**(1), 171–180 (2014)
7. Xie, J., Dai, G., Zhu, F., Wong, E.K., Fang, Y.: DeepShape: deep-learned shape descriptor for 3D shape retrieval. IEEE Trans. Pattern Anal. Mach. Intell. **39**(7), 1335–1345 (2017)
8. Simo-Serra, E., Trulls, E., Ferraz, L., Kokkinos, I., Fua, P., Moreno-Noguer, F.: Discriminative learning of deep convolutional feature point descriptors. In: Proceedings of the IEEE International Conference on Computer Vision, pp. 118–126 (2015)
9. Fang, Y., Xie, J., Dai, G., Wang, M., Zhu, F., Xu, T., Wong, E.: 3D deep shape descriptor. In: Proceedings of the IEEE Conference on Computer Vision and Pattern Recognition, pp. 2319–2328 (2015)
10. Lévy, B.: Laplace-beltrami eigenfunctions towards an algorithm that 'understands' geometry. In: IEEE International Conference on Shape Modeling and Applications, SMI 2006, pp. 13–21. IEEE (2006)

11. Li, C., Hamza, A.B.: Spatially aggregating spectral descriptors for nonrigid 3D shape retrieval: a comparative survey. Multimed. Syst. **20**(3), 253–281 (2014)
12. Bronstein, M.M., Kokkinos, I.: Scale-invariant heat kernel signatures for non-rigid shape recognition. In: 2010 IEEE Conference on Computer Vision and Pattern Recognition (CVPR), pp. 1704–1711. IEEE (2010)
13. Aubry, M., Schlickewei, U., Cremers, D.: Pose-consistent 3D shape segmentation based on a quantum mechanical feature descriptor. In: Joint Pattern Recognition Symposium, pp. 122–131 (2011)
14. Fang, Y., Sun, M., Kim, M., Ramani, K.: Heat-mapping: a robust approach toward perceptually consistent mesh segmentation. In: 2011 IEEE Conference on Computer Vision and Pattern Recognition (CVPR), pp. 2145–2152. IEEE (2011)
15. Ovsjanikov, M., Mérigot, Q., Mémoli, F., Guibas, L.: One point isometric matching with the heat kernel. Comput. Graph. Forum **29**(5), 1555–1564 (2010)
16. Lian, Z., Godil, A., Bustos, B., Daoudi, M., Hermans, J., Kawamura, S., Kurita, Y., Lavoué, G., Van Nguyen, H., Ohbuchi, R.: A comparison of methods for non-rigid 3D shape retrieval. Pattern Recognit. **46**(1), 449–461 (2013)
17. Reuter, M., Wolter, F.E., Peinecke, N.: Laplace-Beltrami spectra as 'Shape-DNA' of surfaces and solids. Comput. Aided Des. **38**(4), 342–366 (2006)
18. Rustamov, R.M.: Laplace-Beltrami Eigenfunctions for Deformation Invariant Shape Representation. In: Proceedings of the Fifth Eurographics Symposium on Geometry Processing, pp. 225–233. Eurographics Association (2007)
19. Sun, J., Ovsjanikov, M., Guibas, L.: A concise and provably informative multi-scale signature based on heat diffusion. Comput. Graph. Forum **28**(5), 1383–1392 (2009)
20. Aubry, M., Schlickewei, U., Cremers, D.: The wave kernel signature: a quantum mechanical approach to shape analysis. In: 2011 IEEE International Conference on Computer Vision Workshops (ICCV Workshops), pp. 1626–1633 (2011)
21. Rosenberg, S.: The Laplacian on a Riemannian Manifold: An Introduction to Analysis on Manifolds. Number 31 in London Mathematical Society Student Texts. Cambridge University Press, Cambridge (1997)
22. Reuter, M., Wolter, F.E., Peinecke, N.: Laplace-spectra as fingerprints for shape matching. In: Proceedings of the 2005 ACM Symposium on Solid and Physical Modeling, pp. 101–106. ACM (2005)
23. Huang, G.B., Zhu, Q.Y., Siew, C.K.: Extreme learning machine: theory and applications. Neurocomputing **70**(1), 489–501 (2006)
24. Lendasse, A., Akusok, A., Simula, O., Corona, F., van Heeswijk, M., Eirola, E., Miche, Y.: Extreme learning machine: a robust modeling technique? yes! In: International Work-Conference on Artificial Neural Networks, pp. 17–35. Springer Heidelberg (2013)
25. Cambria, E.: Extreme learning machines. IEEE Intell. Syst. **28**(6), 30–59 (2013)
26. Akusok, A., Baek, S., Miche, Y., Björk, K.M., Nian, R., Lauren, P., Lendasse, A.: ELMVIS+: fast nonlinear visualization technique based on cosine distance and extreme learning machines. Neurocomputing **205**, 247–263 (2016)
27. Huang, G.B., Chen, L., Siew, C.K.: Universal approximation using incremental constructive feedforward networks with random hidden nodes. IEEE Trans. Neural Netw. **17**(4), 879–892 (2006)
28. Miche, Y., Sorjamaa, A., Bas, P., Simula, O., Jutten, C., Lendasse, A.: OP-ELM: optimally pruned extreme learning machine. IEEE Trans. Neural Netw. **21**(1), 158–162 (2010)

29. Miche, Y., van Heeswijk, M., Bas, P., Simula, O., Lendasse, A.: TROP-ELM: a double-regularized ELM using LARS and tikhonov regularization. Neurocomputing **74**(16), 2413–2421 (2011)

30. Nian, R., He, B., Zheng, B., Van Heeswijk, M., Yu, Q., Miche, Y., Lendasse, A.: Extreme learning machine towards dynamic model hypothesis in fish ethology research. Neurocomputing **128**, 273–284 (2014)

31. He, B., Xu, D., Nian, R., van Heeswijk, M., Yu, Q., Miche, Y., Lendasse, A.: Fast face recognition via sparse coding and extreme learning machine. Cognit. Comput. **6**(2), 264–277 (2014)

32. Zhang, S., He, B., Nian, R., Wang, J., Han, B., Lendasse, A., Yuan, G.: Fast image recognition based on independent component analysis and extreme learning machine. Cognit. Comput. **6**(3), 405–422 (2014)

33. Huang, G.B., Zhou, H., Ding, X., Zhang, R.: Extreme learning machine for regression and multiclass classification. IEEE Trans. Syst. Man Cybern. Part B: Cybern. **42**(2), 513–529 (2012)

34. Moreno, R., Corona, F., Lendasse, A., Graña, M., Galvão, L.S.: Extreme learning machines for soybean classification in remote sensing hyperspectral images. Neurocomputing **128**, 207–216 (2014)

35. Eirola, E., Gritsenko, A., Akusok, A., Björk, K.M., Miche, Y., Sovilj, D., Nian, R., He, B., Lendasse, A.: Extreme learning machines for multiclass classification: refining predictions with gaussian mixture models. In: International Work-Conference on Artificial Neural Networks, pp. 153–164. Springer, Cham (2015)

36. Gritsenko, A., Eirola, E., Schupp, D., Ratner, E., Lendasse, A.: Solve classification tasks with probabilities. statistically-modeled outputs. In: The 12th International Conference on Hybrid Artificial Intelligence Systems. LNAI, pp. 293–305. Springer, Cham (2017)

37. Rao, C.R., Mitra, S.K.: Generalized Inverse of a Matrix and its Applications. Wiley (1971)

38. Pons-Moll, G., Romero, J., Mahmood, N., Black, M.J.: Dyna: a model of dynamic human shape in motion. ACM Trans. Graph. (Proc. SIGGRAPH) **34**(4), 120:1–120:14 (2015)

39. Akusok, A., Björk, K.M., Miche, Y., Lendasse, A.: High-performance extreme learning machines: a complete toolbox for big data applications. IEEE Access **3**, 1011–1025 (2015)

40. Leordeanu, M., Hebert, M.: A spectral technique for correspondence problems using pairwise constraints. In: Tenth IEEE International Conference on Computer Vision, ICCV 2005, vol. 2, pp. 1482–1489. IEEE (2005)

A Highly Efficient Intrusion Detection Method Based on Hierarchical Extreme Learning Machine

Linyuan Yu[1]($\boxtimes$), Yan Liu[1], Wentao Zhao[1], Qiang Liu[1], and Jiaohua Qin[2]

[1] College of Computer, National University of Defense Technology, Changsha 410073, Hunan, People's Republic of China
`809932887@qq.com`, `642293721@qq.com`, {`wtzhao,qiangliu06`}`@nudt.edu.cn`
[2] College of Computer Science and Information Technology, Central South University of Forestry and Technology, Changsha 410004, China
`qinjiaohua@163.com`

Abstract. Cyber security is becoming more and more concerned by people nowadays. Intrusion detection systems (IDSs) is a major approach to ensure the confidentiality, integrity and availability of network system resources. There are many machine learning techniques applied to IDSs. In this paper, we propose a novel and rapid technique based on Hierarchical Extreme Learning Machine (H-ELM) for intrusion detection. We use NSL-KDD 2009 dataset to evaluate our method. Comparing our method with other widely used machine learning methods such as k-Nearest Neighbor (k-NN), Random Forest (RF) and Extreme Learning Machine (ELM), the experimental results show that H-ELM can perform better than or similar to other methods in overall accuracy of 72.87%, while only spends a total time of 2.04 s which is much faster than other methods.

Keywords: Intrusion detection
Hierarchical Extreme Learning Machine (H-ELM)

1 Introduction

With the rapid development of broadband services and social dependence on the network, cyber security issues become increasingly prominent [1]. Especially the wireless network techniques which have recently emerged as a promising technology to provide better services to user terminals are more vulnerable to various attacks [2]. Some techniques, such as user authentication, data encryption, antivirus software and firewalls are used to protect computer security [3]. Intrusion detection systems (IDSs), which use specific analytical techniques to detect attacks have recently been developed [4]. IDSs adopt a certain security strategy, through the hardware, the software, the network and the system operating to monitor the situation, as far as possible to find a variety of attack attempts, attack behavior or attack results to protect the confidentiality, integrity and

© Springer Nature Switzerland AG 2019
J. Cao et al. (Eds.): ELM 2017, PALO 10, pp. 317–326, 2019.
https://doi.org/10.1007/978-3-030-01520-6_29

availability of network system resources. Do an image of the metaphor, if the firewall is a building door lock, then IDS is the monitoring system in the building. Once the thief crossed the door and enter into the building through window, or insiders have cross-border behavior, only the real-time monitoring system can detect the situation and issue a warning.

According to the intrusion detection technology foundation, IDSs can be divided into two categories: signature-based (or misuse-based) and anomaly-based. Some researchers consider a third category: hybrid which combine signature and anomaly detection [5]. But in our opinion, hybrid can be considered as anomaly-based, because there are few pure anomaly-based detection methods.

Long before, IDSs largely used signature-based techniques which were designed to detect known attacks by using signatures of those attacks to detect malicious intrusions [6]. Due to the development of artificial intelligence techniques, especially machine learning and data mining, IDSs mostly use abnormal-based techniques as they can detect novel attacks recent years. Some machine learning classifiers are developed maturely and widely used in IDSs, such as Support Vector Machine (SVM), k-Nearest Neighbor (k-NN), Random Forest (RF), Nave Bayes, Decision Trees, Artificial Neural Networks, etc. In this paper, we select k-NN and RF which are popularly used and have a high performance to compare with our proposed method Hierarchical Extreme Learning Machine (H-ELM). Although SVM is another popular method for supporting IDSs, we do not compare performance between H-ELM and SVM here because the performance of SVM has been demonstrated to be worse than ELM [7,8]. Hence, we alternatively compare H-ELM with ELM in this paper.

The main contributions of this paper are summarized as follows: (1) We propose a novel method based on H-ELM for intrusion detection. (2) The proposed method is much more efficient than other traditional machine leaning methods. The rest of this paper is organized as follows: Sect. 2 introduces the background of ELM and H-ELM. In Sect. 3, we introduce the proposed intrusion detection method in details. Section 4 presents the experimental results and shows the comparable performance. Finally, some conclusions are provided in Sect. 5.

2   Background

2.1   Basics of Extreme Learning Machine

ELM [9] is an effective and efficient solution for the single hidden layer feedforward networks(SLFNs), and has been demonstrated to have a nice learning accuracy and speed in intrusion detection [10]. Unlike some other traditional learning algorithms, e.g., back propagation-based neural networks, or SVM, the parameters of hidden layers of the ELM are randomly generated and do not need to be tuned, thus the hidden nodes can be established before the training samples are acquired [11]. The input weights and the hidden layer threshold are randomly assigned, and the output layer weights are calculated directly by the least squares method. The whole process of learning once completed, without iteration, it can achieve very fast learning speed [11].

Given a training data set with N samples, $\mathbf{D} = \{(\mathbf{x}_i, \mathbf{t}_i) | \mathbf{x}_i \in \mathbf{R}^n, \mathbf{t}_i \in \mathbf{R}^m, i = 1, 2, \cdots, N\}$ where $\mathbf{x}_i = [x_{i1}, x_{i2}, \cdots x_{in}]^{\mathrm{T}}$ is an input vector and $\mathbf{t}_i = [t_{i1}, t_{i2}, \cdots t_{im}]^{\mathrm{T}}$ is an output vector, a SLFN with M hidden nodes and an activation infinitely differentiable function $g(x)$ is modeled as:

$$f(\mathbf{x}_j) = \sum_{i=1}^{M} \boldsymbol{\beta}_i g_i(\mathbf{x}_j) = \sum_{i=1}^{M} \boldsymbol{\beta}_i(\boldsymbol{\omega}_i^{\mathrm{T}} \mathbf{x}_j + b_i) = \mathbf{o}_j \tag{1}$$

where $j = 1, 2, \cdots, N$, $\boldsymbol{\omega}_i = [\omega_{i1}, \omega_{i2}, \cdots \omega_{in}]^{\mathrm{T}}$ is the weight vector connecting the ith hidden node with the input nodes, $\boldsymbol{\beta}_i = [\beta_{i1}, \beta_{i2}, \cdots \beta_{im}]^{\mathrm{T}}$ is the weight vector connecting the ith hidden node with the output nodes, and b_i is the threshold of the ith hidden node. Since the objective of SLFNs with M hidden nodes and activation function $g(\mathbf{x})$ is to approximate all N samples with zero mean square error, The N-dimensional equation can be written as the formula:

$$\mathbf{H}\boldsymbol{\beta} = \mathbf{T} \tag{2}$$

where $\mathbf{H}$ is called the hidden layer output matrix, i.e.,

$$\mathbf{H} = \begin{bmatrix} g(\boldsymbol{\omega}_1^{\mathrm{T}} \mathbf{x}_1 + b_1) & \cdots & g(\boldsymbol{\omega}_M^{\mathrm{T}} \mathbf{x}_1 + b_M) \\ \cdots & \cdots & \cdots \\ g(\boldsymbol{\omega}_1^{\mathrm{T}} \mathbf{x}_N + b_1) & \cdots & g(\boldsymbol{\omega}_M^{\mathrm{T}} \mathbf{x}_N + b_M) \end{bmatrix} \tag{3}$$

$$\boldsymbol{\beta} = \left[\boldsymbol{\beta}_1^{\mathrm{T}} \cdots \boldsymbol{\beta}_M^{\mathrm{T}}\right]_{M \times m}^{\mathrm{T}} \tag{4}$$

$$\mathbf{T} = \left[\mathbf{t}_1^{\mathrm{T}} \cdots \mathbf{t}_M^{\mathrm{T}}\right]_{M \times m}^{\mathrm{T}} \tag{5}$$

Its ith row represents the whole output of the ith input $\mathbf{x}_i$ with respect to the hidden layer. The jth column represents the whole input $\mathbf{x}_1, \mathbf{x}_2, \cdots, \mathbf{x}_n$ respect to the output of the jth hidden node.

The input weight $\boldsymbol{\omega}$ and the threshold $\mathbf{b}$ are subject to a random probability distribution of a continuous probability distribution, so that Eq. 2 is a linear system with variable $\boldsymbol{\beta}$. The solution of the linear system, that is, to find the smallest output weight $\boldsymbol{\beta}$, so that the error $\|\mathbf{H}\boldsymbol{\beta} - \mathbf{T}\|$ tends to a minimum. ELM algorithm uses the least squares method to calculate the output weight matrix $\boldsymbol{\beta}$, the solution can be expressed as:

$$\mathbf{H}\boldsymbol{\beta}^* - \mathbf{T} = \min_{\boldsymbol{\beta}} \mathbf{H}\boldsymbol{\beta} - \mathbf{T} \tag{6}$$

$$\boldsymbol{\beta}^* = \mathbf{H}^\dagger \mathbf{T} \tag{7}$$

where $\mathbf{H}^\dagger$ is the Moore-Penrose generalized inverse of $\mathbf{H}$, i.e.,

$$\mathbf{H}^\dagger = \mathbf{H} \left(\frac{1}{\lambda} + \mathbf{H}\mathbf{H}^{\mathrm{T}}\right)^{-1} \tag{8}$$

where $\frac{1}{\lambda}$ is the positive value to the diagonal on the theoretical basis of the ridge regression.

2.2   Hierarchical Extreme Learning Machine

Typically, the basic ELM and its variants are single-layer learning methods to handle classification problems [12]. However, the proposed H-ELM actually is a multilayered structure of ELM. Each layer of the H-ELM determine the parameters as same as ELM. The inputs of the current layer are come from the resultant outputs of last layer. The relationship between the two neighboring layers is calculated by using the following formula:

$$\mathbf{H}_i = g\left(\mathbf{H}_{i-1} \cdot \boldsymbol{\beta}\right) \tag{9}$$

where $\mathbf{H}_i$ is the output of the ith layer, $\mathbf{H}_{i-1}$ is the output of the $(i-1)$th layer as same as the input of the ith layer.

As every layer of H-ELM is independent of each others, the weights and parameters randomly assigned of the current hidden layer will be fixed. When the layers increase, the resulting feature becomes more compact. The whole system can complete the training without iteration and do not need to be fine-tuned. That is why the training velocity of H-ELM is much faster than other traditional leaning algorithms.

3   The Proposed Intrusion Detection Method

3.1   Overall Framework

At the beginning of the whole work, we need to preprocess the data. Then we use a large number of labeled samples to train the H-ELM classifier for obtaining the weight output matrix of each hidden layer. At last, the output matrix is used to classify the network data to be detected, so as to distinguish the normal data or the attacking ones.

The overall framework of the proposed intrusion detection based on H-ELM is shown in Fig. 1.

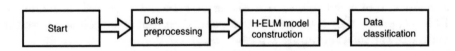

Fig. 1. The overall framework of the proposed method

3.2   Data Preprocessing

As we know, many datasets contain both discrete and continuous features, but as the computational constraints of classifiers, only continuous features are needed. If any discrete feature has only two distinct values like (1 or 0) then we don't need to convert it and can treat it as continuous feature. If it has k (more than two) distinct values, we need to convert it at first. All the discrete features will be represented into continuous ones using 1 of k coding. For this kind of discrete features, we create k different features to represent k distinct values. For example, if any discrete feature has three distinct values like (F0, F1, F2) then it can be represented into three independent continuous features like (001), (010) and (100). These 0 and 1 in represented features are considered as binary values, that means the discrete feature (F0, F1, F2) can also be represented into 1, 2 and 4 in decimal numbers. This step is called numeralization.

Next, we need to normalize the represented features. In this paper, we use Matlab R2014b which contains z-score function to run the experiments. Z-score is the normalization of data based on the mean and standard deviation of the original data. We have quantized the original data in first step, then we need to normalize the quantized value $\mathbf{x}$ to $\mathbf{z}$ using z-score.

$$\mathbf{z}_i = (\mathbf{x}_i - mean(\mathbf{x})) / std(\mathbf{x}) \tag{10}$$

Z-score changes the value of $\mathbf{x}$, but doesn't change the size. When $\mathbf{x}$ is a vector, $\mathbf{z}$ is still a vector. When $\mathbf{x}$ is a matrix, $\mathbf{z}$ is still a matrix. This step is called normalization.

3.3   Construction of the H-ELM Model

In summary, the work flow of the construction of the H-ELM Model is presented in Algorithm 1, And the process of intrusion detection based on the constructed H-ELM model is shown in Algorithm 2.

Algorithm 1. *H-ELM for training data*

Input: The training set of data $\mathbf{D}_{train} = \{(\mathbf{x}_i, \mathbf{t}_i) | \mathbf{x}_i \in \mathbf{R}^n, \mathbf{t}_i \in \mathbf{R}^m, i = 1, 2, \cdots, N\}$, the number of layers M.
Output: The output weight matrix β of each hidden layers.
1. Set up the H-ELM network structure, assign $m = 1$;
2. Random Assign to the input randomly weights $\omega = [\omega_1, \cdots, \omega_N]$ and bias $\mathbf{b} = [b_1, \cdots, b_N]$;
3. Calculate the output matrix $\mathbf{H}_1$ of the hidden layer according to (3);
4. Calculate the output weights β_1 according to (7);
5. **while** $m < M$
6. $\mathbf{H}_{m+1} = g(\mathbf{H}_m \cdot \beta_m)$;
7.   Calculate the output weights β_{t+1} according to (7);
8. $m = m + 1$;
9. **end while**

Algorithm 2. *H-ELM for testing data*

Input: The testing samples $\mathbf{D}_{test} = \{(\mathbf{x}_i, \mathbf{t}_i)|\mathbf{x}_i \in \mathbf{R}^n, \mathbf{t}_i \in \mathbf{R}^m, i = 1, 2, \cdots, N\}$, parameters of each hidden layers $\boldsymbol{\omega}, \mathbf{b}, \boldsymbol{\beta}$.

Output: The multi-class results $\mathbf{Y}$ of intrusion detection.

1. Set up the H-ELM network structure, input the testing samples;

2. While $1 \leqslant m < M$, Calculate the output matrix $\mathbf{H}$ of the hidden layer according to formula (7);

3. Calculate $\mathbf{Y} = \mathbf{H}\boldsymbol{\beta}$;

4   Performance Evaluation

4.1   Datasets

According to the statistical results of literatures, KDD CUP'99 is the mostly widely used dataset in the study of machine learning methods for intrusion detection area [5,13]. Researchers using KDD CUP'99 to evaluate their methods are almost able to achieve a high accuracy. But there are two important issues which highly affects the performance of evaluated systems. Tavallaee et al. have proved that evaluating methods on the basis of accuracy, detection rate and false positive rate on the KDD CUP'99 dataset is not an appropriate option [14]. So in our study, we use NSL-KDD dataset, which consists of selected records of the complete KDD data set and has amended the mentioned shortcomings.

Table 1. The various samples' sizes for both training and testing in NSL-KDD dataset

Type of data	Probing	Dos	U2R	R2L	Normal	Total
Training data	11656	45927	105	942	67343	125973
Testing data	2421	7458	1298	1656	9711	22544

NSL-KDD dataset includes 125,973 and 22,544 records for train and test respectively. It consists of 5 types of data classes, which are, Normal, Denial of Service (DoS), User to Root (U2R), Remote to Local (R2L) and Probing. The dataset has 41 dimensional features in total. The various samples' sizes for both training and testing in NSL-KDD dataset are presented in Table 1.

4.2   Comparative Results and Analysis

We use the Matlab R2014b in the system of Win7 Sp1 64 to run all the experiments. And the testing computer's CPU is Intel®Core™ i7-4790 CPU @ 3.60 GHz, the RAM is 16.0 GB.

To evaluate the proposed method intuitively and objectively, we use full features without selection. As some other learning algorithms e.g. k-NN, it achieves quite different performance within different dimensional features [7]. Moreover, we run independent trials for 10 times to get average results in each group of experiments. We aim to get a both effective and efficient solution applied to IDSs. So we choose the average multi-class accuracy and classification time as the criteria to compare the performance of different methods.

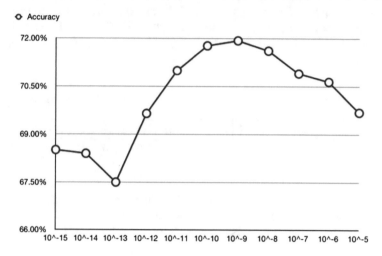

Fig. 2. The overall accuracy with different **C** based on H-ELM

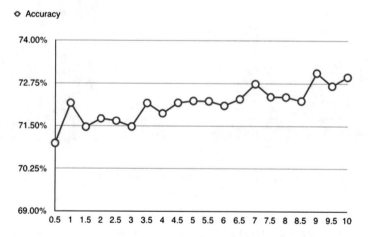

Fig. 3. The overall accuracy with different **s** based on H-ELM

There are two important parameters required before we construct the H-ELM model, i.e., the parameter **C** as the penalty of the last layer, and the parameter **s** as the scaling factor of the activation function. Figures 2 and 3 show the different performance in overall accuracy with different **C** and **s** respectively. According to the experimental results, we choose $\mathbf{C} = 10^{-9}$, and $\mathbf{s} = 9$.

As the number of hidden layers increase, H-ELM spends more time to classify the data. This is contrary to our original intention. So we select 3 layers as it has a higher accuracy and faster speed [11]. To decide the number of nodes of each layer, we run large numbers of experiments. We find out that as the nodes increase, the accuracy increases slowly and the expenditure of time increases fast. The optimal parameter of each layer is 20, 20, 400. Table 2 shows the experimental results of multi-class accuracy of H-ELM with 20, 20 and 400 nodes of each layer.

Table 2. The various accuracy for five types in NSL-KDD dataset

Type of data	Probing	Dos	U2R	R2L	Normal	Total
Accuracy	49.02%	77.50%	0	1.90%	97.11%	72.87%

To evaluate the performance of H-ELM further, we run the experiments with k-NN, RF and ELM respectively. Figure 4 shows the comparative performance in terms of the average multi-class accuracy of these four methods.

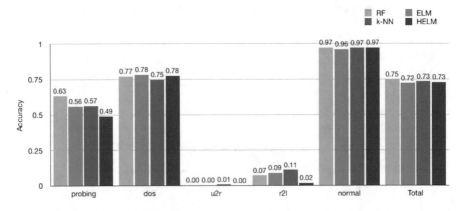

Fig. 4. The comparative results of the average multi-class accuracy of RF, ELM, k-NN and H-ELM.

We can find out that H-ELM acquires a lower accuracy in terms of Probing, and almost all the methods can't detect U2R and R2L which occupy the smallest samples of the dataset. However, in terms of DoS, H-ELM performs better than RF and k-NN and similar to ELM. And in terms of Normal, H-ELM performs better than ELM and similar to RF and k-NN. So we can say H-ELM outperforms the other three methods in terms of the two largest classes: Normal and Dos. The overall accuracy of RF, k-NN, ELM and H-ELM are 74.86%, 73.47%,

72.29% and 72.87% respectively. H-ELM is not the perfect one in accuracy. However, we further examine the classification time, H-ELM is significantly faster than other methods. Table 3 the results obtained by comparing the classification time of these methods.

It is obvious to see that H-ELM takes 2.04 s in total which is 3.5 fold faster than ELM, 57 fold faster than RF and 59 fold faster than k-NN.

Table 3. The training time and testing time of the Random Forest, KNN, ELM and H-ELM, the mean for 10 repeated tests

	Training time	Testing time	Total time
Random Forest	115.29	1.53	116.82
KNN	...	...	121.165
ELM	6.72	0.33	7.05
H-ELM	**1.81**	**0.23**	**2.04**

5 Conclusions

In this paper, we have proposed a highly efficient intrusion detection method based on H-ELM. To evaluate the performance of H-ELM, we run the experiments with NSL-KDD 2009 dataset which is the improved version of KDD Cup'99. After experimental verification, we finally choose the optimal parameters with 3 layers and 20, 20, 400 hidden nodes for each. The proposed method can achieve a overall accuracy of 72.87% which is better than or similar to other methods e.g. RF, k-NN and ELM. On the other hand, H-ELM can easily achieve a 3.5 ~ 59 fold speed without any feature reduction or selection than other previous ones. It is expected to be used in on-line intrusion detection. However, rooms for improvements exist: As to the limitations of this research, H-ELM cannot effectively detect U2L and R2L attacks which are also difficult to detect for other methods. This is an issue that turns into our future main work.

Acknowledgments. This work is supported by the National Natural Science Foundation of China (Grant No. 61772561) and the Key Research & Development Plan of Hunan Province (Grant No. 2018NK2012).

References

1. Singh, R., Kumar, H., Singla, R.K.: An intrusion detection system using network traffic profiling and online sequential extreme learning machine. Expert Syst. Appl. **42**(22), 8609–8624 (2015)
2. Liu, Q., Yin, J., Leung, V.C.M., Cai, Z.: FADE: forwarding assessment based detection of collaborative grey hole attacks in WMNs. IEEE Trans. Wirel. Commun. **12**(10), 5124–5137 (2013)

3. Ambusaidi, M., He, X., Nanda, P., Tan, Z.: Building an intrusion detection system using a filter-based feature selection algorithm. IEEE Trans. Comput. **65**(10), 2986–2998 (2016)
4. Chen, Y., Abraham, A., Yang, B.: Hybrid Flexible Neural-Tree-Based Intrusion Detection Systems: Research Articles. Wiley (2007)
5. Buczak, A.L., Guven, E.: A survey of data mining and machine learning methods for cyber security intrusion detection. IEEE Commun. Surv. Tutor. **18**(2), 1153–1176 (2017)
6. Sarasamma, S.T., Zhu, Q.A., Huff, J.: Hierarchical kohonenen net for anomaly detection in network security. IEEE Trans. Syst. Man Cybern. Part B Cybern. **35**(2), 302–12 (2005)
7. Lin, W.C., Ke, S.W., Tsai, C.F.: CANN: an intrusion detection system based on combining cluster centers and nearest neighbors. Knowl. Based Syst. **78**(1), 13–21 (2015)
8. Zhang, L., Zhang, D., Tian, F.: SVM and ELM: Who Wins? Object Recognition with Deep Convolutional Features from ImageNet. Springer International Publishing (2016)
9. Huang, G.B., Zhu, Q.Y., Siew, C.K.: Extreme learning machine: theory and applications. Neurocomputing **70**(1), 489–501 (2006)
10. Liu, Q., Yin, J., Leung, V.C.M., Zhai, J.H., Cai, Z., Lin, J.: Applying a new localized generalization error model to design neural networks trained with extreme learning machine. Neural Comput. Appl. **27**(1), 59–66 (2016)
11. Tang, J., Deng, C., Huang, G.B.: Extreme learning machine for multilayer perceptron. IEEE Trans. Neural Netw. Learn. Syst. **27**(4), 809–821 (2017)
12. Liu, Q., Zhou, S., Zhu, C., Liu, X., Yin, J.: MI-ELM: highly efficient multi-instance learning based on hierarchical extreme learning machine. Neurocomputing **173**, 1044–1053 (2015)
13. Tsai, C.F., Hsu, Y.F., Lin., Lin, W.Y.: Intrusion detection by machine learning: a review. Expert Syst. Appl. **36**(10), 11994–12000 (2009)
14. Tavallaee, M., Bagheri, E., Lu, W., Ghorbani, A.A.: A detailed analysis of the KDD cup 99 data set. In: IEEE International Conference on Computational Intelligence for Security Defense Applications, pp. 1–6 (2009)

An Extended Extreme Learning Machine with Residual Compensation and Its Application to Device-Free Localization

Jie Zhang[1,2(✉)], Ruofei Gao[1,2], Yanjiao Li[1,2],
and Wendong Xiao[1,2(✉)]

[1] School of Automation and Electrical Engineering,
University of Science and Technology Beijing, Beijing 100083, China
zhangjie2009622@163.com, wdxiao@ustb.edu.cn
[2] Beijing Engineering Research Center of Industrial Spectrum Imaging,
Beijing 100083, China

Abstract. Extreme learning machine (ELM) was proposed for training single hidden layer feedforward neural networks (SLFNs), and can provide an efficient learning solution for regression problem. However, the prediction error is unavoidable when the underlying regression problem is nonlinear and stochastic, due to the limited modeling capability of a given ELM. In this paper, an extended ELM, named as ELM-RC, is proposed for regression problem with residual compensation induced by the prediction error. ELM-RC employs a two-layer structure with the baseline layer for building the feature mapping relationship between the input and the output, and another layer for residual compensation to improve the accuracy further. The final output of the proposed ELM-RC is the weighted accumulation of the outputs of the two layers. In order to verify the validity of the proposed ELM-RC, device-free localization (DFL) is used for experimental testing, by comparing it with the original ELM. Experimental results show that ELM-RC has better generalization performance and robustness.

Keywords: Extreme learning machine · Regression problem
Residual compensation · Device-free localization

1 Introduction

In the past decades, machine learning approaches have been widely studied and successfully applied to solve many engineering problems and puzzles in our daily life. However, the traditional machine learning approaches, such as back propagation neural network (BPNN) and support vector machine (SVM), often suffer the problems such as trapping in local optimum and insufficiency in learning speed. In order to tackle these issues, Huang et al. [1, 2] proposed extreme learning machine (ELM) as the extension of these traditional machine learning approaches. In contrast to other approaches, the hidden layer learning parameters of ELM are assigned randomly without tuning, and the output weights are determined through the least square method. Due to its faster learning speed and better generalization performance, ELM has been receiving

© Springer Nature Switzerland AG 2019
J. Cao et al. (Eds.): ELM 2017, PALO 10, pp. 327–337, 2019.
https://doi.org/10.1007/978-3-030-01520-6_30

increasing research interests, and many progresses have been made. Huang et al. proposed the incremental ELM (I-ELM) and its variants [3–6], by adopting an incremental construction approach to adjust the number of hidden nodes. Zong et al. [7] and Xiao et al. [8] proposed the weighted ELM and class-specific cost regulation ELM for imbalanced data distribution, respectively, which widely exists in many fields. Liang et al. [9] proposed the online sequential ELM (OS-ELM) for online learning problems when samples come sequentially. Comparing with its faster speed and better performance, the robustness of ELM may be relatively weak. The mainly effects of the robustness of ELM come from the randomly assigned hidden layer parameters, which also can result in the relatively large change of the output weights [10]. The regularization factor is often used in ELM to balance and reduce the structural risk and the empirical risk, but with little impacts on the robustness of the algorithm, as the large changes of the output weights will increase largely both the two risks [11, 12]. In order to improve the robustness of ELM, some modifications have been performed. Man et al. [10] proposed FIR-ELM to reduce the structural risk and the empirical risk, and removed some undesired frequency components through the FIR filtering. In [13], discrete Fourier transform-based ELM (DFT-ELM) was proposed to improve the robustness of ELM, which has achieved better performance than FIR-ELM. Zhang et al. [14] proposed denoising Laplacian multi-layer ELM (D-Lap-ML-ELM), but its computational complexity is relatively high. In addition, ELM has been widely used in many fields, such as localization [15, 16], industrial production [17], etc.

The above approaches improve the robustness of ELM mainly by reducing the input disturbance or remove the undesired input samples. To our best knowledge, it rarely involves in making use of these input disturbance or the undesired input samples in ELM field. However, the residual of the input samples still contain useful information, if we can make fully use of it, the accuracy of ELM will be improved further. In order to tackle this problem, we will propose an extended ELM with residual compensation, named ELM-RC. ELM-RC employs a two-layer structure with the baseline layer for building the feature mapping relationship between the input and the output, and another layer for residual compensation to improve the accuracy further. The final output of the proposed ELM-RC is the accumulation of the outputs of the two layers. In order to verify the validity of the proposed ELM-RC, device-free localization (DFL) is used for experimental testing, by comparing it with the original ELM.

The paper is organized as follows. Some preliminaries are given in Sect. 2, including the brief introduction to ELM, and how errors are generated in regression problem. The proposed ELM-RC is detailed in Sect. 3. ELM-RC based DFL is introduced in Sect. 4. Experimental results and further analysis are reported in Sect. 5. Finally, conclusions are given in Sect. 6.

2   Preliminaries

In this section, ELM and how errors are generated in regression problem are introduced to facilitate the understanding of the problem to be addressed.

2.1   Extreme Learning Machine

ELM was originally proposed for the single hidden layer feedforward neural networks (SLFNs) and was extended to the generalized SLFNs where the hidden layer need not be neuron alike [4]. Different from other traditional machine learning approaches, the hidden layer parameters need not be tuned in ELM.

In this paper, we denote the number of hidden nodes as L. The output of hidden node i can be written as $g(x; a_i, b_i)$, where $g(\bullet)$ is the activation function, a_i and b_i are the corresponding hidden layer parameters, $i = 1, \ldots, L$. For a given dataset $\{(x_i, t_i)\}_{i=1}^{N} \subset R^n \times R^m$, where x_i is a n-dimensional input vector and t_i is the corresponding m-dimensional observation vector. Its mapped feature vector can be represented as

$$h(x) = [g(x; a_1, b_1), g(x; a_2, b_2), \ldots, g(x; a_i, b_i)] \tag{1}$$

From the network architecture point of view, the output function of ELM for the generalized SLFNs with L hidden nodes can be represented by

$$f_L(x) = \sum_{i=1}^{L} \beta_i h_i(x) = h(x)\beta \tag{2}$$

where $\beta = [\beta_1, \beta_2, \ldots, \beta_L]^T$ is the vector of the output weights connecting the hidden layer and the output.

From the learning point of view, ELM aims to reach the smallest training error and also the smallest norm of the output weights:

$$Minimize : \|\beta\|_p^{\sigma_1} + C\|H\beta - T\|_q^{\sigma_2} \tag{3}$$

where $\sigma_1 > 0$, $\sigma_2 > 0$, $p, q = 0, (1/2), 1, 2, \ldots, +\infty$, C is a user specified parameter and provides a tradeoff between the minimization of the training errors and the maximization of the marginal distance, H is the hidden layer output matrix (randomized matrix):

$$H = \begin{bmatrix} h(x_1) \\ \vdots \\ h(x_N) \end{bmatrix} = \begin{bmatrix} h_1(x_1) & \cdots & h_L(x_1) \\ \vdots & \ddots & \vdots \\ h_1(x_N) & \cdots & h_L(x_N) \end{bmatrix} \tag{4}$$

and T is the training data target matrix:

$$T = \begin{bmatrix} t_1^T \\ \vdots \\ t_N^T \end{bmatrix} = \begin{bmatrix} t_{11} & \cdots & t_{1m} \\ \vdots & \ddots & \vdots \\ t_{N1} & \cdots & t_{Nm} \end{bmatrix} \tag{5}$$

The three-step learning process of ELM can be summarized as

(1) Randomly assign the hidden node parameters, e.g., the input weights a_i and biases b_i for additive hidden nodes.
(2) Calculate the hidden layer output matrix H.
(3) Obtain the output weight vector

$$\beta = H^{\dagger} T \tag{6}$$

where $H^{\dagger}$ is the Moore-Penrose generalized inverse of the matrix H.

Usually, ELM is more stable and has better generalization performance with $\sigma_1 = \sigma_2 = p = q = 2$, thus, the optimization problem can be mathematically written as

$$Min : L_{P_{ELM}} = \frac{1}{2}\|\beta\|^2 + \frac{1}{2}C\sum_{i=1}^{N}\xi_i^2$$

$$s.t., \ h(x_i)\beta = t_i - \xi_i, i = 1,\dots,N \tag{7}$$

where $\xi_i = [\xi_{1,m},\dots,\xi_{i,m}]$ is the training error vector of the m output nodes with respect to the training sample x_i.

Then, based on the KKT theorem, we have

$$\beta = \begin{cases} H^T(\frac{I}{C} + H^T H)^{-1}T, & N < L \\ (\frac{I}{C} + H^T H)^{-1}H^T T, & N > L \end{cases} \tag{8}$$

where I is the unit matrix.

2.2 Errors in Regression Problem

In this subsection, we will introduce the main sources of errors in the regression problem. Usually, there are three kinds of errors:

(1) Noisy error. This kind of error is caused by the abrupt noise, when the output of the model deviates from the normal range. Such errors should be removed.
(2) System error. This kind of error can reflect the true relationship of the input and the output, but we cannot remove this kind of error directly.
(3) Uncontrollable error. This kind of error is caused by missing some key input attributes.

Unfortunately, there are no strategies to distinguish the three kinds of error, thus, on one hand, if we directly remove the undesired input samples, the model will not obtain the useful information completely, which will degrade the performance of the model; on the other hand, if we reserve all these undesired input samples, especially those with noise, it will lead to the overfitting of the model. So, we should extract the useful information from the undesired input samples, but not directly remove them. In this paper, we will extract the useful information and make fully use of them through the residual compensation mechanism.

3   Extreme Learning Machine with Residual Compensation

As we know, regression is an important task in machine learning. We can find that some of the approaches for regression can obtain good performance when the underlying regression problem is stable or the input disturbance is small. However, its performance may drop significantly under the unstable conditions of the underlying regression problem, or the input disturbance is relatively large. A possible reason is that much useful information existing in the input disturbance is removed or not modeled, that is to say, we cannot make fully use of the useful information existing in the residual. In such case, it is always difficult to obtain accurate model.

For the underlying regression problem, it is difficult to build the corresponding mechanism model, because there must be measuring error or system error. For machine learning approaches, all of them only can approximate the targeted process or system, so it must exist some kinds of errors in the machine learning-based models. Likewise, unless all the prediction value can equal to the true value, if not, there must be useful information in the residuals. Thus, making fully use of the useful information from the input disturbance is one of the key issues in the machine learning field.

Considering the following original underlying regression problem:

$$y = f(x) + \varepsilon \tag{9}$$

where $f(\bullet)$ stands for the real process or system, x is the input, y is the output, and ε denotes the error.

When building the model of the above process or system, we can obtain the following relationship

$$y = \tilde{f}(x) + \tilde{\varepsilon} \tag{10}$$

where $\tilde{f}(\bullet)$ stands for the created model, $\tilde{\varepsilon}$ stands for the corresponding error, x and y are the same input and output of (9). As analyzed above, $\tilde{\varepsilon}$ not only contains some kinds of errors, but also useful information.

From the view of ELM, when we consider the input disturbance, we have

$$H\beta + \varepsilon = T \tag{11}$$

where $\varepsilon \in N(0, \sigma^2)$ to facilitate the following analysis.

After considering the interference of external disturbance, the solution of ELM without regularization factor can be represented as

$$\beta_\varepsilon = \frac{\sum_{i=1}^{L} H_i^T (H_i \beta_i + \varepsilon_i)}{\sum_{i=1}^{L} H_i^T H_i} = \beta + \frac{\sum_{i=1}^{L} H_i^T \varepsilon_i}{\sum_{i=1}^{L} H_i^T H_i} \tag{12}$$

Thus, we can calculate the corresponding expectation (E), variance (V) and mean square error (MSE) of ELM:

$$E(\beta_\varepsilon) = \beta \tag{13}$$

$$V(\beta_\varepsilon) = E(\beta_\varepsilon^2) - E(\beta_\varepsilon)^2 = \frac{\sigma^2}{\sum_{i=1}^{L} H_i^T H_i} \tag{14}$$

$$MSE(\beta_\varepsilon) = \frac{1}{L}E[(\beta_\varepsilon - \beta)^T(\beta_\varepsilon - \beta)^T] = \frac{1}{L}\frac{\sigma^2}{\sum_{i=1}^{L} H_i^T H_i} \tag{15}$$

According to (12), (14) and (15), we can find that the external noise has effects on the output weight, V and MSE. Furthermore, Bartlett [18] pointed out that for SLFNs the smaller the norm of output weight and the training error are, the better generalization performance the networks tend to have. Therefore, one of the most effective ways to improve the generalization performance of ELM is to make the output weight and the training error smaller.

In order to address this issue, we will propose an extended two-layer ELM with residual compensation (ELM-RC). ELM-RC employs a two-layer structure with the baseline layer for building the feature mapping relationship between the input and the output, and another layer for residual compensation to improve the accuracy further. As shown in Fig. 1, x is the input, $\tilde{y}$ is the output of the baseline ELM, i.e., the predicted value, y is the real output, here $\tilde{\varepsilon}$ is the predicted value of the residual, $\tilde{\tilde{y}}$ is the final output of ELM-RC. According to Fig. 1, the final output of ELM-RC is

$$\tilde{\tilde{y}} = \tilde{y} + \tilde{\varepsilon} \tag{16}$$

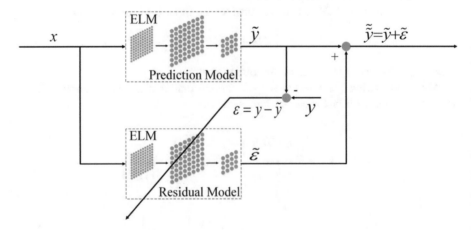

Fig. 1. Main structure of ELM-RC

Remark 3.1. It should note that the final output of ELM-RC highly correlates with the baseline ELM, i.e., the prediction model in Fig. 1. If the prediction model is exactly similar to the real model, it means the residual contains little useful information, so the performance of the residual model will be relatively poor. On the contrary, if the prediction model is different from the real model significantly, the residual will contain more useful information, so the performance of the residual model will be better.

4   ELM-RC Based Device-Free Localization

Device-free localization (DFL) was introduced based on wireless sensor network radio-frequency signal measurements, where the target does not need to have any attached electronic device [19]. As shown Fig. 2, a DFL system can estimate the location of the target by sensing the differential received signal strength (ΔRSS) measurements of the affected links. DFL can be applied to many applications, such as human health and medical care with the help of location detection and behavior analysis.

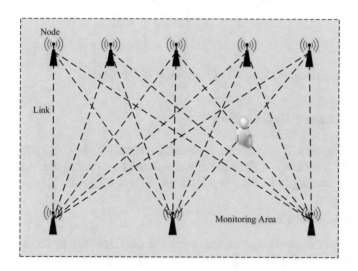

Fig. 2. Schematic diagram of device-free localization

The ELM-RC based DFL involves two phases, i.e., the offline training phase and the online localization phase. In the offline training phase, each reference point (RP) is associated with a number of affected links, and the corresponding affected links of all the RPs can be used for providing the ELM-RC training samples. In the online localization phase, the trained ELM-RC model outputs the location of the target.

Assuming that there are l RPs, and each RP corresponds to n affected links, so the input matrix of prediction model in ELM-RC can be written as

$$Input = x = \begin{bmatrix} x_1 \\ x_2 \\ \vdots \\ x_n \end{bmatrix} = \begin{bmatrix} \Delta RSS_1^1 & \Delta RSS_1^2 & \cdots & \Delta RSS_1^n \\ \Delta RSS_2^1 & \Delta RSS_2^2 & \cdots & \Delta RSS_2^n \\ \vdots & \vdots & \ddots & \vdots \\ \Delta RSS_l^1 & \Delta RSS_l^2 & \cdots & \Delta RSS_l^n \end{bmatrix} \tag{17}$$

and the input weights with L hidden nodes are

$$\omega = \begin{bmatrix} \omega_1 \\ \omega_2 \\ \vdots \\ \omega_L \end{bmatrix} = \begin{bmatrix} \omega_1^1 & \omega_1^2 & \cdots & \omega_1^n \\ \omega_2^1 & \omega_2^2 & \cdots & \omega_2^n \\ \vdots & \vdots & \ddots & \vdots \\ \omega_L^1 & \omega_L^2 & \cdots & \omega_L^n \end{bmatrix} \tag{18}$$

The output can be expressed as

$$T = \begin{bmatrix} t_1 & t_1^v \\ t_2 & t_2^v \\ \vdots & \vdots \\ t_l & t_l^v \end{bmatrix} \tag{19}$$

where $\{(t_l, t_l^v)\}$ denotes the location coordinate of the target.

5   Performance Verification

In this section, we will evaluate the performance of the proposed ELM-RC using DFL compared with the original ELM. All the experiments are carried out in Matlab 2012a environment running in an Inter i5 3.2 GHz CPU and 4G RAM.

5.1   Experimental Environment

We performed the experiment on the campus of the University of Science and Technology Beijing. The monitoring area is 6 m × 6 m square, with 16 wireless nodes placed along its boundary and the adjacent node distance of 1.5 m. We use the IEEE 802.15.4 ZigBee device for communication, each device operates in 2.4 GHz frequency band. In the experiment, we have calculated average error with the distance between each reference point of 0.3 m.

5.2   Experimental Results and Analysis

Figure 3 shows the results of ELM-RC (sigmoid function as the activation function) with the changing of the numbers of links from 2 to 10, and all links. According to Fig. 3, we can find that ELM-RC obtains the best accuracy with $L = 40$ and all links, which equals to 1.36 m.

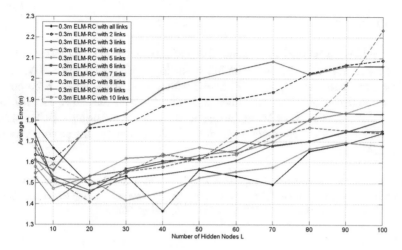

Fig. 3. Results of ELM-RC with the increasing of links from 2 to 10, and all links

Figure 4 illustrates the results of the original ELM (sigmoid function as the activation function) with the changing of the number of links from 2 to 10 and all links. According to Fig. 4, we can find that ELM obtains the best accuracy with $L = 10$ and 10 links, which equals to 1.71 m.

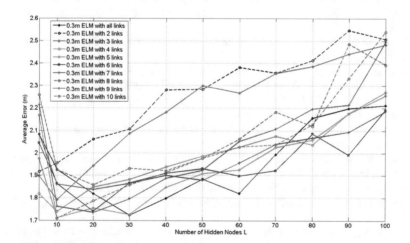

Fig. 4. Results of ELM with the increasing of links from 2 to 10, and all links

According to the above two figures, ELM-RC and ELM obtain their own corresponding best localization accuracy when all links and 10 links are performed, respectively, and the result of ELM-RC is about 0.4 m better than ELM. In addition, most of the localization accuracies of ELM-RC with different number of links are better than ELM, which indicates the good generalization performance of the proposed ELM-RC.

6   Conclusions

For the underlying regression problem, the prediction error is unavoidable when it is nonlinear and stochastic, due to the limited modeling capacity of a given ELM. In order to address this issue, in this paper, ELM-RC with a two-layer structure is proposed by residual compensation. The experimental results indicate that ELM-EC has better accuracy than the original ELM. In this paper, we only use the original ELMs in ELM-RC, so we are thinking to use other variants of ELM to improve the performance, such as OS-ELM, etc.

Acknowledgement. This work is supported by the National Key Research and Development Program of China under Grant 2017YFB1401203 and the National Natural Science Foundation of China (NSFC) under Grants 61673055, 61673056, and 61773056.

References

1. Huang, G.B., Zhu, Q.Y., Siew, C.K.: Extreme learning machine: theory and applications. Neurocomptuing **70**, 489–501 (2006)
2. Huang, G.B., Zhou, H., Ding, X., et al.: Extreme learning machine for regression and multiclass classification. IEEE Trans. Syst. Man Cybernetics. Part B: Cybern. **42**(2), 513–529 (2012)
3. Huang, G.B., Chen, L., Siew, C.K.: Universal approximation using incremental constructive feedforward networks with random hidden nodes. IEEE Trans. Neural Netw. **17**(4), 879–892 (2006)
4. Huang, G.B., Chen, L.: Convex incremental extreme learning machine. Neurocomptuing **70**, 3056–3062 (2007)
5. Huang, G.B., Li, M.B., Chen, L., et al.: Incremental extreme learning machine with fully complex hidden nodes. Neurocomptuing **71**, 576–583 (2008)
6. Huang, G.B., Chen, L.: Enchanced random search based incremental extreme learning machine. Neurocomptuing **71**, 3460–3468 (2008)
7. Zong, W.W., Huang, G.B., Chen, Y.Q.: Weighted extreme learning machine for imbalance learning. Neurocomptuing **101**, 229–242 (2013)
8. Xiao, W.D., Zhang, J., Li, Y.J., Zhang, S., Yang, W.D.: Class-specific cost regulation extreme learning machine for imbalanced classification. Neurocomptuing **261**, 70–82 (2017)
9. Liang, N.Y., Huang, G.B., Saratchandran, P., et al.: A fast and accurate online sequential learning algorithm for feedforward networks. IEEE Trans. Neural Netw. **17**(6), 1411–1423 (2006)
10. Han, Z.H., Lee, K., Wang, D.H., et al.: A new robust training algorithm for a class of single-hidden layer feedforward neural networks. Neurocomptuing **74**, 2491–2501 (2011)
11. Vapnik, V.: Statistical Learning Theory. Wiley, New York (1998)
12. Anthony, M., Bartlett, P.L.: Neural network learning: theoretical foundations. Cambridge University Press, Cambridege (1999)
13. Man, Z.H., Lee, K., Wang, D.H., et al.: Robust single-hidden layer feedforward network-based pattern classifier. IEEE Trans. Neural Netw. Learn. Syst. **23**(12), 1974–1986 (2012)
14. Zhang, N., Ding, S.F., Shi, Z.Z.: Denoising Laplacian multi-layer extreme learning machine. Neurocomptuing **171**, 1066–1074 (2016)

15. Zhang, J., Sun, J., Wang, H.L., Xiao, W.D., Tan, L.: Large-scale WiFi indoor localization via extreme learning machine. In: Proceedings of the 36th Chinese Control Conference, Dalian, China, 26–28 July, pp. 4115–4120 (2017)
16. Zhang, J., Xiao, W.D., Zhang, S., Huang, S.D.: Device-free localization via an extreme learning machine with parameterized geometrical feature extraction. Sensors **17**(4), 879–890 (2017)
17. Li, Y.J., Zhang, S., Yin, Y.X., Xiao, W.D., Zhang, J.: A novel online sequential extreme learning machine for gas utilization ratio prediction in blast furnaces. Sensors **17**(8), 1847–1870 (2017)
18. Bartlett, P.L.: The sample complexity of pattern classification with neural networks: the size of the weight is more important than the size of the network. IEEE Trans. Inf. **44**(2), 525–536 (1998)
19. Youssef, M., Mah, M., Agrawala, A.: Challenges: device-free passive localization for wireless environments. In: Proceedings of the 13th Annual ACM International Conference on Mobile Computing and Networking, Montreal, QC, Canada, 9–14 September, pp. 222–229 (2007)

Author Index

© Springer Nature Switzerland AG 2019
J. Cao et al. (Eds.): ELM 2017, PALO 10, pp. 339–340, 2019.
https://doi.org/10.1007/978-3-030-01520-6

Printed in the United States
By Bookmasters